Quantitative Analysis Methods for Substantive Analysts

QUANTITATIVE ANALYSIS METHODS FOR SUBSTANTIVE ANALYSTS

HENRY F. DeFRANCESCO
Consultant,
Advanced Technology Systems, Inc.

Published by **Melville Publishing Company,**
a Division of John Wiley & Sons, Inc.

Library of Congress Cataloging in Publication Data:

DeFrancesco, Henry F
Quantitative analysis methods for substantive analysts.

Includes bibliographies and index.
1. Mathematical statistics. 2. Logic, Symbolic and mathematical. 3. Set theory. 4. Reasoning.
I. Title.
QA276.D42 519.5 74-20886
ISBN 0-471-20529-X

Printed in the United States of America

10 9 8 7 6 5 4 3 2 1

To my
Parents,
Wife,
and Family
They did so much
with so little

Preface

This text has been prepared to satisfy the needs of professional analysts, be they in private practice, industry, or in the service of their country. It is hoped that all substantive analysts who must make analyses, test hypotheses, draw conclusions, and make important decisions from information extracted from or supplied by others where the information may be replete with inaccuracies, uncertainties, and deception will find this presentation useful. The text is especially designed for analysts whose formal education in their collegiate careers did not provide the skills in science and mathematics which adequately enable them to communicate and progress in their professions. Today and tomorrow all analysts will become increasingly involved with information systems, statistics, probability, logical processing, and "systems" thinking.

Specifically, this text provides an introduction and some elementary concepts in sets, logic, plausible reasoning, statistics, and probability. Concepts and their interpretation and utilization in the analysis of information have been emphasized, and no attempt has been made to stress computational procedures or proofs.

The first three chapters bridge the language and analysis methods gap which exists between analysts such as political or news analysts, who analyze subjective information and analysts such as mathematicians and physicists, who analyze objective information. It is demonstrated through examples how the language and analysis methods used in objective analytic situations can be applied to analysis situations which confront analysts who normally deal with subjective information. These three chapters, and the chapters which follow, provide the analyst with the information required by him to choose intelligently the language and analytic methods best suited for his analysis situation.

The remaining chapters provide an introduction to the concepts of statistics, probability, and inferential statistics. Special emphasis is placed on giving step-by-step procedures to follow in the application of techniques such as hypothesis testing and Bayes' rule.

It is my hope that this text, developed on the basis of many classroom and user experiences, will provide the student and professional analyst with a satisfying, enriching, and useful understanding of the new analytic methods applicable to relevant areas of substantive analysis.

To the many professionals and students who have in one way or another contributed so greatly to the development of the material between the covers of this book, may I express my deepest appreciation and thanks. Especially, I wish to single out Capt. R. W. Bates, USN, Lt. Col. T. H. Murray, USA, Dr. Gordon T. Shahin and Dr. Reed B. Dawson for their encouragement and contributions to this text.

Finally, I wish to express my thanks to my wife, Bobbie, and to my former secretary, Mrs. Adelyne Godfrey, for their patience and untiring effort in preparing and editing the many drafts of this text.

McLean, Virginia *Henry F. DeFrancesco*

Contents

List of Examples

Chapter 3

Introduction

The primary responsibility of an analyst is to examine information, make assessments and predictions, and communicate findings on the possible manifestations of the substantive content of the information. An analyst is concerned with questions of substantive content and with analytic processes performed on this content. For example, the analyst uses all substantive information which include incomplete and uncertain information (or intelligence) of the political, social, military, diplomatic geographical, and historical, aspects of his areas of concern. He assesses his substantive information by employing analytical tools which include elements of the deductive, inductive, and plausible reasoning processes. The methods and techniques of analysis employed by the analyst normally determine the accuracy, reliability, and completeness of his findings. Good information poorly analyzed leaves much to be desired. But poor information properly evaluated and analyzed can contribute much to the understanding of the substantive content and its implications, even if only to determine the areas wherein more information is required. Every analyst should be aware of, and should strive to achieve, a reasonable level of skill in as many methods and techniques of analysis as possible. The chapters which follow provide analysts with some of the basic methods and techniques which have proven useful in the analysis of subjectively and objectively derived information.

Proper use of analysis will achieve one or a combination of any of the following objectives:

1. Enhance understanding of the past and present.
2. Provide suggestions and rationale on predictions about the future.
3. Align reader with the analyst's or established normative views of the past, present, and future.
4. Prescribe certain goals and provide plans or means by which these goals can be achieved.

These four objectives of analysis are best achieved by the descriptive, predictive, normative, and prescriptive forms of analysis, respectively. These forms of analysis will now be described.

DESCRIPTIVE ANALYSIS

The type of analysis most frequently used by analysts is termed "descriptive." The objective is to describe the characteristics of things, statements, or events on the basis of historical or current observations. A description usually consists of two components: One describes the event, statement, or thing; the other rationalizes the existence of the thing, the truthhood of the statement, or the occurrence of the event.

Descriptive analysis uses a spectrum of individualized styles and a modicum of techniques and methods. For example, intuition and heuristics (educated guesses) play an important role in the development of ideas which lead to more accurate descriptions. Several pieces of information may be tied together deductively or inductively, yielding an input to a better description of the thing, statement, or event. In the worst case, reams of information will be studied in an attempt to make some useful generalizations about the properties or characteristics of the thing, statements, or events under observation.

Descriptive analysis normally takes the form of a verbal description, a statistical description, or a combination of the two. Nonscientifically oriented analysts use verbal descriptive analysis most frequently in their work. Training in this technique is acquired mainly through experience or courses in writing and communicating. This aspect of descriptive analysis will always be an important output of any analyst attempting to communicate his values, thoughts, or findings to the interested but not professionally prepared public. The scientific analyst would do well in learning the techniques of good verbal descriptive analysis.

PREDICTIVE ANALYSIS

The analyst must of necessity be anticipatory or predictive of things developing or events occurring. For example, to answer questions such as, "what positions will the Russians assume if we supply additional arms material to the Israeli armed forces?" or "what are the chances of the Russians developing a laser weapon of significant destructive capability before the Chicoms?," the analyst will employ predictive analysis. Here the analyst normally attempts to describe the thing or event as it might appear to exist in the future.

Descriptions of things or events as they existed in the past, and as they currently exist, and the rate of change in key properties from the past to the present guides the analyst in deciding whether the same rate of change will continue into the future. Most of this analysis is descriptive and yet it does possess a predictive component. In some instances, it is difficult to separate the descriptive analysis component from the predictive component. In other instances, the predictive components are utilized to test the value and efficacy of the descriptive component and methods of prediction. For example, an analyst can have information on the period from 1920 to 1970. He may wish to set up and check the validity of some methods whereby he can predict or forecast events out to 1975. To do this the analyst uses only the information from the period 1920-1965, restricting the use of information for the period 1966-1970. His descriptive analysis then covers the period 1920-1965 and on the basis of his description and the predictive analysis to be employed, he can make assertions about the period 1966-1970. If his description and methods yield predictions about the period 1966-1970 which are credible (even if less than completely accurate) in that they predict the events which actually occurred in the 1966-1970 period, then his confidence in making predictions out to the year 1975 is enhanced proportionately.

NORMATIVE ANALYSIS

Another type of analysis used by almost everyone, and hopefully to a lesser degree by trained analysts, is called normative analysis. It is analysis based on one's own value system. In normative analysis, explicit or implicit judgments are formed from one's personal view of what exists or what will exist. Normative analysis reflects in varying degrees personal preferences be they legal, moral, political, or social. Normative analysis reflects an individual's view of reality tempered by his notions of right and wrong, good and bad, or acceptable and unacceptable. These notions become the main theme of his or her reasoning processes.

For example, someone who believes strongly that communism is the greatest evil confronting the free world would have a very difficult time objectively evaluating any benevolent act by a communist nation. Most descriptions of the act and reasons for the act would imply, if not directly state, that the acting nation had some devious or ulterior motive other than the one of pure benevolence. Normative judgments are generally made with respect to the analyst's value system in which he assumes that the recipient of his judgments has the same value system. One indicator of an analyst's objectivity is his effort to employ value systems other than his own in his analysis.

No individual is void of normative analytic techniques in his reasoning processes. Each of us has formed sets of values. How these values are acquired is of utmost importance. If the value system reflects a lifetime of objective analysis, then future analysis of events utilizing values already formed will provide a working value base more akin to the real world and should therefore result in better descriptions of the present or future state of affairs. However, if the current value system does not reflect reality, then rational and objective thought processes based on these value systems are subject to question. Briefly then, as stated in Chapter 2, if the hypothesis (our value system) does not reflect reality, then even the best logical thought processes employing this value system cannot guarantee a rational let alone a true description or prediction.

PRESCRIPTIVE ANALYSIS

Another type of analysis is called "prescriptive." Prescriptive analysis is used to demonstrate how and why future goals, policies, values, and strategies should assume the forms that the analyst considers to be desirable. Normally, prescriptive analysis includes a suggested plan of action on how to accomplish the analyst's desired objectives. Descriptive, normative, and predictive analytic components are also present in the plan and analysis to indicate why these forms are desirable and should be sought.

There are several techniques used in prescriptive analysis. The techniques include implication, suggestion, and generalization and are best illustrated by means of an example. The United States, at the end of World War II, set for itself the following aims:

Goal: Containment of communism to existing boundaries.
Value: Self-determination and individual freedom.
Policy: Support and development of needy noncommunist nations.
Strategy: World peace through alliances of "free world" nations.

Initially the U.S. used the implication form of prescriptive analysis to illustrate how foreign aid would serve the U.S. in achieving the goal, value, policy, and strategy stated above. The foreign aid program was sold to the public (and Congress) solely on the implication that if the stated aims of our nation were achieved, then the world would be a better place to live in. And the quickest way to achieve these aims was through an extensive foreign aid program. In other words, the U.S. prescribed a better world and implied that its plan for foreign aid was the best one to achieve this major aim.

After foreign aid had been in effect for several years, its continuance to nations not cooperating fully was used as a means to suggest to these nations the desirability of undertaking specific actions favorable to the goals of U.S. This constituted the suggestive form of prescriptive analysis.

After foreign aid had been in effect for a considerable period of time, the U.S. felt itself burdened most heavily and made its general prescriptive analysis which stated that if the free world were to achieve the four aims previously stated, all free nations should share in the aid to other free nations requiring assistance.

RESPONSIBILITY OF THE ANALYST

Every analyst should be familiar with each of these analysis forms and their special attributes. Further, the analyst should be aware of the type of analysis he is performing. To mix or confuse forms of analyses results in a misunderstanding of one's stated position, predictions, values, and prescriptions. It is most important that the analyst understand whether his statements are descriptive, predictive, normative, or prescriptive. By stating the form the analyst achieves greater objectivity in his analysis and he impresses others by clarifying the method.

For example, the above paragraph is prescriptive. It prescribes desirable goals to be achieved by analysts. It describes the differences or attributes of analytic types, and it describes the benefits to be derived as determined by the writer's value system. However, it does not indicate plans or methods for accomplishing these goals. The remainder of the text is dedicated to that purpose.

The analyst will find the techniques and methods described in this text to be generally applicable to a wide class of problems. Whether the methods and techniques are applicable to any specific problem is basically a function of the structure of the problem and the availability of pertinent data. However, real life factors also enter into the analytic picture. For example, the methods and techniques of analysis may be available but the data needed to complete the analysis may not be. In the real world real people must respond in real time to real problems. Delays cannot be tolerated. Decisions are made with the information available. Thus, by default good analyses are not applicable unless the time and data availability factors are handled satisfactorily.

Sometimes the substantive analyst will find that the "structure" of the problem under consideration does not fit or is not analogous to one solved previously. The analytical tools may not be precisely adequate, i.e., nonlinearities, subjective factors, and questionable data may have to be handled and accounted for. In these cases simplifying assumptions are normally made in order to make the analysis "work." The problem now becomes one of determining whether the problem "solved" resembles reality. Does it take into account all important factors? Or have some of these factors been "simplified" out of the problem? Most simplifications deal with time constraints, power structures, political sensitivities, people problems, and data credibility and availability. These form the substance on which most subjectively oriented problems are based. Without them technically beautiful but empty and useless solutions are achieved. The substantive analyst or any analyst for that matter must be aware of the existence and importance of these factors. Further, these factors should be considered where and when they are part of the problem. How to take account of them is a main objective of this text. But this text or the texts to follow will never provide the analyst with the most important asset–the ability to sense those factors that are important and those that are not. Only experience in the real world with real problems can provide that ingredient.

Chapter 1 / Sets: Concepts and Applications

1.1 INTRODUCTION

Basic to the methods and techniques of objective analysis is the Theory of Sets. In this chapter sets will be discussed from the viewpoint of a nonscientific substantive analyst whose interest lies in the applications of sets to the field of information processing and analysis.

The meanings of the term "set," and its synonyms–collection, aggregate, class, ensemble, or family–are frequently used but seldom questioned by nonscientifically trained analysts. If we look up the meaning of the above synonyms in a dictionary, we find statements such as "an accumulation of objects," "a group constituting an organic whole...," or "a number of things of the same kind...." All of these definitions are only descriptions and merely add to our intuitive or normative understanding of the concept of a collection or set.

To avoid problems of definition which do not interest applications-oriented readers, the notion of a set will be accepted as an intuitive term at this time. With experience in the applications of sets the set concept will take on new meaning just as the terms "point" and "line" (undefined terms in geometry) assume meaning through usage in the study and application of geometry.

This chapter demonstrates how the concept of sets and their proper use in nonscientific fields can be a useful aid to the professional analyst who wishes to analyze information and to present his findings in a clear, concise, and unambiguous manner.

1.2 SET CONCEPTS

The terms "set," "collection," "aggregate," "ensemble," "class," "family," occur in common everyday usage. Automobiles are divided into sets by year, make, model, price, or color. Airplanes are divided into classes on the basis of payload, range, manufacturer, country of origin, etc. Ethnic groups and countries are aggregated as separate entities, and nations are classified as friend, foe, or neutral. When communicating with others, discussions are classified by focusing our attention to a specific ensemble (or set) of items. For purposes of clarity, discussions are restricted to classes of items having the same characteristics which allow them to be considered as parts of the same or different sets. Efficient and systematic use of set concepts introduced in the next section helps reduce the amount to be written and establishes singularity of meaning in the written statements or the characteristics under consideration.

1.3 SET NOTATION

In working with sets, the following notation convention is typical and will be employed in this text. Lowercase letters $(a,b,c,\ldots)$ will be used to denote elements* of sets. Uppercase English $(A,B,C,\ldots)$ and Greek $(\mathrm{A},\mathrm{B},\Gamma,\Delta,\ldots)$ letters will be used to designate sets. The special symbol $\in$ will be used to express the notion "is an element (member) of." A slash through the symbol ($\notin$) denotes "is not an element (member) of."

Example 1.1 (Notation). Information Storage and Retrieval Systems (IS&R) are important applications in the field of information processing and analysis. A major problem which arises in IS&R is that of retrieving only those documents containing subject matters of interest. This problem reduces to identifying a "set" of documents which satisfies the need of the requester. Documents are identified by properties which include author, title, publisher, or subject content. As will be seen in later examples on IS&R, documents are identified mainly by index terms. Indexing creates concise descriptions through which documents of like characteristics can be classified into sets for retrieval. If we were interested in identifying all textbooks which contain information on the theory of sets, then we could identify many such texts by listing all textbooks which have the words "set theory" in their title. The resulting list would identify a set designated by the symbol S, and a textbook x would be an element of the set S, i.e.,

$$x \in S$$

if x is a textbook which possessed the words "set theory" in its title.■■

*The terms "elements" or "members" as used in set theory are the generic terms of the "things" that belong to a set. We call the elements by proper names (diplomat, chinese national, tank) when it is ascertained that the elements possess identifiable properties which earn them the right to be so named.

Developing this idea a little further, suppose that from a set of textbooks R we request specific textbooks through the index term I. The index term I separates the set of textbooks R into two sets or classes of textbooks. One set of textbooks, S, is the set addressed and recalled by index term I; the other is the set of textbooks T not addressed by index term I. As we will see in the discussions which follow, indexing for retrieval purposes is a set-forming process. Knowledge of the operations on sets, as they apply to indexing, is of considerable value to both the requester and the indexer. Through the set approach to indexing and retrieving of documents, the requester can state his document needs to coincide with the indexer's interpretation of his needs, resulting in more efficient and relevant IS&R systems. These points will be clarified in the discussions which follow.

Exercise 1.1

(a) Suppose that you had misplaced one of your textbooks. How would you describe the book in a lost and found ad?

(b) Does your description of the misplaced textbook remove any possibility of having the wrong textbook returned to you?

(c) Suppose that you forget the color, author, or title of the textbook. Could you describe the content of the book in a simple, concise manner and still not have the wrong textbook returned to you?

(d) Discuss the above problem with the change that you are attempting to find a book you require in a library. How would you inform the librarian of your needs? ■■

Life becomes a little more complex when it is necessary to define sets whose elements have several properties. For example, we may wish to review all textbooks having the terms "set," "theory," *or* "set theory" in their titles. Obviously, the resulting set of textbooks is a larger set than set S discussed in Example 1.1. To handle this situation effectively we need to introduce additional set definitions and notations.

1.4 SPECIFYING SETS

Sets can be specified in the following two commonly used ways:

1. By *tabulation* or *listing* by names, e.g., John, Mary, etc.
2. By *description* of all properties which any object must have in order to be included as a member of the set. This method is also called the "*rule*" method, i.e., a rule for specifying which elements can be

selected as members of a set. A set S defined by description or rule P is expressed by the following notation:

$$S = \{x | x \text{ has property } P\}$$

This notation is read "S is the set of all elements x such as that x has property or follows rule P."

The following examples illustrate the specification of sets by listing and description (or rule).

Example 1.2 (Listing, Description).

1. The set S consisting of all vowels of the English language
   $S = \{a,e,i,o,u\}$ (listing)
   $S = \{x | x \text{ is an English vowel}\}$ (description)

   The description reads "S is the set of all x such that x is an English vowel."

2. The set M of American Mothers
   $M = \{\text{Mary Smith, Joan Dyer, ..., Toni Zype}\}$ (listing)
   $M = \{x | x \text{ is an American Mother}\}$ (description)

■■

Note that the complete listing of M would involve the writing of approximately 50 million names. Hence the need for set definition by description.

1.5 SET DEFINITIONS AND RELATIONS

First, we consider the notion of a set of elements *being contained* in another set. We say that a set A is a *subset* of or is *contained* in a set B if every element of A is also an element of B. The relation is expressed symbolically as $\subset$:

$$A \subset B \quad \text{or} \quad B \supset A$$

which is read A is contained in B or B contains A. Other terms in common usage which are synonyms for subset but which are normally used to indicate special types of subsets are:

Subdivision:	For a neighborhood, organization, etc.
Task Force:	Groups of people, equipment, etc., on special assignment; part of a larger organization.
Subcontract:	Part of the total contractual effort.
Directorate, Agency, Department:	Subsets of a larger organization.

Another way of expressing this relationship is: $A \subset B$ means that whenever x is a member of A $(x \in A)$, then x is also a member of B $(x \in B)$.

From the definition, $A \subset A$ for every set A; i.e., every set is a subset itself. It is important to know when considering subsets whether the subset is or is not the whole set. We want to know, when A is a subset of B, whether there are elements of B that are not elements of its subset A; i.e., $x \in B$ but $x \notin A$. When B contains elements which are not contained in A, then A is called a proper subset of B. We denote this relationship by

$$A \subsetneqq B$$

i.e., the set A is contained in B but is not equal to B. If a set (or subset) contains no elements, it is called an *empty* set (or subset).

A family consisting of a mother, father, and children is a well-defined set. If both parents are living, then two *proper* subsets can be defined; the subset of females, and the subset of males. If the father dies and there are no male children, then the subset of females is no longer a *proper* subset of the family. Also the subset of males becomes an *empty* set.

Defining Sets by Description

Next, we indicate how sets are defined by describing the properties its elements must possess in order to qualify for membership in the set.

The method used in forming sets can be expressed in a simple, concise manner. The symbol $\{\,|\,\}$ is used to define sets as follows:

$$A = \{x | x \text{ has the property } P\}$$

We would read the above as "A is the set of all things or elements x such that x has property P." The braces $\{\ \}$ are used to enclose the listing of all elements defined by the statement contained in the braces. The symbol $|$ may be read "such that."

For example, a Russian soldier s is a member of the Russian Army A $(s \in A)$. If N is used to denote the Russian Navy, the fact that the soldier is not a member of the Navy is denoted $s \notin N$. If M stands for the set of all Russian military personnel $[M = \{x | x \text{ is a Russian military service man}\}]$, then both Army A and Navy N are proper subsets of M, i.e.,

$$(A \subsetneqq M, N \subsetneqq M)$$

M contains elements (Naval and Air Force personnel) which are not part of the Army, and there are Army and Air Force personnel which are not part of the Navy.

When defining sets by property or description, the property or description must be clear, concise, and not subject to misinterpretation. For example, the set,

$$T = \{x | x \text{ is a tall male}\}$$

is not a well-defined set. It is not clear what the word *tall* means in terms of height. To pygmies, the word *tall* may mean 5 ft. 8 in., to a basketball player, the word *tall* may mean 6 ft. 10 in. However, if the set is redefined as

$$T = \{x | x \text{ is a male over 6 ft. tall}\}$$

then the ambiguity is removed. Briefly, in defining sets by property or description, avoid the use of subjective terms such as some, tall, many, few, heavy, wide, etc. Make a concerted effort to express in unambiguous terms the property or characteristic required to gain membership in the set.

Universal Set

Normally it is advisable to limit discussion to members of a particular set of elements; e.g., the set T of Russian tanks, the set A of Russian aircraft, or the set B of Russian aircraft operating in East Germany. When a set of elements under discussion is restricted in membership by prior agreement, then the set of elements to which discourse is limited is called the *universal set.* For example, the set of all Russian military men can be considered as a universal set. Subsets of this universal set can be identified as military men stationed in East Germany and attached to a numbered army. Other examples of universal sets are: (*a*) the set of all Russian soldiers, (*b*) the set of all U.S. battleships, and (*c*) the set of all \$100 bills. Subsets of these universal sets are, respectively: (*a*) the set of all Russian Army officers, (*b*) the New Jersey and the Missouri, and (*c*) the set of \$100 bills with prespecified serial numbers.

Venn Diagram

A pictorial device which is used to describe universal sets and their subsets is the Venn diagram (sometimes called an Euler diagram). In a Venn diagram a rectangle and its interior represent the elements of a (universal) set S. The subsets of S are then represented by sets of distinct points or areas interior to the rectangle. For example in Figure 1.1, the rectangle S and its interior constitute the universal set S. Subsets of S are illustrated as circles A and B and the areas enclosed by the circles.

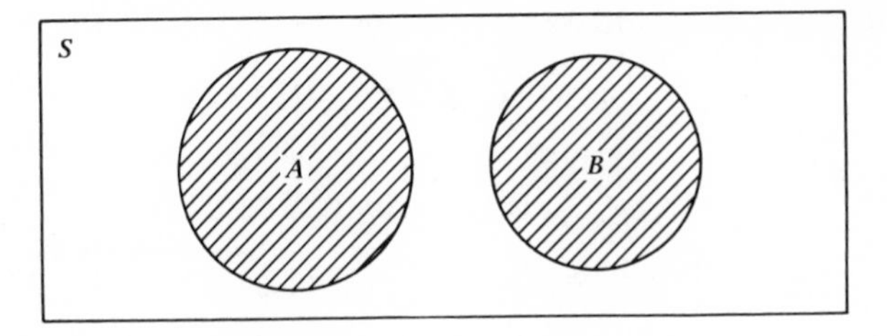

Figure 1.1 Venn diagram.

Another example is the universal set S consisting of the nine points contained in the rectangle illustrated in Figure 1.2. Subsets of S are rectangles A and B containing six and four points, respectively. Note that the subsets A and B of S are proper subsets of S. In each instance, S contains points (or elements) not contained in A or B.

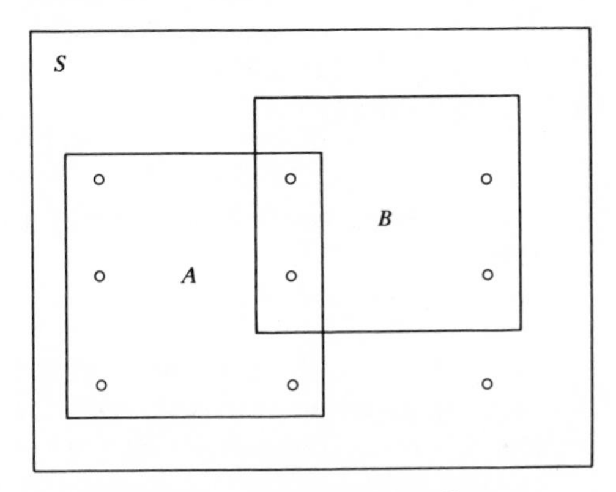

Figure 1.2 Discrete Venn diagram.

Exercise 1.2. Consider the following four sets defined by listing and describing their elements:

A: Contains the tanks with serial numbers 0, 1, 2, 3, 4, 5, 6, 7, 8, 9. (listing)

$A = \{x|x$ is a tank and x has serial number y, $0 \leqslant y \leqslant 9$, y is an integer$\}$ (description)

The symbol $\leqslant$ reads "is less than or equal to." Thus y is a serial number which takes on any integral value from 0 to 9.

B: Contains the tanks with serial numbers 0, 2, 4, 6, 8. (listing)

$B = \{x|x$ is a tank and x has serial number y, $0 \leqslant y \leqslant 9$, y is an even integer$\}$ (description)

C: Contains the tanks with serial numbers 1, 3, 5, 7, 9. (listing)

$C = \{x|x$ is a tank and x has serial number y, $0 \leqslant y \leqslant 9$, y is an odd integer$\}$ (description)

D: Contains the tanks with serial numbers 1, 2, 4, 5, 7, 8. (listing)

(no description possible)

(a) Is B a subset of A?
(b) Is B a proper subset of A?
(c) Do B and C have any elements in common?
(d) Does D have any elements in common with C?
(e) Is there any element of A which is not an element of B or C?
(f) What elements of A are not elements of D? ▪▪

Identical Sets

Two sets A and B having the same elements are called *identical sets*. We express this by the equality sign

$$A = B$$

Saying that two sets A and B are *identical* is equivalent to saying that the two sets contain each other, i.e.,

$$A \subset B \quad \text{and} \quad B \subset A$$

Note that when two sets are defined by description (identification of properties) the two sets are identical only with respect to the properties defining them. For example, if two sets of tanks A and B are described by

$$A = \{x|x \text{ is a tank}\}$$
$$B = \{y|y \text{ is a tank}\}$$

then by the set definition, the two sets A and B are identical ($A = B$). If we wish to be more precise and express the fact that A may be composed of light tanks and B of medium and heavy tanks, then these differences must enter into consideration. We thus change the definition of A and B to read

$$A = \{x|x \text{ is a light tank}\}$$
$$B = \{y|y \text{ is a medium or heavy tank}\}$$

It then becomes obvious that even though A and B might possibly contain the same number of elements, they would not under these defining properties be considered equal.

Null Set

A set having no elements is called the *empty* or *null* set. The empty set is usually denoted by 0 or $\emptyset$. In any collection of sets there can be at most one empty set. The empty set is a subset of every set. However, the empty set is not an *element* of every set. For example, the set consisting of all *subsets* (8) of three books, B_1, B_2, and B_3 has as its elements the subsets $\{B_1\}$, $\{B_2\}$, $\{B_3\}$, $\{B_1,B_2\}$, $\{B_1,B_3\}$, $\{B_2,B_3\}$, $\{B_1,B_2,B_3\}$, and $\emptyset$, and hence contains the empty subset (the subset containing no books) as an element. The set whose elements consist of three books B_1, B_2, and B_3 does not contain the empty subset as an element. As interesting distinction exists between the empty set and the numerical value of 0. To illustrate the difference between the empty set and the numerical concept of 0, consider the difference between receiving the grade of 0 (an element of value 0) for a submitted exam against receiving no grade (the null set) for not submitting an exam.

Set Properties

Before proceeding to the discussion of set properties and operations, it will be instructive to review how sets defined by listing or description are employed in a specific application. The following example and exercise utilizes sets of textbooks identified by key words in their titles, and shows how the retrieval of such sets of textbooks is dependent on one's knowledge of how the sets were originally identified.

Example 1.3 (IS&R Application). Let the set S consist of six textbooks labeled a through f and titled as follows:

a: *A System of Axiomatic Set Theory*

b: *A Set of Postulates for Plane Geometry*

c: *Notions on Orders of Sets*

d: *The Theory of Axiom-Systems*

e: *Partially Ordered Sets*

f: *A Set of Postulates for the Theory of Betweenness*

The set S can be defined as consisting of those texts which have the property that they are labeled a through f. In set notation this would be

$$S = \{x | x \text{ has label } a \text{ through } f\}$$

From the set S let us arbitrarily define subsets A, B, C, D, and E of S as follows:

$A = \{x|x$ has the word "set" in its title$\}$
$B = \{x|x$ has the word "theory" in its title$\}$
$C = \{x|x$ has the words "theory" *or* "set" in its title$\}$
$D = \{x|x$ has the words "theory" *and* "set" in its title$\}$
$E = \{x|x$ has the word "postulate" in its title$\}$

where it is understood that x is a member of the universal set S of six textbooks.

In terms of the text labels, the sets A, B, C, D, and E are given by

$A = \{a, b, c, e, f\}$
$B = \{a, d, f\}$
$C = \{a, b, c, d, e, f\} = S$
$D = \{a, f\}$
$E = \{b, f\}$

The Venn diagram for the sets A-E is given in Figure 1.3. In the Venn diagram only the points labeled a-f are in the set S. Other areas in the rectangle S are not to be considered as part of S or the subsets of S. ■■

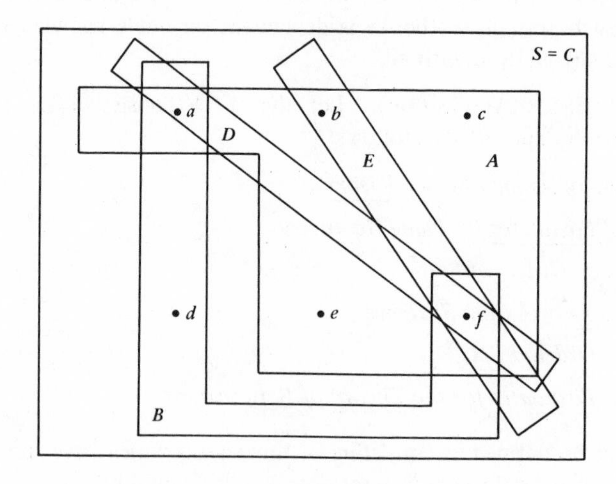

Figure 1.3 Venn diagram of document sets.

Exercise 1.3. Using the set S and the subsets A, B, C, D, and E defined in Example 1.3, answer the following questions:

(a) Does B contain D as a proper subset?

(b) Is C a subset of S?

(c) Is C a proper subset of S?

(d) How is subset E related to A?

(e) How is subset A related to C?

(f) What are the textbooks of the subset G of S?
$G = \{x|x$ has "order" *or* "axiom" in its title$\}$

(g) What are the textbooks of the subset H of S?
$H = \{x|x$ does *not* have "theory" in its title$\}$ ■■

The number of properties used to identify the elements of sets and how these properties are *conjoined* determine the "size" of, or the number of elements in, the resulting set. For example, the sets A and B were defined to contain textbooks which possess only one property. The textbooks of the sets C and D are defined in terms of two properties. The textbooks of D, by definition, were required to have both properties and resulted in set D having fewer textbooks than sets A, B, or C. More will be said about the conjoining operation as it applies to selecting elements of sets after we define the fundamental conjoining operations and illustrate their application to specific sets.

Example 1.4 (Intersection). In Example 1.3 and Exercise 1.3 we employed subsets of a set of three terms (set, theory, postulate) to select textbooks from the set of six textbooks. A natural question to ask here is as follows: If we use one of the three identifying terms, say theory, to select the texts in which we may have interest, how many textbooks would we be selecting? Going back to Example 1.3, we would note that we would be selecting three textbooks identified by the set $B = \{a,d,f\}$, i.e., we have determined that the titles of the textbooks of set B have the word "theory" in common. Further, no other textbook in the list of six has the word "theory" in its title.

In discovering that three textbooks are retrieved (selected through employing the word "theory"), we might decide three textbooks are too many to study or peruse. We may also reconsider and determine that our primary interest lies in set theory, and not in just the theory of anything. With this enlightenment we may decide to use the two terms set and theory which more closely identifies our interest. However, on closer consideration it becomes apparent that we must question what happens in our selection process (retrieval) when we assert that *either or both* terms "set" and "theory" appear in the textbook titles, or when we insist that *both* terms must appear in the titles. In the former, the set C of Example 1.3 is selected. In the latter, set D is selected. It is obvious that to reduce the number of textbooks to be considered, use of both terms is preferable to use of either or both terms. The following discussion on set intersections and unions highlights the "selection" properties of the operations "union" and "intersection" as they apply to the selection of elements from sets which may or may not have elements in common.■■

When only two sets are involved in the intersection, the symbols $\cdot$ or $\cap$ are most frequently used. When more than two sets are involved, it is permissible to use the same notation for their intersection, i.e.,

$$A_1 \cap A_2 \cap \cdots \cap A_n \quad \text{or} \quad A_1 \cdot A_2 \cdots A_n$$

This notation proves to be unwieldy in practice. A simplified notation will now be described. Let I be the set of integers, called an *index set,* defined as follows:

$$I = \{i | i \text{ is a positive integer} \leqslant n\} \qquad \text{(description)}$$

or

$$I = \{1,2,3,\ldots,n\} \qquad \text{(listing)}$$

Using the symbols $\cap$ or $\top$ to denote a continued intersection of sets, either the notation

$$A_1 \cap A_2 \cap \cdots \cap A_n = \bigcap_{i \in I} A_i$$

or

$$A_1 \cdot A_2 \cdots A_n = \top_{i \in I} A_i$$

can be employed in place of the longhand notation for the intersection of n sets.

While the definition of intersection appears to be formal and precise, it is used in this same precise manner (as is its synonyms) in ordinary and intuitive discourse. For example, we use the term "overlap" to identify an area common to two items. We have a common understanding of intersection when applied to the meeting of two streets or roadways. We also talk of "joining" two edges or the conjunction (overlap, intersection) of events in space or time or both. Briefly, the concept of the intersection operation is a simple and basic one which conforms with the experience and intuition of most educated people.

Example 1.5 (Intersection).

1. Consider the three sets A_1, A_2, and A_3. The index set is $I = \{i | i = 1,2,3\}$. The intersection of the three sets is notationally given by

$$A_1 \cap A_2 \cap A_3 \quad \text{or} \quad \prod_{i \in I} A_i \quad \text{or} \quad \bigcap_{i \in I} A_i$$

The intersection can be illustrated by means of the Venn diagram in Figure 1.4.

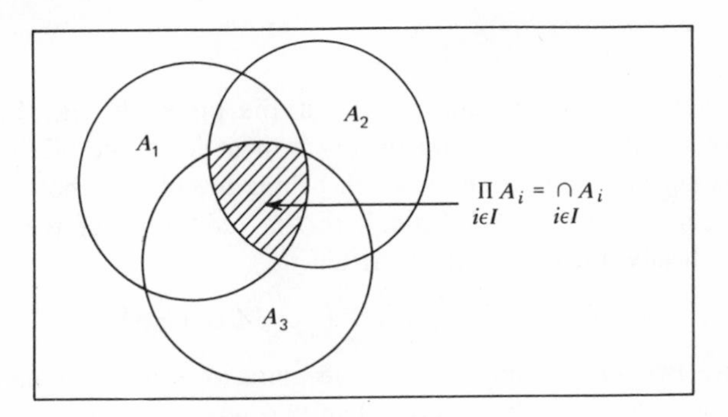

Figure 1.4 Venn diagram of intersection of three sets.

2. From Example 1.3 and Figure 1.3 on the indexing of textbooks, the intersection of some of the subsets of the set S can be written:

$$A \cdot B = A \cap B = \{a,f\} = D$$
$$D \cap E = \{a,f\} \cap \{b,f\} = \{f\}$$
$$A \cap B \cap C = \{a,f\} = D$$
$$C \cap D \cap E = \{f\}$$

3. The concept of set intersection arises in many interesting ways in applications as is illustrated in the following example:

let A = the set of Russian light tanks
let B = the set of Russian medium tanks
let C = the set of Russian heavy tanks

Note the tanks are classified according to weight. Each tank class is well-defined if the classes light, medium, and heavy are well-defined. However, the analyst may be asked the questions

i. What Russian tanks carry 20 mm cannons?
ii. What Russian tanks require a personnel complement of three or more?

Note that to answer these questions the analyst seeks specific intersections of the three sets A, B, and C defined by the properties "carry

20 mm cannons," "crew of three or more." Only specific subclasses or subsets of A, B, and C may qualify. Those subsets that do appear in the intersection form the sets which answers affirmatively the questions asked. ■■

Set Union

Another notion in set theory is that of the union of sets. Elements of the union of a number of sets are members of at least one of the sets. The set C, consisting of those elements which belong to either or both of two sets A and B is called the "union," "sum," or the "addition" of the two sets A and B. Symbolically, the definition is written

$$C = A + B = A \cup B = \{x | x \in A \text{ or } x \in B\}$$

When only two sets are involved in the union or sum, the symbols + or $\cup$ are most frequently used. When more than two sets are involved, the notation for the union is

$$A_1 \cup A_2 \cup \cdots \cup A_n \quad \text{or} \quad A_1 + A_2 + \cdots + A_n$$

This notation is further simplified by use of the symbols: Σ and $\cup$ to denote the union of more than two sets as below:

$$\sum_{i \in I} A_i = \bigcup_{i \in I} A_i = A_1 + A_2 + \cdots + A_n$$

where I denotes the index set $I = \{1,2,...,n\}$. Figure 1.5 is a Venn diagram for the union of six sets $A_1, A_2, ..., A_6$.

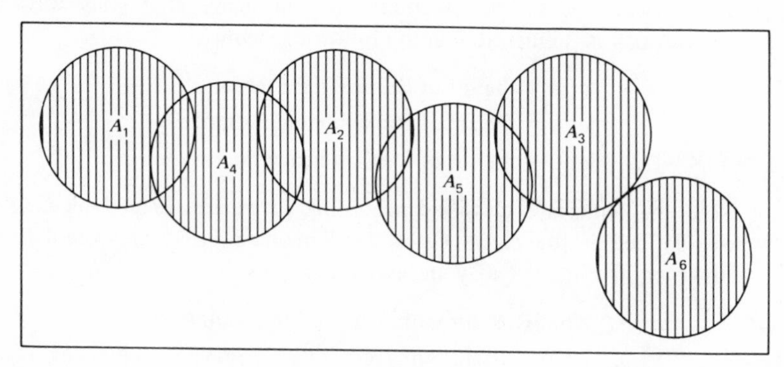

Figure 1.5 Venn diagram of union of six sets.

As with the set operation intersection, the set operation union is used in ordinary discourse in an intuitive yet precise manner. For example, we use the term "union" in the sense of joining two or more things into one. We speak of the combination or coalition of separate political entities or nations. And "unions" of workers are formed to provide strength in bargaining power. Another term synonymous with unions is confederation where the confederation of independent nations or persons means the bringing together for some common purpose.

In terms of the set S of textbooks, the sets

$$D = \{x | x \text{ has "theory" and "set" in title}\}$$
$$E = \{x | x \text{ has "postulate" in title}\}$$

has a union defined as

$$\begin{aligned} L &= D \cup E \\ &= \{x | x \text{ has "theory" and "set," or "postulate" in title}\} \\ &= \{a,f\} \cup \{b,f\} \\ &= \{a,b,f\} \end{aligned}$$

A word on common usage of the symbols $\cap$, $\cup$, $\cdot$, and $+$. When the symbols $\cap$ and $\cup$ are used, it is clear that the operations are applied to sets. For example, if the union $A \cup B$ or intersection $A \cap B$ is used in text, it is clear that A and B are both sets. (In logic the symbols are used to denote the connectives "and," "or.") If the symbols $A + B$ and $A \cdot B$ are used in text, it is not clear whether A and B are sets or numbers. In either case the meanings of $A + B$ and $A \cdot B$ should be clear from the context of the statements. One should be able to determine whether the symbols $+$, $\cdot$ are applied to sets or numbers. Since both pairs of symbols $(+, \cdot)$; $(\cup, \cap)$ are in common usage, readers must be prepared to use either and accept with understanding the confusion caused by their interchangeability.

Set Difference

The *difference* of two sets A and B, denoted $A - B$, is defined as the set of elements belonging to A but not to B. Symbolically,

$$A - B = \{x | x \in A \text{ and } x \notin B\}$$

and

$$B - A = \{x | x \in B \text{ and } x \notin A\}$$

Again, the negative sign $-$ can mean the difference of two sets or of two numbers. As previously stated, the meaning will normally be clear from the context of the discussion.

Example 1.6 (Union of Sets). In terms of the subsets A, B, C, D, E defined previously, we can define a number of sets obtained from the unions or differences of the subsets taken in pairs. Six such sets are as follows:

$$D + E = D \cup E = \{a, b, f\} \qquad D - E = \{a\}$$
$$B + D = B \cup D = \{a, d, f\} = B \qquad B - D = \{d\}$$
$$B + E = B \cup E = \{a, b, d, f\} \qquad B - E = \{a, d\}$$

Figure 1.6 provides a graphical illustration of the union, difference, and intersection of sets. The two sets A and B can be represented by the plane figures given in Figure 1.6. The intersection $A \cap B$ is represented by the area which contains *both* the vertical and horizontal lines. The union $A \cup B$ is represented by the area which contains the vertical *or* the horizontal lines. The difference $A - B$ is represented by the area which contains vertical lines only. The difference $B - A$ is represented by the area which contains horizontal lines only. Normally, the difference operator $-$ is used to delimit a set to a specific class or subset of elements or properties requiring closer attention. The term "but not" is a close approximation to the meaning of the difference operator. For example, we may say that we are interested in studying the elements or properties of the set and we want to count the elements of A *but not* the elements of B. These statements can be expressed as

$$A - B \quad \text{or} \quad A - A \cap B$$

In the first instance $A - B$, if B has no elements or properties in common with A, then restricting one's attention to A avoids consideration of elements of B.

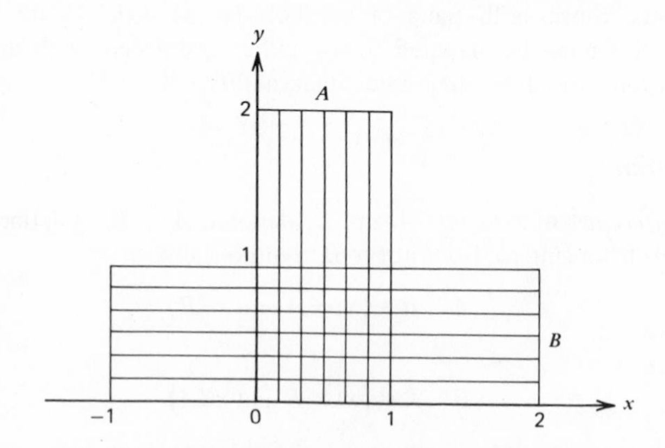

Figure 1.6 Venn diagram of the union and difference of sets where A is the vertical rectangle and B is the horizontal rectangle.

However, if A and B have elements in common, then the common elements of A and B, given by $A \cap B$, must be removed from consideration. Hence the notation

$$A - A \cap B$$

can be read "A, but not those elements of B which are also elements of A." ■■

Disjoint Sets

When two sets A and B have no elements in common, we can express this fact as $A \cap B = \emptyset$. When the intersection of two sets is empty (null set), the two sets are said to be *disjoint* or *mutually exclusive.*

Other words and phrases in common usage which have essentially the same meaning as the words disjoint and mutually exclusive are "separate," "apart," "free from each other," "detached," and "different." Can you list others?

Indexing a set of documents D in terms of their content or other characteristic is an operation which partitions the collection of documents D into a set of subsets. When a set of documents D is indexed by one term I, two subsets are formed. One subset D_1 consists of those documents of D which are labeled by the index term I. The remaining subset D_2 consists of those documents which are not labeled by the index term I. The subsets D_1 and D_2 partition D into mutually exclusive subsets.

Example 1.7 (Disjoint Sets). Consider the set S of six texts previously listed. If B denotes the set of texts of S which have the word "theory" in their title, then

$$B = \{a, d, f\}$$

The subset B_1 of texts not having the word "theory" in their title is disjoint from the set B, and is given by

$$B_1 = \{b, c, e\}$$

Note that

$$S = B \cup B_1 \qquad B = S - B_1$$
$$B_1 = S - B \qquad \emptyset = B \cap B_1$$

Set Complement

When a set D is partitioned into two mutually exclusive sets, D_1 and D_2, as in Figure 1.7, the following relationships hold:

$$D = D_1 \cup D_2 \qquad D_2 = D - D_1$$
$$D_1 = D - D_2 \qquad \emptyset = D_1 \cap D_2$$

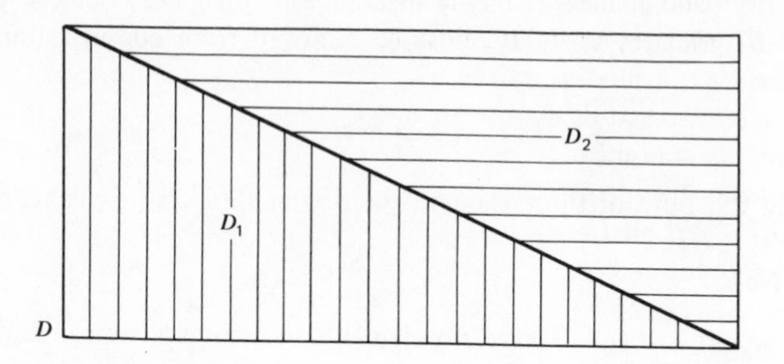

Figure 1.7 Venn diagram of the union, difference, and complement of sets.

In general, for any sets D, D_1, and D_2 for which the above relations hold, the set D_1 is called the *complement* of D_2 relative to D. Notationally, this relationship is denoted

$$D_1 = \bar{D}_2$$

and is illustrated in Figure 1.7. The set D_2, similarly is called the complement of D_1 relative to D, i.e., $D_2 = \bar{D}_1$.

Whenever a set of elements is under consideration, the elements of the set are usually selected from a larger class or set of elements. For example, specific classes of people (white, black, male, female, professionals, etc.) are selected from the class or set of all people. Light, medium, or heavy tanks are selected from the set of all tanks. The set which contains all elements under consideration as previously indicated, is called the *universal* set U. Thus every set is defined relative to a universal set. The identity of the universal set is usually specified or implied at the beginning of discourse. When considering the complement of a set, it is necessary to have a precise definition of the universal set relative to which the complement is taken.

The complement of a set A is the set of elements in the universe which are excluded from A. For example, if A is given by

$$A = \{X | X \text{ is a man and } X \text{ is over 6 ft. tall}\}$$

then the complement of A, denoted $\bar{A}$, relative to the universe of human beings, is the set of all women and the set of all men under 6 ft. tall. Thus, if a set A is determined by having its elements possess property P, then the complement of A is the set of elements which do not have property P. Any set A and its complement $\bar{A}$ are disjoint. Their union $A \cup \bar{A}$ forms the universe of elements under discussion.

Example 1.8 (Complement and Difference Set). Let S be the set of radio and radar transmitters operating in a given area. Let D be the set of receivers assigned the task of monitoring the output of the transmitters of S. Receivers in D are to be selected so as to cover specific portions of the frequency spectrum utilized by the transmitters of S. The receivers of D can be selected so that the frequencies they cover form a mutually exclusive set. For example, one receiver D_1 covers all frequencies below 10 MHz (MHz = megahertz or million cycles per second) and a second receiver D_2 covers frequencies above 10 MHz. Of course, in practice more receivers are normally utilized, each covering a narrower frequency range. For the set D of two classes of receivers D_1 and D_2, the partition is

$$D_1 = \{x | x \text{ covers frequencies below 10 MHz}\}$$
$$D_2 = \{y | y \text{ covers frequencies equal to or above 10 MHz}\}$$

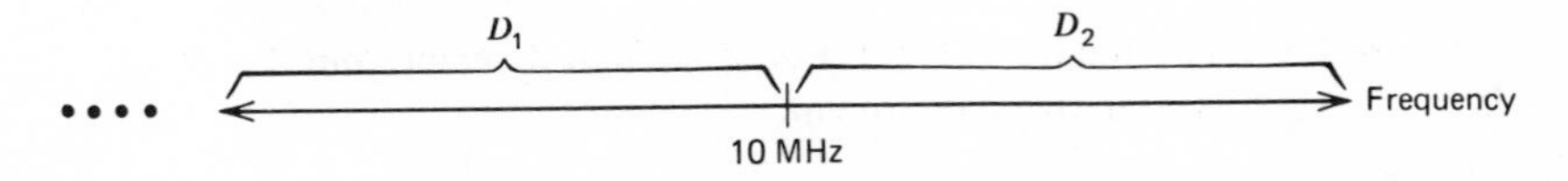

where $D = D_1 \cup D_2$ and D_1 and D_2 are complements of each other with $D_1 \cap D_2 = \emptyset$, $D_2 = D - D_1$ and $D_1 = D - D_2$.

The receivers can be selected to have overlapping or intersecting frequency bands. For example, to get complete coverage, a set of receivers $R = \{R_1, R_2, R_3, R_4\}$ can be selected as follows:

R_1: 30 kHz to 30 MHz
R_2: 20-400 MHz
R_3: 300-8000 MHz
R_4: 6000 MHz and above

For the set of receivers R, there is no disjoint partition. Note that coverage by R_1 and R_2 overlap in the frequency range 20-30 MHz. Hence dual coverage is provided in that frequency range. As an exercise, identify the other frequency ranges wherein dual coverage is provided. The diagram below should help in identifying frequency ranges of dual coverage:

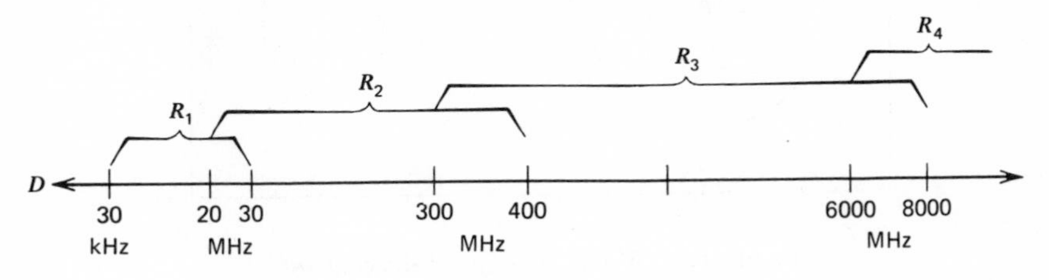

Formally the definition for the *complement* of a set A (relative to D) is

$$\bar{A} = \{x | x \in D \text{ and } x \notin A\}$$

where the set D is the set of all elements under consideration. Again, using a Venn diagram, the complement of A relative to D is illustrated in Figure 1.8.

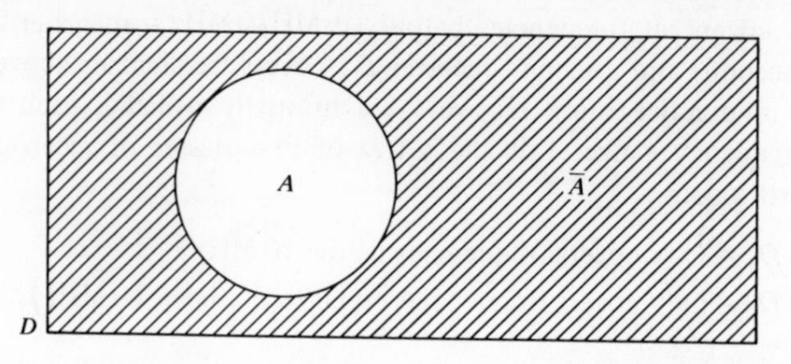

Figure 1.8 The relation $D = \bar{A} + A$ is a Venn diagram, and $\bar{A} = D - A$ is a relative component.

The following example uses a Venn diagram to illustrate the set operations of difference and complement when two sets are disjoint or overlap or when one set is contained in the other.

Example 1.9 (Complement and Difference). In Figure 1.9 the sets B and C are disjoint, i.e., $B \cap C = \emptyset$. The two set differences are $B - C = B$ and $C - B = C$. Note that $A - B \supset C$ and $A - C \supset B$.

The set A in Figure 1.9 can be identified as all U.S. personnel in Viet Nam. The set B can represent all Army personnel, and the set C can represent all Navy personnel. The remaining U.S. personnel (civilian contractors, State Department, Air Force, etc.) are represented by $A - (B \cup C)$.

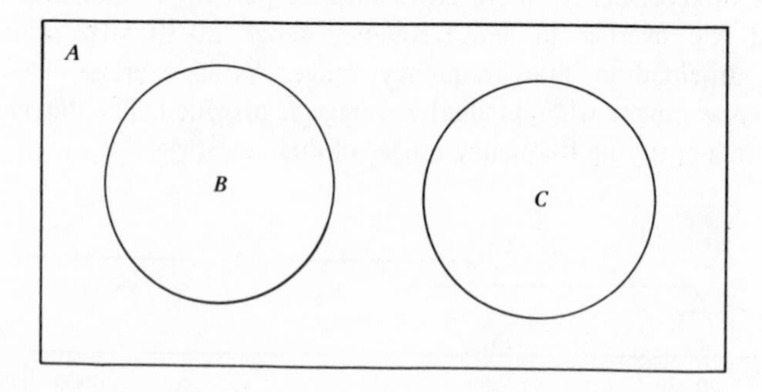

Figure 1.9 Venn diagram of disjoint sets.

In Figure 1.10 the sets B and C are not disjoint, i.e., $B \cap C \neq \emptyset$. The difference $B - C$ is the portion of B covered with horizontal lines. The difference $C - B$ is the portion of C covered with vertical lines. The differences $A - B$ and $A - C$ do not contain B and C, respectively, in their entirety.

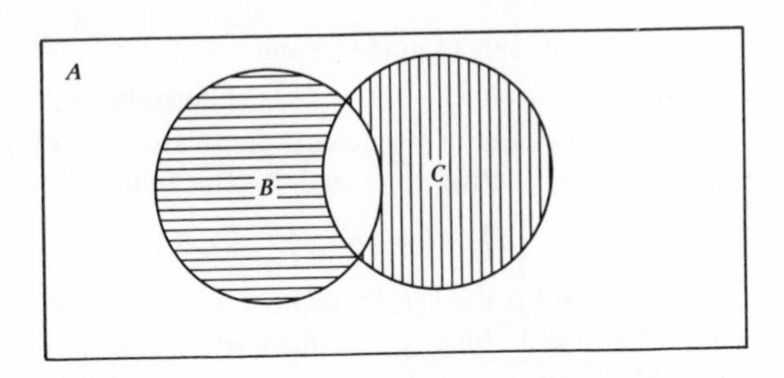

Figure 1.10 Venn diagram of set analysis.

In Figure 1.10 the set A can represent all Army personnel, the set B can represent all Army personnel on active military duty, and the set C can represent all Army military reservists. The intersection $B \cap C$ hence represents all Army military reservists on active duty. The set $A - (B \cap C)$ represents all Army personnel except reservists on active duty. Following the same line of reasoning, Army civilian employees, advisors, etc., are in the set $A - (B \cup C)$.

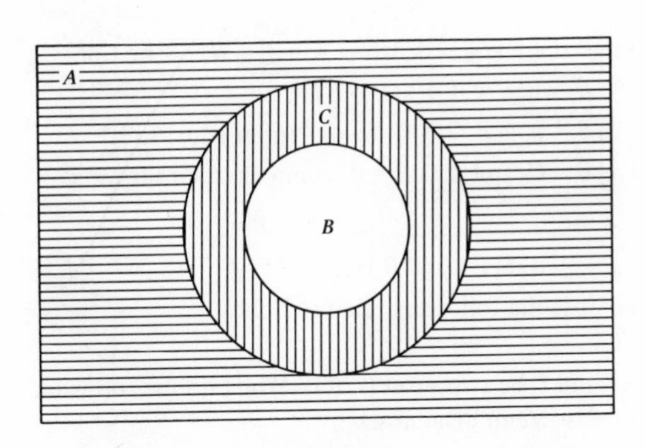

Figure 1.11 Venn diagram of set analysis.

In Figure 1.11 the sets B and C are not disjoint. In fact, $B \subset C \subset A$ and $B \cap C = B$. The difference $A - C$ is the portion of A covered by horizontal lines; $C - B$ is the portion of C covered by vertical lines. Relative to Figure 1.11, let the sets A, B, C be defined as follows:

A = all Army personnel

B = all Commissioned Army officers of grade 04 and above

C = all Commissioned Army officers

The derived sets $C - B$ and $A - C$ then take on the following meanings:

$C - B$ = Army officers of grade 03 and below

$A - C$ = Army personnel who are not commissioned officers

Let S, A, and B be the sets of documents considered in Example 1.3. The set S is the universal set relative to A and B. The following relations hold:

$$\bar{A} = \{d\}, \quad \bar{B} = \{b,c,e\}, \quad \bar{S} = \emptyset$$

Note that the sets A and B used in Example 1.3 consist of those documents having the words "set" and "theory" in their titles, respectively. The intersection consists of those documents having *both* the words "set" and "theory" in their title. The resulting set (the intersection of A and B) is the set D. The intersection $D \cap E$ identifies the one document whose title contains the words "set," "postulate," and "theory" in its title. Note that the order of the three words does not enter into consideration. Thus it cannot be assumed that the document deals with set theory or the postulates for set theory. The only conclusions that can be drawn are that the document considers some "postulates," discusses "sets," and deals with "theory." ▪▪

Exercise 1.4.

(a) If A and B are any sets, when will $A - B = A$ be true?

(b) If $A \subset B$, when will $A \cap B = B$?

(c) If $A \cap B \neq \emptyset$, is $A \subset B$ ($\neq$ reads "is not equal to")?

(d) If $A \cup B = C$, and $A \cap B = \emptyset$, complete the following:

$C - A =$ ____________, and $C - B =$ ____________.

(e) For any two sets A and B is $A \cap B \subset A$?

(f) When does the following relation hold?
$A \subset (A - B)$

(g) Let A and B be two subsets of C. When do the following relations hold? (*Hint:* Use Venn diagrams.)

$(A - B) \cap C = A$ $\qquad$ $(A - B) \cap C = B - A$

$(A - B) \cap C = \emptyset$ $\qquad$ $(A - B) \cap C = A - B$ ▪▪

1.6 APPLICATIONS OF SET THEORY

Many practical problems can be solved with a knowledge of set operations and of the number of elements contained in any finite collection of finite sets.

Let X, Y, and Z be any three sets containing only a finite number of elements. Let $N(X)$, $N(Y)$, and $N(Z)$ denote the number of elements in X, Y, and Z, respectively.

First, let us consider the element counting relationship for any two sets X and Y.

$$N(X \cup Y) = N(X) + N(Y) - N(X \cap Y)$$

The relation states that the number of elements $N(X \cup Y)$ in the union $X \cup Y$ of two sets X and Y is equal to the sum of number of elements $N(X)$ in X and $N(Y)$ in Y, diminished by the number of elements $N(X \cap Y)$ in the intersection $X \cap Y$ of X and Y. The number $N(X \cap Y)$ is subtracted from the sum $N(X) + N(Y)$ when the two sets intersect, i.e., $X \cap Y \neq \emptyset$, because the elements in the intersection $X \cap Y$ are accounted for in both $N(X)$ and $N(Y)$. The element-counting relationship can be extended to any three sets X, Y, and Z. In the case of three sets, we have

$$N(X \cup Y \cup Z) = \\ N(X) + N(Y) + N(Z) - N(X \cap Y) - N(X \cap Z) - N(Y \cap Z) + N(X \cap Y \cap Z)$$

where again the last four terms account for duplication in counting elements which belong to two or more intersecting sets.

Example 1.10 (Counting Formula). Let us consider the set S as the lattice of points as depicted in Figure 1.12. Let the subsets A, B, and C be defined as the points of S contained in the rectangles labeled by A, B, or C. Note that $A \cup B \cup C \neq S$, i.e., one point of S is not in A, B, or C. By inspection we have

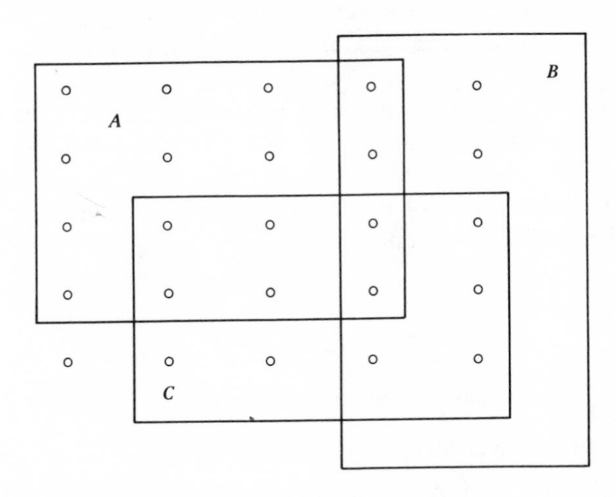

Figure 1.12 Venn diagram of counting formula analysis.

$$N(A) = 16 \qquad N(B) = 10 \qquad N(C) = 12$$
$$N(A \cap B) = 4 \qquad N(A \cap C) = 6 \qquad N(B \cap C) = 6$$
$$N(A \cap B \cap C) = 2$$

$$\begin{aligned} N(A \cup B \cup C) &= \\ &N(A) + N(B) + N(C) - N(A \cap B) - N(A \cap C) - N(B \cap C) + N(A \cap B \cap C) \\ &= 16 + 10 + 12 - 4 - 6 - 6 + 2 \\ &= 24 \end{aligned}$$

■■

The following example illustrates how sets and counting of set elements can be used in the discussion of effectiveness of IS&R.

Example 1.11 (IS&R Application). Let D represent the set of all documents indexed in an IS&R. Let A represent the set of documents that should be retrieved for a retrieval request if the retrieval system were perfect. Let B represent the set of documents actually retrieved by means of the retrieval request. The Venn diagram of the retrieval sets is

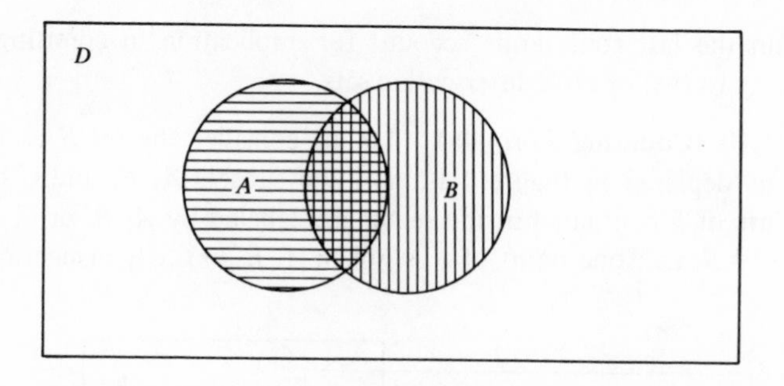

The set $A \cap B$ is a subset of the sets A (documents desired to be retrieved) and B (documents actually retrieved). If $N(A)$, $N(B)$, and $N(A \cap B)$ represent the number of documents in the sets A, B, and $A \cap B$, respectively, then a measure of the effectiveness of the indexing scheme for the retrieval system is given by the following two ratios:

$$\frac{N(A \cap B)}{N(B)} \quad \text{and} \quad \frac{N(A \cap B)}{N(A)}$$

The ratio $\dfrac{N(A \cap B)}{N(A)}$ relates the number $N(A \cap B)$ of desired or relevant documents retrieved (recalled) to the number $N(A)$ of documents in the system that should have been retrieved (recalled). Hence $\dfrac{N(A \cap B)}{N(A)}$ is called the *recall ratio.* It provides a measure of the effectiveness of the indexing system of the retrieval system in the recall of documents existing in the system which were intended to be retrieved by the requester.

The ratio $\frac{N(A \cap B)}{N(B)}$ relates the number $N(A \cap B)$ of desired or relevant documents retrieved to the total number $N(B)$ of documents actually retrieved. Hence $\frac{N(A \cap B)}{N(B)}$ is called the *precision ratio,* i.e., of the total number $N(B)$ of documents retrieved, what number $N(A \cap B)$ is actually desired by or relevant to the requester and his needs.

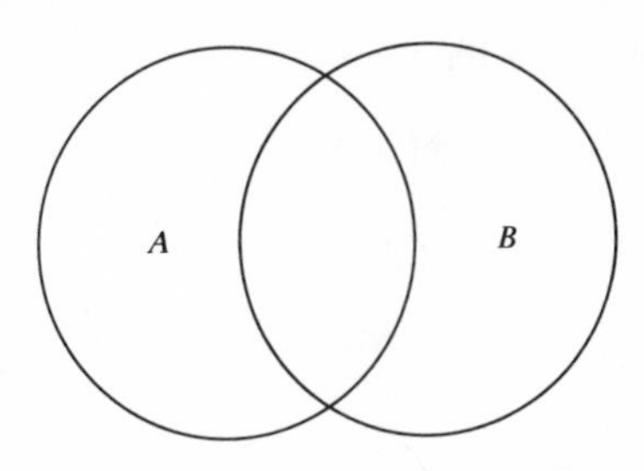

The set $A - A \cap B$ is the set of documents of interest to the user which were missed by his retrieval request.

$$N(A - A \cap B) = N(A) - N(A \cap B)$$

since $A \cap B \subset A$. Hence $N(A - A \cap B)$ represents the number of documents missed by the retrieval request.

The ratio $\frac{N(A - A \cap B)}{N(A)}$ is called the *miss ratio,* i.e., the ratio of the numbers of documents missed to the number of documents that a perfect request would have retrieved. In a similar manner, the ratio $\frac{N(B - A \cap B)}{N(B)}$ is called the *overage ratio* and is the ratio of the number of excess or irrelevant documents retrieved to the total number of documents retrieved.

In all four measures it is apparent that an objective of an IS&R system is to maximize the quantity $N(A \cap B)$ over all possible retrieval requests.■■

Next we will examine how the element-counting relationship of set theory can be applied to identify and isolate contradictions in the analysis of seemingly related and consistent data. The following example illustrates how a knowledge of set operations and the set-counting relation can be used to determine whether information is incomplete, inconsistent, or contradictory.

Example 1.12 (Information Analysis). An analyst attempts to determine the capability of three types of weapons denoted A, B, and C in the enemy's arsenal. The analyst reports that she surveyed 42 users and found

20 users considered weapon A as best
17 users considered weapon B as best
23 users considered weapon C as best
10 users considered weapons A and B as best
12 users considered weapons A and C as best
10 users considered weapons B and C as best
5 users considered all three as equals
7 users did not like any of the three

The analyst's supervisor used set theory to identify a significant discrepancy in the data. The supervisor drew Figure 1.13 and labeled the square as the set S of all sources contacted. He noted that

$$N(S) = 42$$

Continuing further he noted,

$$N(A) = 20 \qquad N(B) = 17 \qquad N(C) = 23$$
$$N(A \cap B) = 10 \qquad N(A \cap C) = 12 \qquad N(B \cap C) = 10$$
$$N(A \cap B \cap C) = 5$$
$$N(\overline{A \cup B \cup C}) = 7$$

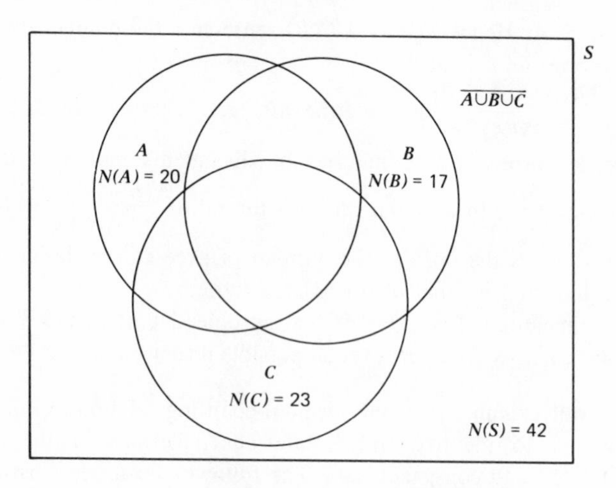

Figure 1.13 Venn diagram of counting formula.

where $\overline{A \cup B \cup C}$ is the subset of users who did not like any of the three weapons. Using the counting formula, he noted

$$N(A \cup B \cup C) =$$
$$N(A) + N(B) + N(C) - N(A \cap B) - N(A \cap C) - N(B \cap C) + N(A \cap B \cap C)$$
$$= 20 + 17 + 23 - 10 - 12 - 10 + 5$$
$$= 33$$

This gives the number of analysts who considered at least one of the weapons as best. The total does not include those who did not like any of the three. Counting in the seven users who did not like any of the weapons, the supervisor pointed out that the reports showed that the survey covered but 40 users, whereas the analyst indicated that 42 users were surveyed. He suggested that the discrepancy be removed. ■■

Exercise 1.5. Identify and describe fully an application of any set relation or operation that you consider to be (or has been) useful to you in performing your job as an analyst. For example, where you employed the concept of set intersection or complement, etc., in your work? Are there analysis situations where the set-counting relation could help you in making equipment or people counts, retrieval, etc.? ■■

Exercise 1.6. Find answers to the following exercise using the counting formula, Venn diagram, or other technique. Watch for inconsistencies in results obtained. Note that the information is incomplete.

Of 100 persons asked:

40 favored the Viet Nam war

60 supported federal aid to education

70 thought the United Nations should be made stronger

25 favored both the Viet Nam war and federal aid to education

20 favored the Viet Nam war and also believed the U.N. should be strengthened

55 advocated federal aid to education and also a stronger U.N.

How many at most supported the Viet Nam war, federal aid to education, and also a stronger U.N.? ■■

Exercise 1.7. Suppose that you are retrieving documents from an IS&R system which contains 100 documents. Assume that all documents are indexed by at least one term A, B, or C and that

65 documents are indexed by term A

55 documents are indexed by term B

40 documents are indexed by term C
30 documents are indexed by terms A and B
25 documents are indexed by terms A and C
20 documents are indexed by terms B and C

Determine the number of documents you would retrieve if your retrieval request asked for

(a) A and B and C
(b) B or C
(c) B and C
(d) A or (B and C)
(e) Not (A and B)
(f) Not (B or C) ■■

Example 1.13 (IS&R). When someone requests documents from an IS&R system, it is important for the requester to be aware of the manner in which the documents were indexed, i.e., the terms, attributes, or words used by the indexer to originally classify the complete set of documents. The requester should also be aware of the admissable ways in which the index terms can be conjoined through union and intersect. For example, a naval aviator may request information on landing aids thinking only in terms of radar landing aids for naval carrier aircraft. The information system, on receiving the request for "landing aid," provides the requester with a document list containing references for landing aids for aircraft of all types, and for vessels of other types. Hence the requester is inundated with documents, only a small percentage of which are relevant to his information needs. His predicament can be illustrated by the accompanying Venn diagram.

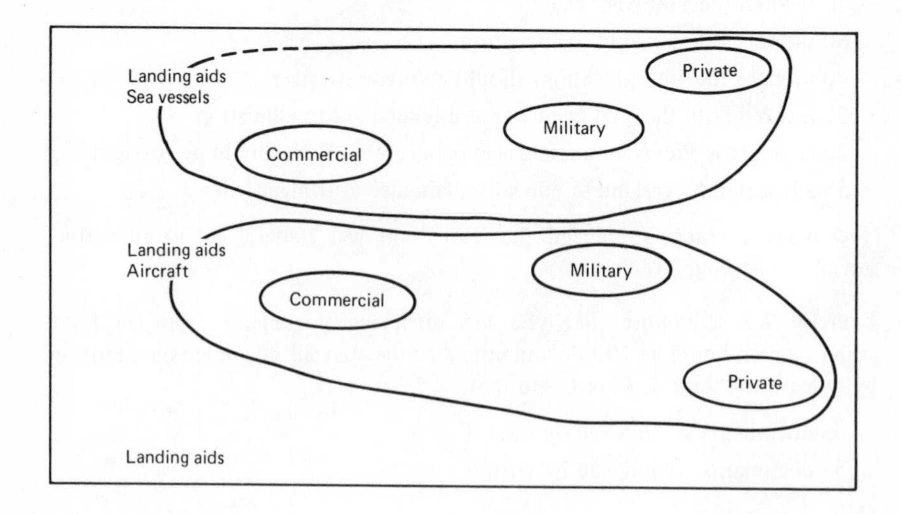

The aviator, somewhat smarter, reenters his request for documents on landing aids, but this time he asks for Landing Aids–Aircraft. The document list he receives from this request is smaller than previously obtained but is still too large. For it includes documents covering all aircraft landing aids. Looking over the document list the aviator eventually sees that if he wants a relevant, complete, but not excessive set of documents to peruse, his request should state that his interests lie in landing aids for military aircraft. But on receiving the documents so requested, the naval aviator will find that he has recalled also references on Army and Air Force types of landing aids. These may or may not be of interest to him. Figure 1.14 indicates that to refine his need further the aviator should request "Navy-carrier-based radar landing aids."

Note that in the above discussion a hierarchical structure of index terms or attributes has been assumed. A tree or graph, Figure 1.14, can be employed to illustrate the relationships which obtain in this hierarchical structure. From the tree structure one also obtains relationship of inclusion which exists among the branches of the tree. For example, the set of radars is a subset of the set of all carrier-based landing aid radars, which in turn is a subset of both naval

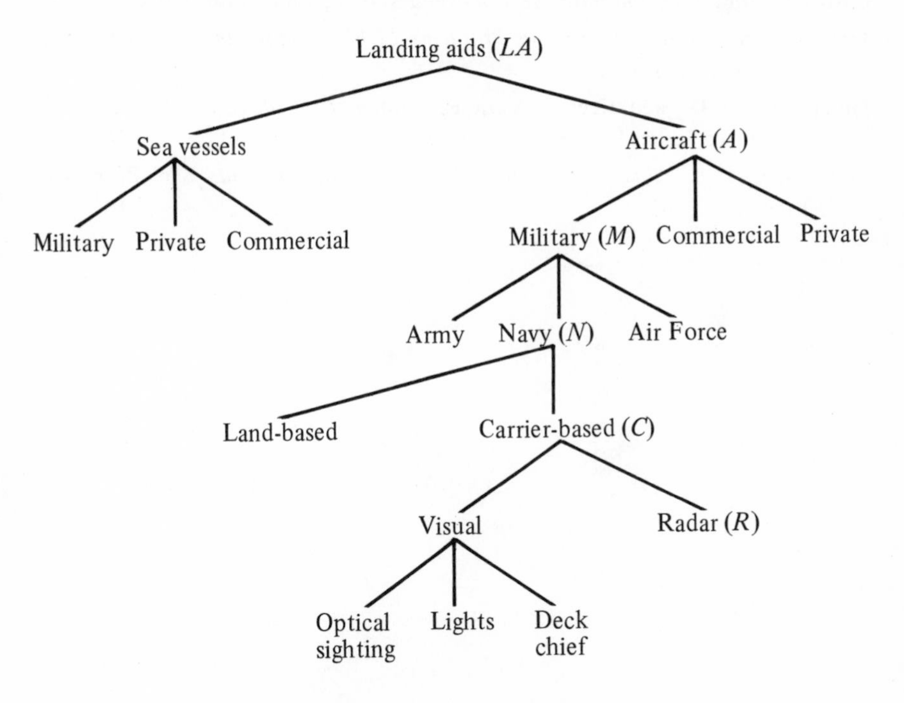

Figure 1.14 Indexing tree.

carrier and possibly commercial aircraft landing aids. Using the symbols applied to each class in the figure, we can indicate the class inclusion as

$$R \subset C \subset N \subset M \subset A \subset LA$$

If a requester knows the relationships which exist among document sets defined through the indexing system employed in the IS&R system and he uses these relationships, his requests will result in more efficient and relevant responses, hence saving his valuable time, while reducing the load on the information system.

Note that requests can be made to cut across sets of documents identified in the separate branches of the hierarchical structure. For example, if one searched for "commercial landing aids," the documents retrieved would cross over the two branches (sea vessels, aircraft) in the landing aids tree. Such retrieval schemes are called associative.■■

REFERENCES

Christian, Robert. *Introduction to Logic and Sets.* Boston: Ginn, 1958.

Lipschutz, Seymour. *Theory and Problems of Finite Mathematics.* New York: McGraw-Hill, 1966.

Miller, Charles D., and Heeren, Vern E. *Mathematical Ideas: An Introduction.* Glenview: Scott, Foresman, 1968.

Stoll, Robert R. *Sets, Logic and the Axiomatic Method.* San Francisco: Freeman, 1961.

Chapter 2 / Logic, Truth Tables, and the Validity of Arguments

2.1 INTRODUCTION

Logic is often described as the science of necessary inference. It studies how logical structure and quantification of simple statements govern the truth and falsehood of derived complex statements. In this sense logic studies the theory of statement composition. It is concerned with logical structures which emerge when compound statements are constructed from simple statements by means of the basic locutions (particles) "and," "or," "not," "if, then," "unless." However, in this chapter we will cover only a few of the elementary concepts of logic leaving the formidable studies to .specialists in logic, mathematics, and philosophy.

A natural question a reader of this chapter should ask is why should a substantive analyst study some of the elementary concepts of logic? There are three answers to this question. First, logic is basic to any field of intellectual endeavor. Once you become familiar with some elementary concepts of logic you will find that you have been using (or misusing) the principles of logic in your everyday speaking, writing, and analysis. With a knowledge of the basic concepts of logic, you should be able to surface and understand your inductive thought processes which you intuitively felt were correct.

The second reason for studying logic is its importance in statement formation and quantification. Fundamental to the correct analysis of information and the clear and concise communication of information is the necessity for correct IS&R, statement structure, and interpretations. Knowledge of logic provides a basis for the formal structuring and hence the storing of precise information. Equally important, logic provides a means for minimizing or avoiding errors of commission or omission in statement preparation and interpretation. This fact is especially true where quantification is involved.

The third reason for studying logic (perhaps the most important one for substantive analysts) is that formal logic provides a standard against which the value of plausible reasoning processes can be measured. As will be demonstrated in Sections 3.2-3.15, it is possible to consider plausible reasoning processes when the premises and conclusions are neither absolutely true nor absolutely false; i.e., assume some level of credibility between true and false, and the argument assumes some level of plausibility between the values of invalid and valid. Plausible reasoning allows a rational approach to decision-making when working with uncertain information or questionable argument structures. Through a knowledge of plausible reasoning techniques, the analyst can make use of the advantages, values, and limitations associated with this form of logical reasoning. Therefore, his analytic abilities are considerably extended when he is faced with uncertainties. Thus everyone, and in particular analysts, programmers, writers, or experimenters should be familiar with the content of this chapter. Application of the basic principles of elementary logic to everyday analysis will result in clearer statements, better understanding and interpretation, and fewer arguments over problems of definition and implication.

As you proceed in your study of the logic of statements, you will find many similarities with the theory of sets. These similarities will be highlighted in Section 2.10; for readers already expert in the theory of sets, it is suggested that Section 2.10 be read first. Others should start with Section 2.2 and proceed in natural order through the chapter. The remaining sections are to be considered as preparatory to a better understanding of Chapter 3, where applications of logical and plausible reasoning to analysis will be discussed.

2.2 STATEMENTS

Statements are declarative sentences to which a truth value can be assigned. A statement is called a *simple* statement if it expresses a single fact. If it expresses more than a single fact, it is called a *compound* statement. The *truth value* of a statement is either truth or falsehood according to whether the statement is true or false. A statement cannot be both true and false. Statements are sometimes referred to as propositions. Figure 2.1 illustrates the relationships which exist among the various types of sentences and statements. For example, a sentence such as "x is a Russian diplomat" is called an *open sentence.* When a real name is substituted for x, the open sentence becomes a closed simple statement and can assume a truth value of truth or falsehood. From Figure 2.1 it should be noted that not all sentences are statements. Imperative, exclamatory (not shown in Fig. 2.1), and interrogative sentences are not statements. Following the tree from top to bottom, it is noted that not all declarative sentences are statements. Open and nonsense declarative sentences are not to be considered as statements until ambiguity is removed.

Open sentences can be converted to nonsense sentences, indeterminant sentences, or closed statements. If terms which yield meaningful sentences are substituted for the variables in an open sentence, a closed statement results. Otherwise either an indeterminant or an incomprehensible nonsense sentence results. When a closed meaningful statement results, the statement is either simple or compound. If the statement is simple, then the truth value "truth" or "falsehood" can be directly assigned to its factual content. If the statement is compound, truth tables (to be described) are employed to determine the truth value of the compound statement.

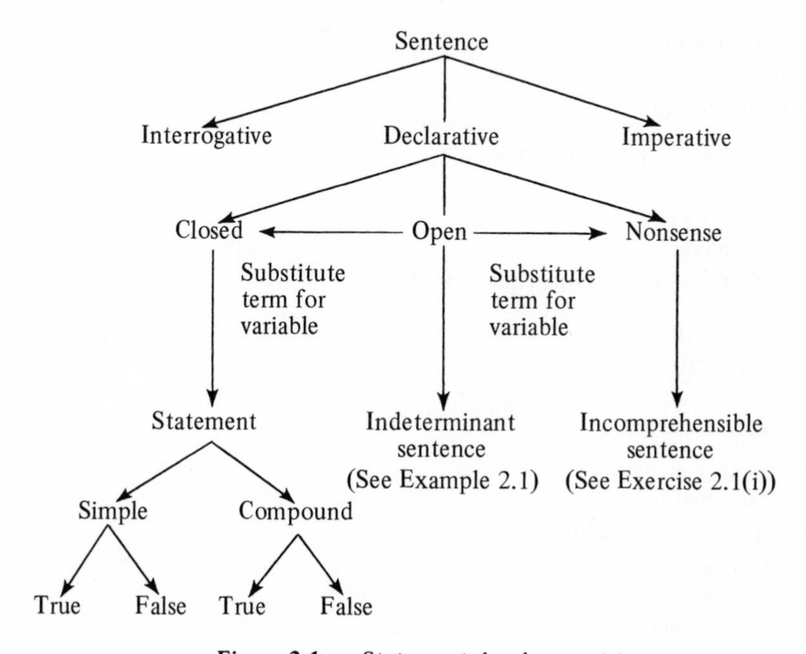

Figure 2.1. Statement development tree

When evaluating the truth or falsehood of a simple statement, it is important that differences between relative and absolute truth value be understood. For example, the utterance, "I am a Russian," may be true for one speaker but false for another. The statement, "Democracy is the best form of government," is true for most Americans but false for many communists. The statement, "Richmond is a large city," may be true to a Virginian but false to a New Yorker. In the sequel we adopt a convention (for discussion purposes) which states that a statement will be assigned a truth value which, when assigned, will be invariant with time, place, context, or speaker. Thus extra care will be taken to avoid ambiguous words or phrases in forming statements. Statements must in their own right have unambiguous truth values. The following examples illustrate some of the different types of sentences considered in the sequel.

Example 2.1 (Statements and Sentences).

The following are open sentences:

The *x*'s are attacking.
John is not ________________ .
His name is *x*.

The following are simple statements:

Russians are not shy.
Joe is his name.
Clouds are white.

The following are not statements:

Run!
Can't you walk?
Go fly a kite.
I now pronounce you man and wife. ■■

Exercise 2.1. State whether each of the following sentences is an (1) open sentence, (2) statement, (3) neither.

(a) The object *x* is larger than a Sherman tank.
(b) John is taller than Bill.
(c) Yankee go home.
(d) Tote that bale.
(e) Americans are softhearted.
(f) Leave well enough alone.
(g) April is a winter month.
(h) He likes Mary.
(i) All mumboos are jumboos. ■■

The *logical structure* of a statement is the pattern assumed by the basic logical locutions of a sentence formed from other nonlogical phrases. For example, the following two sentences have the same logical structure:

Every Turk is a European *or* an Asian.
Every American is an Indian *or* a foreigner.

The logical structure of the two sentences is identified by the form

Every ________________ is ________ or ________________ .

The word "every" is a quantifying particle while the word "or" is a connecting particle. Only the quantifying and connecting particles every and or are important in determining the logical structure of the two sentences. The verb "is" does not contribute to the logical structure. It is the logical structure and quantification of statements which determine their truth or falsehood. However, the implied or understood meanings of sentence subjects, verbs, predicate nominatives, direct objects, and modifiers are very important in assigning truth values to statements. Usage should be restricted to those words which are least ambiguous within the often difficult semantic constraints imposed by the English language.

Some statements are true solely by virtue of their logical structure. These are called *tautologies.* An example of a tautology is "*Every* Turk is sick *or not* sick." In some instances two statements will be found to be logically *equivalent*–that is, they agree in truth or falsehood solely by virtue of their logical structure. An example of such a pair of statements follows:

> *Every* Turk is *either* a European *or* an Asian.
> If *someone* is *neither* an Asian *nor* a European, he is *not* a Turk.

In the section on truth tables (2.4) more will be said about *tautologies* and *logical equivalence.* Techniques will be developed to allow analysis of the truth or falsehood of statements as related to their quantification and logical structure. First, however, let us consider the formation of compound statements.

2.3 COMPOUND STATEMENTS

Of primary importance in logic is the construction of *compound statements* from simple statements. The basic logical locutions and, or, either...or, if, then, neither...nor, are called *logical connectives* and are employed in combining simple statements to form compound statements. Some examples of combining statements follow.

Example 2.2 (Connectives). From the following simple statements:

> John is smart.
> Harry is dumb.
> Bill is average.

the logical connectives and, or, not, if, then, neither...nor can be used to form the compound statements:

> John is smart *or* Harry is dumb.
> John is smart *and* Harry is dumb.
> John is *not* smart. (It is false that John is smart.)

If John is smart, *then* Bill is average.
If John is *not* smart, *then* Bill is average.
Neither is John smart *nor* is Harry dumb. ■■

Exercise 2.2. State whether the following statements are (1) simple, (2) compound, (3) neither:

(a) Some statements are simple and some compound.
(b) John and Henry are small.
(c) Sauce is good for the goose and the gander.
(d) Can Bill and Bob excel?
(e) Laugh and the world laughs with you.
(f) What is good for the goose is good for the gander.
(g) Can neither Bill nor Bob do it?
(h) What can you expect from Bill or Bob?
(i) If John does his job, will Bill do his?
(j) If John does his job, so also will Bill.
(k) Try John or Bill. ■■

If each of the simple statements in Example 2.2 is assigned a value of truth or falsehood, the question arises as to the determination of the truth or falsehood of the compound statements formed from them. The answer to this question is obtained through the use of *truth tables.*

2.4 TRUTH TABLES

In an earlier discussion we noted that the truth value of a statement was determined solely by its logical structure and quantification. For what truth values of its simple statements is a compound statement true? A convenient format used in answering this question is the *truth table.* A truth table is a tabular compilation of truth values which identifies the connectives employed in a compound statement and lists the truth values of the compound statement for all combinations of truth values of its constituent simple statements. There are four fundamental truth tables which are standard in logic and are employed to structure truth tables for more complex compound statements. The four truth tables relate to the connectives and, or, not, and if, then. These will be discussed first and then the technique for applying these to structure more complex truth tables will be described.

Consider the case where the connective and is used to form the compound statement, "John is smart and Harry is dumb." Let us analyze this compound statement intuitively. If "John is smart" is true and "Harry is dumb" is true, we can intuitively expect that the compound statement is also true. If either or both statements are false, we can expect the compound statement to be false. All of this can be summarized with a truth table where we use P to represent "John is smart" and Q to represent "Harry is dumb." The following is the truth table for P and Q:

P	Q	P and Q
T	T	T
T	F	F
F	T	F
F	F	F

The entries of T, F in the first two columns headed P, Q denote all of the possible combinations of values of truth and falsehood which the simple statements P, Q might assume. The entries in the third column headed "P and Q" are the truth values of the compound statement "P and Q" for the four possible truth value combinations of P and Q. Thus every compound statement of the structure "P *and* Q" is true only if both P and Q are true; otherwise it is false.

Example 2.3 (And). Consider the compound statement

John is tall and Bill is short.

Let P be the statement "John is tall," and Q the statement "Bill is short." When P is true and Q is false, the reference is to the second row of the "and" truth table from which it is read that the compound statement "P and Q" is false. If P and Q are both true, this corresponds to the first row of the truth table which yields that the compound statement "P and Q" is true.■■

Exercise 2.3. Let P, Q, R, and S be simple statements with truth value, T, T, F, F, respectively. What is the truth value of

(a) P and Q
(b) P and R
(c) Q and S
(d) R and S ■■

Next consider the case when the connective "or" is used. In ordinary discourse or has two meanings. The statement "P or Q" can mean "P or Q or both," (inclusive or), or it can mean "P or Q but not both" (exclusive or). The connective or as used in the discussions to follow will mean "P or Q or both," unless otherwise stated. If we denote P by "John is smart" and Q by "Harry is dumb," the compound statement (P or Q), "John is smart or Harry is dumb," will have a truth value depending on the truth values of the simple statements P and Q and the meaning attached to the connective or. The truth table for the connective or is

P	Q	P or Q
T	T	T
T	F	T
F	T	T
F	F	F

From the truth table it is easy to see that the compound statement "P or Q" is true if either P or Q is true, and false only if both P and Q are false.

Example 2.4 (Or). Consider the compound statement "John is smart or Harry is dumb." If P ("John is smart") is assumed to be false and Q ("Harry is dumb") is assumed to be true, these conditions are identified by the third row in the truth table. The resulting statement "P or Q" by the truth table then has a truth value of true. ■■

Exercise 2.4. Let P, Q, R, and S be simple statements with truth value of T, T, F, F, respectively. What is the truth value of

(a) P or Q
(b) P or R
(c) Q or S
(d) R or S

The third connective of importance in ordinary discourse and logic is the *negation* or *denial* "not." From any statement it is possible to form another by negating or denying the given statement. If the original statement is true or false, then its negation is accordingly false or true. The denial of the statement "John is smart" is "John is not smart." There are other ways of forming the denial of statements. For example, the statement, "John fails to be smart" denies "John is smart." A preferred form is "It is not the case that John is smart" or "It is false that John is smart." To simplify the writing of the denial the symbol $\sim$ is used to signify not. For example, "$\sim$ John is smart" replaces "It is false that John is smart." The truth table for the denial of a statement is

P	$\sim P$
T	F
F	T

Example 2.5 (Truth Tables). This example illustrates how the truth tables for the connectives and, or, and not are employed in determining truth tables for compound statements. Let P, Q, and R denote the statements "John is smart," "Bill is strong," and "Bob is tall," respectively. Let the symbols $\wedge$, $\vee$, and $\sim$ designate the connectives and, or, and not, respectively. Then

$P \wedge Q$	replaces	P and Q
$P \vee Q$	replaces	P or Q
$\sim R$	replaces	not R

Compound statements can be formed using one or more of these connectives; for example,

$$\sim(P \wedge Q) \vee R$$
$$(P \vee Q) \wedge (\sim P \vee R)$$

A typical question asked is, for what truth values of P, Q, and R is the compound statement $\sim(P \wedge Q) \vee R$ true? To answer this question use is made of the truth tables for the connectives $\wedge$, $\vee$, $\sim$.

First consider the truth value of $P \wedge Q$. From the "and" truth table $P \wedge Q$ is true only if P and Q are both true and is false otherwise. Hence, the negation $\sim(P \wedge Q)$ is true except when P and Q are true. The statement $\sim(P \wedge Q) \vee R$ is true whenever $\sim(P \wedge Q)$ is true or R is true. Hence the truth values of P, Q, and R for which the compound statement is true includes all possible truth structures except the one where P and Q are both true and R is false. All of this effort can be systematized by the composite truth table given here.

P	Q	R	$P \wedge Q$	$\sim(P \wedge Q)$	$\sim(P \wedge Q) \vee R$
T	T	T	T	F	T
T	T	F	T	F	F
T	F	T	F	T	T
T	F	F	F	T	T
F	T	T	F	T	T
F	T	F	F	T	T
F	F	T	F	T	T
F	F	F	F	T	T

In the composite truth table, the three columns on the left contain the eight possible truth state combinations the three simple statements can assume. The truth state combinations for the variables P, Q, and R are entered in the following fashion. Starting with the extreme right variables column, truth values for R are alternated T, F from top to bottom until the column is completed. For 3 variables, there are $2^3 = 8$ entries or rows. (For n variables, there are 2^n rows.) In the next column to the left of R, truth values for Q are alternated in runs of two, T, T, F, F, from top to bottom, until the column is completed. In the next column to the left of Q, truth values from P are alternated in runs of four, T, T, T, T, F, F, F, F, from top to bottom until the column is completed. If more variables were involved, this systematic method of assigning truth values to all the variables would be continued until all the columns were completed. Simply stated, for n variables the truth values are alternated one at a time, two at a time, up to 2^{n-1} at a time starting from the rightmost column and proceeding to the leftmost column.

After completion, the truth state combinations for P, Q, and R, are headed by the simplest compound statements. In this case $P \wedge Q$ is the simplest compound statement. Higher order compound statements are selected as column headings as the truth table is built up from left to right. The last column contains the compound statement for which we desired to determine the truth value. In this form it is a simple matter to determine the truth value of the statements selected from column headings by observing the truth values of their constituent statements contained in the columns to the left. The process continues from left to right until the last column is filled with truth values. The columns containing the truth values for the constituent simple statements (P, Q, R) of the compound statement and the completed last column form the truth table for the compound statement. The truth table for $\sim(P \wedge Q) \vee R$ reduces to the following:

P	Q	R	$\sim(P \wedge Q) \vee R$
T	T	T	T
T	T	F	F
T	F	T	T
T	F	F	T
F	T	T	T
F	T	F	T
F	F	T	T
F	F	F	T

■■

Exercise 2.5. Construct the truth tables for the following compound statements:

(a) $(\sim P \vee Q)$
(b) $\sim P \wedge \sim Q$
(c) $(P \vee \sim Q) \wedge R$
(d) $\sim(P \vee Q)$
(e) $\sim(P \wedge Q)$
(f) $\sim(P \vee Q \vee R)$
(g) $\sim(P \wedge Q) \vee R$

■■

2.5 OTHER CONNECTIVES

In ordinary discourse connectives other than and, or, and not are employed. Most common are the connectives "but," "although," and "unless." All three are used to form compound sentences from simple sentences; for example,

John is ill but Mary is well.

John is ill although Mary is well.

John will be ill unless Mary cares for him.

In the first two sentences the connectives but and although are used in place of the connective and to emphasize a contrast between the two simple statements they connect. The connective although also attempts to indicate an element of surprise in the contrast. However, in the sequel, the connectives but and although will be used interchangeably with the connective and. The only difference in usage will be rhetorical in nature and not relevant to the resultant truth value of the statement.

When considering the connective unless, the meaning is more difficult to construe. Consider the four truth states that the third sentence can assume where P is "John will be ill," and Q is "Mary cares for him."

$\sim P \wedge Q$: John will not be ill and Mary cares for him.

$P \wedge Q$: John will be ill and Mary cares for him.

$\sim P \wedge \sim Q$: John will not be ill and Mary does not care for him.

$P \wedge \sim Q$: John will be ill and Mary does not care for him.

The meaning of the connective unless in this sentence is rendered as the inclusive or if the user is willing to accept the fact that "Mary cares for John yet John becomes ill" as a possible truth. On the other hand, if the user does not accept the above possibility, then the connective unless is employed as the exclusive or, i.e., one and only one of the component statements can assume a truth value of true. In this case one agrees that only the first and fourth states of the compound sentence are acceptable consequences of the connective unless.

In ordinary discourse the connective unless is most often employed in place of an inclusive or. Its rhetorical connotation is that the first statement is more likely to be true or deserves more emphasis than the second.

Exercise 2.6. Determine the truth table for the following compound statements:

(a) P unless Q

(b) P but not Q

(c) Not P although Q

(d) P or Q but not both

(e) $(P \vee Q) \wedge \sim (P \wedge Q)$

(f) What can be concluded from the answers obtained in (d) and (e)? ▪▪

Up to this point we have been considering various logical combinations of simple statements. We have assumed that we know with *certainty* the truth value of each of the simple statements involved. Through the use of truth tables we determine with *certainty* the truth values of the compound statements. Note the emphasis on the word *certainty.* In the real world we are not afforded the luxury of making all our statements with certainty. Rather, statements are often qualified such as: "it is sometimes true that;" "with a high degree of probability it is true that;" "the probability is 0.7 that;" etc. When we admit statements whose truth values are not certain, we enter the realm of probability theory.

The calculus of statements is used in probability theory to make determinations on the probability of truth of compound statements based on the logical connectives employed and the probability of truth of the simple statements involved.

For example, let us consider two statements which might be derived from an attempt to determine the effectiveness of two means of communications.

P: The people of Zork read the magazine *POT.*

Q: The people of Zork listen to radio station BULL.

Using simple logic, we can construct two compound statements to indicate the level of media contact with the people of Zork.

$P \wedge Q$: The people of Zork read *POT* and listen to BULL.

$P \vee Q$: The people of Zork read *POT* or listen to BULL.

Our attache stationed in Zork tells us that only 20% of the population of Zork read *POT*–i.e., the probability that a randomly selected citizen of Zork reads *POT* is 0.2. Thus we can make the statement that the probability that statement P is true (denoted $p(P)$) is 0.2–i.e., $p(P) = 0.2$. Similarly, we are told that the probability that some citizen of Zork listens to BULL is 0.4–i.e., $p(Q) = 0.4$.

From these statements we can obtain the probability that citizens of Zork both read *POT* and listen to BULL. This probability is denoted $p(P \wedge Q)$ and is calculated as

$$\begin{aligned} p(P \wedge Q) &= p(P) \times p(Q) \\ &= 0.2 \times 0.4 \\ &= 0.08 \end{aligned}$$

Thus we can expect only 8% of the citizens of Zork to do both.

However, a natural question for USIA to ask is, "What is the probability of reaching the people of Zork through at least one of these media? Stated otherwise, what is the probability that the statement $P \vee Q$ is true? If we denote this probability as $p(P \vee Q)$, we can write

$$p(P \vee Q) = p(P) + p(Q) - p(P \wedge Q)$$

where $p(P \wedge Q)$ accounts for those citizens of Zork who are both readers and listeners and hence are counted twice in $p(P) + p(Q)$. Using the probabilities given above, we can compute

$$\begin{aligned} p(P \vee Q) &= p(P) + p(Q) - p(P \wedge Q) \\ &= 0.2 + 0.4 - 0.08 \\ &= 0.52 \end{aligned}$$

Thus the statement, "The people of Zork read *POT* or listen to BULL," has a 0.52 probability of being true. Stated in another way, we could expect to reach 52% of the people of Zork through the two media. The moral of this story is as follows:

> If we want to gain a 52% coverage for any given message, repeat that message over *POT* and BULL. Don't arrange message so that both *POT* and BULL must be received in order to assemble the full message. If the latter is the case, then only an average of 8% will receive the message.

Now let us return to a consideration of additional connectives used in forming compound statements. The next connective to be discussed is one used in most forms of good arguments.

2.6 IMPLICATION

A compound statement of the form "if P, then Q," where P and Q are statements, is called a *conditional statement* or an implication. It is important to notice that a cause-and-effect relation need not exist between the statements P and Q. One is free to conjoin any two statements to form a conditional statement. For example, the conditional statement, "If Mary goes swimming, then the tide will change," conjoins the unrelated statements

P: Mary goes swimming.

Q: The tide will change.

Notice that the statement, "The tide will change," is a true statement independent of whether Mary goes swimming.

The notation for the *implication connective* is an arrow. The notation "$P \rightarrow Q$" reads: The premise P implies the conclusion Q. Another commonly used form is "If P, then Q." Occasionally the then of "if, then" is dropped: "If it rains tonight, the football game will be called off." Occasionally, the connective "only if" is used in place of "if, then." For example, the last statement can be written: "It will rain tonight only if the football game is called off." Care should be taken in using "only if" rather than "if, then." In a statement of the form

if P, then Q

the equivalent statement using "only if" is

P only if Q

Prefixing "only if" to one statement of an implication is equivalent to prefixing if of the "if, then" to the remaining statement of the implication. The truth table for the conditional connective $\rightarrow$ is

P	Q	$P \rightarrow Q$
T	T	T
T	F	F
F	T	T
F	F	T

To illustrate the rationale behind the entries to this table, consider the following conditional statement: "If it rains tonight, then the game will be called off." It is possible to combine the two simple statements in four ways using and as the connective.

It rained tonight and the game was called off.

It rained tonight and the game was not called off.

It did not rain tonight and the game was called off.

It did not rain tonight and the game was not called off.

The first combination agrees with the first row of the truth table. If it is assumed that it rained and the game was called off, realization of the occurrence, regardless of cause, is enough to accept the entire statement as true. However, if it rained and the game was not called off, this would contradict the intended meaning of the compound statement. Hence, the compound statement would be accepted as false. This is in agreement with the second row of the truth table. If it did not rain, the nonoccurrence of rain could have no affect on whether the game was played or not. Hence, the compound statement would be accepted as true in both cases, corresponding to the third and fourth row of the truth table.

In an implication a true premise cannot yield a false conclusion. It is this fact which leads to the statement: In implication a true hypothesis always leads to a true conclusion. Stated in another way, in implication it is false that the premise may be true and the conclusion false. In other words, "$P \rightarrow Q$" and "not (P and not Q)" are equivalent. Hence, the truth tables for the two statements are the same. This fact is illustrated in Example 2.6.

Example 2.6 (Equivalence). It is possible to demonstrate the fact that the two statements $P \rightarrow Q$, $\sim(P \wedge \sim Q)$ are equivalent. To do this, consider the truth table constructed as follows:

P	Q	$P \rightarrow Q$	not Q	P and not Q	$\sim(P \wedge \sim Q)$
T	T	T	F	F	T
T	F	F	T	T	F
F	T	T	F	F	T
F	F	T	T	F	T

In the first two columns we list all the possible combinations of truth values assumed by the statements P and Q. In the next column we copy the truth values of $P \rightarrow Q$ from the truth table for implication. In the fourth column we negate Q and list the truth values of "not Q." In the fifth column are listed the truth values for the conjunction "P and not Q" using the truth values of columns 1 and 4. Negating the fifth column yields the sixth column which contains the truth values for $\sim(P \wedge \sim Q)$. Comparing column 6 with column 3 we note that their truth values agree for each row entry of truth values for P and Q. Hence, the statements heading columns 3 and 6 are equivalent.

The implication statement $P \rightarrow Q$ is usually associated with three other statements:

$$Q \rightarrow P \qquad \text{(converse)}$$

$$\sim P \rightarrow \sim Q \qquad \text{(inverse)}$$

$$\sim Q \rightarrow \sim P \qquad \text{(contrapositive)}$$

The relationships which exist among these four compound statements are identified best through use of a truth table.

Variable Statements		Direct Implication	Converse	Inverse	Contrapositive
P	Q	$P \rightarrow Q$	$Q \rightarrow P$	$\sim P \rightarrow \sim Q$	$\sim Q \rightarrow \sim P$
T	T	T	T	T	T
T	F	F	T	T	F
F	T	T	F	F	T
F	F	T	T	T	T

Examples of these relationships would be:

If I have ten planes, then I can make an airstrike. (direct implication $P \rightarrow Q$)
If I can make an airstrike, then I have ten planes. (converse $Q \rightarrow P$)
If I do not have ten planes, then I cannot make an airstrike. (inverse $\sim P \rightarrow \sim Q$)
If I cannot make an airstrike, then I do not have ten planes. (contrapositive $\sim Q \rightarrow \sim P$)

Note that the truth values listed in the third and sixth columns are identical, as are those listed in the fourth and fifth columns. Thus the statements $P \rightarrow Q$ and $\sim Q \rightarrow \sim P$ are *equivalent,* as are $Q \rightarrow P$ and $\sim P \rightarrow \sim Q$. When two statements have identical truth tables, one of the statements can be substituted for the other in any discourse containing it. Hence, the *inverse* and *converse* are interchangeable as are the *direct implication* and the *contrapositive.* The latter exchange is widely used in proving theorems in mathematics and is called an indirect proof or *reductio ad absurdum.* The truth table for two additional equivalents is shown in Example 2.7.

Example 2.7 (Further Equivalences). The following truth table shows that the three statements $P \rightarrow Q$, $\sim P \vee Q$, and $\sim (P \wedge \sim Q)$ are equivalent:

P	Q	$P \rightarrow Q$	$\sim P \vee Q$	$\sim (P \wedge \sim Q)$
T	T	T	T	T
T	F	F	F	F
F	T	T	T	T
F	F	T	T	T

The last two columns of the truth table illustrate an important relationship between the connectives and and or when statements containing them are negated. From the table, we can write

$$\sim (P \wedge \sim Q) = \sim P \vee Q$$

where the symbol = means "equivalent to." From this it can be observed that when statements are negated, and transforms to or, and vice versa, and the negation is applied to all statements; for example,

$$\sim (P \wedge Q \wedge R) = \sim P \vee \sim Q \vee \sim R$$

$$\begin{aligned} \sim [P \wedge (Q \vee R)] &= \sim P \vee \sim (Q \vee R) \\ &= \sim P \vee (\sim Q \wedge \sim R) \end{aligned}$$

Exercise 2.7. For each of the following statements write the converse, inverse, and contrapositive statements.

(a) All cats have fat tails.
(b) People who throw rocks should not live in glass houses.
(c) Smile and you will not smile along.
(d) Smoke pot and you'll land in jail.
(e) Girls have it easier than boys.
(f) What's good for the goose is good for the gander.
(g) I will drive if you are tired.
(h) I will drive only if it rains.
(i) Decreased pay is a necessary* condition for workers to quit.
(j) My boss saying so is a sufficient* condition for me to work harder.
(k) If you want to pass this course, you had better complete all the exercises.

2.7 BICONDITIONAL

Another connective of importance arises when the direct implication $P \to Q$ and its converse $Q \to P$ are simultaneously true. From the previous truth table, this occurs when P and Q are both true or are both false. The new connective, called the *biconditional,* and denoted by $\leftrightarrow$ has the following truth table:

P	Q	$P \leftrightarrow Q$	$(P \to Q) \wedge (Q \to P)$
T	T	T	T
T	F	F	F
F	T	F	F
F	F	T	T

Normally the direct implication is written "John is well if Mary cares for him." The converse statement is "John is well only if Mary cares for him." The biconditional is written "John is well if and only if Mary cares for him." Another written form of the biconditional is "In order that John be well it is necessary and sufficient that Mary care for him." In a biconditional statement $P \leftrightarrow Q$, the statement $P \to Q$ is translated as "P is sufficient for Q" or "Q is necessary for P." The converse statement $Q \to P$ is read "P is necessary for Q" and "Q is sufficient for P."

*See Section 2.7.

Example 2.8 (Necessary and Sufficient Conditions). A condition is considered to be necessary for the occurrence of a specified event if the event will not occur in the absence of the condition. For example, a fire will not occur if oxygen is not present. Hence the presence of oxygen is considered necessary for the occurrence of a fire.

A condition is considered to be sufficient for the occurrence of a specified event if the event must occur in the presence of the condition. For example, for most substances a sufficient condition is the attaining of a specific temperature in the presence of oxygen. Note that the presence of oxygen or the attainment of a specific temperature alone does not guarantee combustion. Each is a necessary condition. Together they constitute a sufficient condition. Generally, there will be several necessary conditions which in part or in total will constitute a sufficient condition.▪▪

Review

Exercise 2.8. Express the following compound statements in terms of simple statements, P: John will go; Q: Peter will go; R: Mary will go and any of the connectives defined in the text.

(a) Neither John nor Mary will go.
(b) John and Mary will go.
(c) John will go unless Mary goes.
(d) John will go only if Mary goes.
(e) John will go if Mary goes.
(f) John will go although Mary goes.
(g) John and Mary will go if Peter goes.
(h) If Peter and John go, Mary will go.
(i) If Peter goes or Mary goes, John will not go.
(j) John will go if and only if Peter or Mary go.
(k) Only John or Mary, but not both will go.
(l) When Mary goes John will remain.▪▪

Exercise 2.9. Express the following statements in words, where P: John will go; Q: Peter will go; and R: Mary will go.

(a) $\sim P \wedge \sim R$
(b) $P \wedge R$
(c) $\sim R \rightarrow P$ or $P \vee R$
(d) $P \rightarrow R$
(e) $R \rightarrow P$
(f) $P \wedge R$
(g) $Q \rightarrow (P \wedge R)$
(h) $(P \wedge Q) \rightarrow R$
(i) $(Q \vee R) \rightarrow \sim P$
(j) $P \leftrightarrow (Q \vee R)$
(k) $(P \vee R) \wedge \sim (P \wedge R)$
(l) $R \rightarrow \sim P$ ▪▪

Exercise 2.10. Construct truth tables for each of the following compound statements recalling that a statement is called a *tautology* if its truth table contains all Ts, two statements are *equivalent* if their truth tables are identical (agree in truth or falsehood for each truth state of the table), and a statement is *self-contradictory* if its truth table contains all Fs. Use the truth tables to determine which of the three terms apply to each (or pairs of) compound statement in this exercise.

(a) $\sim(P \wedge \sim P)$ Law of excluded middle
(b) $\sim(P \wedge \sim Q)$ Implication
(c) $(\sim P \vee Q) \wedge (\sim Q \vee P)$ Equivalence
(d) $[(P \to Q) \wedge P] \to Q$
(e) $P \vee \sim P$
(f) $(P \to Q) \wedge (Q \to P)$
(g) $[(P \to Q) \wedge \sim Q] \to \sim P$
(h) $\sim P \vee Q$
(i) $P \leftrightarrow Q$
(j) $[(P \to Q) \wedge (Q \to R)] \to (P \to R)$
(k) $[(P \to Q) \wedge Q] \to P$
(l) $[(P \to Q) \wedge \sim P] \to \sim Q$ ■■

2.8 VALID ARGUMENTS

This section discusses one of the more important applications of logic, i.e., its use to determine the validity of reasoning or arguments. Up to this section, statements have been considered as abstract elements which assume values of truth or falsehood. In this section the emphasis will be on examining some relationships between pairs of statements. The relationships of implication and equivalence between statements receive the greatest use in these applications and hence will be given the most emphasis in the discussion to follow.

The relationship of implication $(P \to Q)$ between two statements P and Q is characterized by its truth table. Recall that P implies Q means that Q must be true whenever P is true. The implication relationship between P and Q is also stated by "Q is a consequence of, follows from, or is deducible from P."

Example 2.9 (Form of Arguments). Consider the implication "If Mary studies, then she will pass." If we also assume that Mary does study–i.e., "Mary studies" is a true statement, then from the truth table for the implication it is concluded that Mary will pass. The structure of the argument or reasoning assumes the following form:

If Mary studies, then she will pass.	$P \to Q$
Mary studies.	P is true
Therefore (∴) Mary will pass.	$\therefore Q$

Note that the argument contained in the example consists of two hypotheses and a conclusion. Symbolically the argument assumes the form of the compound statement:

$$[(P \to Q) \wedge P] \to Q$$

The truth table for the above statement is

P	Q	$(P \to Q)$	$(P \to Q) \wedge P$	$[(P \to Q) \wedge P] \to Q$
T	T	T	T	T
T	F	F	F	T
F	T	T	F	T
F	F	T	F	T

Since the last column of the truth table contains all Ts, the compound statement is true regardless of the truth values of the statements P and Q. This states that, if Q is assumed to be a consequence of P, and P is assumed true, then the statement Q must be true. Any argument of this form is a *valid argument.* ■■

Example 2.10 (Arguments). Suppose it is stated that P implies Q and that Q is false. Can it be concluded the P *must be* false? The pattern of the argument may be written as

$P \to Q$	P implies Q
$\sim Q$	Q is false
$\therefore \sim P$	∴ P is false

or in the form of a compound statement

$$[(P \to Q) \wedge \sim Q] \to \sim P$$

To answer the question, the truth table for the above compound statement is constructed.

P	Q	$(P \to Q)$	$(P \to Q) \wedge \sim Q$	$[(P \to Q) \wedge \sim Q] \to \sim P$
T	T	T	F	T
T	F	F	F	T
F	T	T	F	T
F	F	T	T	T

The truth table contains all Ts for the compound statement and hence the argument or reasoning contained in the compound statement is valid. It is concluded that P must be false.■■

Example 2.11 (Checking Validity of Arguments). The following two examples illustrate how texts, which appear in every form of communication can be viewed as arguments and their validity checked by the methods considered in this chapter. The two examples and the two exercises which follow were taken from published accounts of statements by national figures. Notation is added for analysis purposes.

1. If N.V. agrees to a cease fire (P), then the U.S. will win world favor (Q) or Russians will not agree to peace terms ($\sim R$). However, it can be stated unequivocally that the U.S. will not win world favor ($\sim Q$) and that the N.V. will agree to a cease fire (P). Hence Russia will not agree to peace terms ($\sim R$).

   The first statement in the paragraph can be written $P \rightarrow (Q \vee \sim R)$. The second statement is $P \wedge \sim Q$. The third statement and ultimate conclusion in the argument is $\sim R$.

   To check the validity of the argument we list and combine the premises and conclusion in argument form

	$P \rightarrow (Q \vee \sim R)$	first statement
	and $P \wedge \sim Q$	second statement
therefore $\sim R$		conclusion

   The truth table indicates that the argument presented is a valid one.

P	Q	R	$P \rightarrow (Q \vee \sim R)$	$(P \wedge \sim Q)$	$\{[P \rightarrow (Q \vee \sim R)] \wedge (P \wedge \sim Q)\}$	$\{\} \rightarrow \sim R$
T	T	T	T	F	F	T
T	T	F	T	F	F	T
T	F	T	F	T	F	T
T	F	F	T	T	T	T
F	T	T	T	F	F	T
F	T	F	T	F	F	T
F	F	T	T	F	F	T
F	F	F	T	F	F	T

2. If the N.V. launch an offensive, then Russia must give approval and China must provide arms support $[P \rightarrow (Q \wedge R)]$. However, China has not provided arms support ($\sim R$). Therefore, the N.V. will not launch an offensive ($\sim P$).

To check the validity of the argument, we list the premises and conclusions in argument form:

$$\begin{array}{ll} P \to (Q \wedge R) & \text{first statement} \\ \sim R & \text{second statement} \\ \hline \therefore \sim P & \text{conclusion} \end{array}$$

The truth table shows that this argument is valid.

P	Q	R	$P \to (Q \wedge R)$	$\sim R$	$\{[P \to (Q \wedge R)] \wedge \sim R\}$	$\{\} \to \sim P$
T	T	T	T	F	F	T
T	T	F	F	T	F	T
T	F	T	F	F	F	T
T	F	F	F	T	F	T
F	T	T	T	F	F	T
F	T	F	T	T	T	T
F	F	T	T	F	F	T
F	F	F	T	T	T	T

■■

Exercise 2.11. If the N.V. launch an offensive, then the Russians will support them and Communist China will be unhappy. It has been determined that the Russians will support the N.V. Hence the Red Chinese will be unhappy.■■

Exercise 2.12. If the Russians enter the Vietnamese conflict, S.V. will lose its independence. Also, if the Red Chinese enter the Vietnamese conflict, S.V. will lose its independence. Russia or Communist China will enter the Vietnamese conflict. Therefore, S.V. will lose its independence.■■

Exercise 2.13. Check whether the following arguments are valid:

(a) $$\begin{array}{l} P \to Q \\ Q \\ \hline \therefore P \end{array} \qquad [(P \to Q) \wedge Q] \to P$$

(b) $$\begin{array}{l} P \to Q \\ \sim P \\ \hline \therefore \sim Q \end{array} \qquad [(P \to Q) \wedge \sim P] \to \sim Q$$

(c) $$\begin{array}{l} (P \to Q) \wedge (Q \to P) \\ P \\ \hline \therefore P \vee Q \end{array}$$

(d) $$\begin{array}{l} P \to Q \\ Q \to P \\ \hline \therefore P \wedge Q \end{array}$$

(e) All students are demonstrators.
Demonstrators are Hippies.
____
All students are Hippies. ■■

2.9 DIAGRAMMING ARGUMENTS

In Exercise 2.10(j) it was shown that the truth table for the compound statement

$$[(P \rightarrow Q) \wedge (Q \rightarrow R)] \rightarrow (P \rightarrow R)$$

contained all Ts. Hence this argument is valid. Employing meaningless statements, P, Q, and R are defined as follows:

$P \rightarrow Q$: All Mumboos are Jumboos.
$Q \rightarrow R$: All Jumboos are Cuckoos.

$P \rightarrow R$: All Mumboos are Cuckoos.

The final statement is true if, regardless of meaning, it is assumed that the first two implications are true–i.e.,

$P \rightarrow Q$ is true
and $Q \rightarrow R$ is true
---
then $P \rightarrow R$ is true

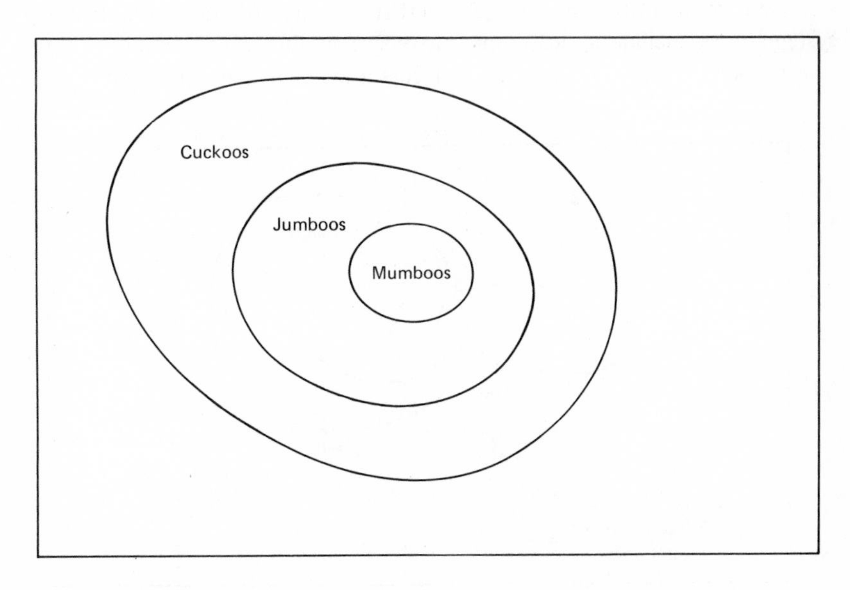

Figure 2.2

We can employ Venn diagrams from set theory to illustrate the validity of this argument. Let the rectangle in Figure 2.2 represent the universal set of elements under discussion. The first premise $P \to Q$ states that the set of Mumboos is contained in the set of Jumboos. The second premise $Q \to R$ states that the set of Jumboos is contained in the set of Cuckoos. From the diagram it is obvious that the set of Mumboos is contained in the set of Cuckoos–i.e., all Mumboos are Cuckoos.

Example 2.12 (Diagram of Argument). It is not easy to acquire the ability to assess the validity of a given argument. Even more difficult is the ability to infer valid conclusions from stated premises. Venn diagrams help in making inferences and checking their validity. Suppose that the following statements were made in the course of some discussion on weapon systems.

Weapons systems that cost a lot of money are effective.

Some weapons systems that cost a lot of money are transportable.

What would you, as an analyst, conclude from these two statements?

To answer this question use can be made of the Venn diagram of Figure 2.3. Note that the first statement is a direct implication and commits costly weapon systems to be effective (not necessarily true in the real world). The second statement is quantified by "some" so that the class of transportable weapon systems must include at least one costly system. One cannot imply that some transportable weapon systems do not cost a lot of money, although the Venn

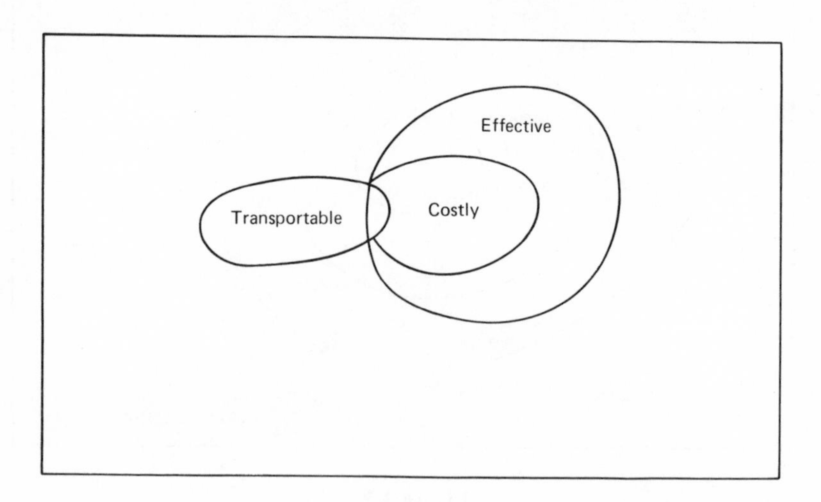

Figure 2.3.

diagram indicates otherwise. From the Venn diagram we can deduce two valid conclusions:

Some transportable weapon systems are effective.

Some transportable weapon systems cost a lot of money. ■■

Exercise 2.14. What conclusions can be drawn from the following groups of premises? Illustrate each argument with Venn diagram.

(a) All sheep are black.
Some sheep make good mutton.

(b) Some dogs have four legs.
Not all four-legged animals are pets.

(c) To be a pet is to be loved.
To be loved entails some work.

(d) All Americans love a good comedy.
All rich people love a good comedy.
Some rich people attend plays. ■■

The following example and its solution techniques illustrate how logic and its truth tables can be employed effectively in uncovering inconsistencies or purposeful deception in supplied information.

Example 2.13 (The Agent Problem). Two foreign agents A and B supply contradictory information on the whereabouts of an important diplomat C. Each agent states that the other agent's information is purposefully false. Agent A states that diplomat C is in North Korea, while agent B states that diplomat C is not in North Korea. Collateral information from a reliable source provides us with the following facts:

1. One of the two agents is a plant and hence always lies while the other always tells the truth. Not having the identities of A and B at hand, there is no way of knowing which is the plant.
2. Either diplomat C is not in North Korea or agent A is not the plant.
3. Either diplomat C is not in North Korea or agent B is not the plant.

Which agent is the plant and is diplomat C in North Korea?

Solution: We are concerned with three statements for which we must decide whether each is true or false. To do so we adopt the following notation:

P: Agent A is the plant.
Q: Agent B is the plant.
R: Diplomat C is in North Korea.

An examination of the problem indicates that there are five statements to be investigated for conflicts or inconsistency. The statements are:

1. The plant lies and the other agent tells the truth.
2. Agent A states that diplomat C is in North Korea.
3. Agent B states that diplomat C is not in North Korea.
4. Diplomat C is not in North Korea or agent A is not the plant.
5. Diplomat C is not in North Korea or agent B is not the plant.

Using the notation introduced above, the five statements can be restated symbolically as follows:

1. Statements 2 and 3 below result from the application of the relationship between being a plant and the truthfulness of his statements.
2. (a) $\overline{P \cap R}$, i.e., it is false that agent A is the plant and diplomat C is in North Korea.
   (b) $\overline{\bar{P} \cap \bar{R}}$, i.e., it is false that agent A is not the plant and diplomat C is not in North Korea.
3. (a) $\overline{Q \cap \bar{R}}$, i.e., it is false that agent B is the plant and diplomat C is not in North Korea.
   (b) $\overline{\bar{Q} \cap R}$, i.e., it is false that agent B is not the plant and diplomat C is in North Korea.
4. $\bar{R} \cup \bar{P}$, i.e., it is true that diplomat C is not in North Korea or agent A is not the plant.
5. $\bar{R} \cup \bar{Q}$, i.e., it is true that diplomat C is not in North Korea or agent B is not the plant.

To check each of these statements for consistency, we construct the following truth table:

P	Q	R	$\overline{P \cap R}$	$\overline{\bar{P} \cap \bar{R}}$	$\overline{Q \cap \bar{R}}$	$\overline{\bar{Q} \cap R}$	$\bar{P} \cup \bar{R}$	$\bar{Q} \cup \bar{R}$
T	T	T	2(a)				4	5
T	T	F			3			
T	F	T	2(a)			3(b)	4	
T	F	F						
F	T	T						5
F	T	F		2(b)	3			
F	F	T				3(b)		
F	F	F		2(b)				

The table lists all possible combinations of truth values for the three original statements, P, Q, and R. We seek a unique solution, i.e., truth values for P, Q, and R which are consistent with the statements 1-5. Statement 2, i.e., $\overline{P \cap R}$ indicates that rows 1 and 3 of the table yield truth values of P, Q, and R which yield inconsistencies. For example, in row 1 statements P and R are true. Hence the statement $P \cap R$ is true for these values of P and R. However, statement 2(a) indicates that $P \cap R$ must be false, i.e., $\overline{P \cap R}$. Thus the truth combinations of rows 1 and 3 are excluded from further consideration by the statement $\overline{P \cap R}$.

Continuing through the statements and applying them against the possible combinations of truth values for P, Q, and R, we eliminate those combinations of truth values which contradict or are inconsistent with any of the five statements. The application of the statements, by number, and the rows of the truth table each eliminates is indicated in the table. The row (row 4) which is not eliminated has that P is true, i.e., agent A is the plant, Q is false, i.e., agent B is not the plant, and R is false, i.e., diplomat C is not in North Korea. This is the solution.

Note that statement 4 is not required in the solution of this problem. Why is this the case?

Alternate Solution: A second solution to the agent problem is to construct a truth table to determine the validity of the arguments offered by the two suspected foreign agents. We again adopt the following notation:

P: Agent A is the plant.
Q: Agent B is the plant.
R: Diplomat C is in North Korea.

Note that $P \wedge R$ and $Q \wedge \sim R$ are statements by agents A and B, respectively. Collateral information yields the two statements, $\sim R \vee \sim P$ and $\sim R \vee \sim Q$ which are relevant to agent A and B, respectively. Thus the two implications are

$$(P \wedge R) \rightarrow (\sim P \vee \sim R)\text{: Agent } A$$
$$(Q \wedge \sim R) \rightarrow (\sim Q \vee \sim R)\text{: Agent } B$$

The requirement now is to check the validity of the two arguments. The following truth table yields that agent B's argument is valid but agent A's argument contains some inconsistencies. Hence agent A is the plant and diplomat C is not in North Korea.

P	Q	R	$\sim P$	$\sim Q$	$\sim R$	$P \wedge R$	$Q \wedge \sim R$	$\sim R \vee \sim P$	$\sim R \vee \sim Q$	$(P \wedge R) \rightarrow (\sim P \vee \sim R)$	$(Q \wedge \sim R) \rightarrow (\sim R \vee \sim Q)$
T	T	T	F	F	F	T	F	F	F	F	T
T	T	F	F	F	T	F	T	T	T	T	T
T	F	T	F	T	F	T	F	F	T	F	T
T	F	F	F	T	T	F	F	T	T	T	T
F	T	T	T	F	F	F	F	T	F	T	T
F	T	F	T	F	T	F	T	T	T	T	T
F	F	T	T	T	F	F	F	T	T	T	T
F	F	F	T	T	T	F	F	T	T	T	T

2.10 RELATIONSHIP OF FORMAL LOGIC TO SET THEORY

One way to define a set is to specify the properties its elements must possess, i.e., a set S is defined as

$$S = \{x | x \text{ has property } P\}$$

In formal logic it is possible to associate a special set with every statement. Let X be a compound statement made up of constituent statements $P, Q, R, \ldots$. We can define the set S_X, called the *truth* set of statement X, as follows:

$$S_X = \{\text{truth states of } P, Q, R, \ldots, | X \text{ is true}\}$$

For example, let X be the statement "P or Q." The S_X is the set consisting of the truth values for P and Q which make "P or Q" true.

$$S_X = \begin{Bmatrix} P = T, Q = T \\ P = T, \bar{Q} = F \\ \bar{P} = F, Q = T \end{Bmatrix}$$

The truth set S_X for any statement X is readily identified from the truth table for the statement X. Briefly, it consists of those truth values of the simple constituent statements for which the compound statement is true. In the truth table for $X = P$ or Q, the first three rows qualify for membership in S_X:

P	Q	P or Q	
T	T	T	} S_X
T	F	T	} S_X
F	T	T	} S_X
F	F	F	

In the truth table for $X = \sim [(P \wedge Q) \vee R]$ depicted below, rows 4, 6, and 8 identify elements of S_X:

P	Q	R	$\sim [(P \wedge Q) \vee R]$	
T	T	T	F	
T	T	F	F	
T	F	T	F	
T	F	F	T*	
F	T	T	F	
F	T	F	T*	S_X
F	F	T	F	
F	F	F	T*	

From the above it becomes clear that set theory and the logic of statements are closely related in structure and order. The relationships which exist between set operations and the conjoining of statements are summarized in Table 2.1.

Table 2.1

Sets	Statements	Remarks
$\in$	T	Membership ↔ truth state T
$\notin$	F	Nonmembership ↔ truth state F
$A \cup B$	P or Q	See Figure 2.4
$A \cap B$	P and Q	See Figure 2.5
$\overline{A}$	$\sim P$	Complementation ↔ negation
$\overline{A} \cup B$	$P \rightarrow Q$	See Figure 2.6
$\overline{A \cap \overline{B}}$	$\sim P \cup Q$	Equivalent to above
$\overline{A - B}$	$\sim (P \cap \sim Q)$	Equivalent to above

The table is explained best through the use of examples. Again let $X = P$ or Q and $X_s = A \cup B$. Figure 2.4 gives the truth table and the Venn diagram for the statement and its set equivalent.

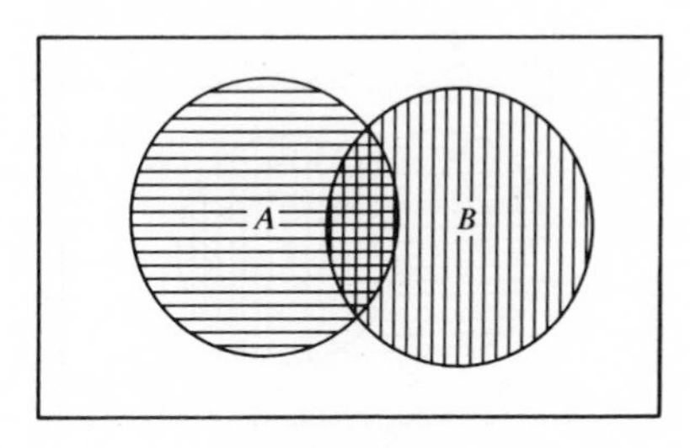

Figure 2.4

P	Q	P or Q	
T	T	T	S_X
T	F	T	S_X
F	T	T	S_X
F	F	F	

The truth set S_X of X contains those rows of the truth table for which "P or Q" is true. The set A is defined by the points residing in the circle containing horizontal lines. The set B is defined as the vertically lined circle. The set $A \cup B$ is the lined area within the square. Note that membership in set A corresponds to truth value T for statement P. Membership in set B corresponds to truth value T for statement Q. Membership in $A \cup B$ corresponds to truth state T for "P or Q" and hence membership in S_X.

Figure 2.5 depicts the truth table and the Venn diagram for the statement $X = P$ and Q and its equivalent set representation $X_s = A \cap B$.

P	Q	P and Q	
T	T	T	S_X
T	F	F	
F	T	F	
F	F	F	

For X = "P and Q," S_X consists of the first row of the truth table; X_s consists of points residing in cross-hatched portion of the square. Note again that membership in $A \cap B$ corresponds to membership in the truth set S_X or X = "P and Q."

Figure 2.6 portrays the truth table and the Venn diagram for the statement $X = P + Q$ and its equivalent set representation $X_s = \sim A \cup B$ or $X_s = \bar{A} \cup B$.

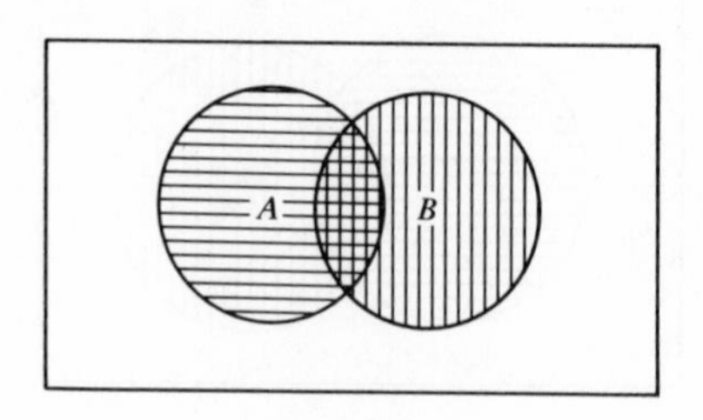

Figure 2.5

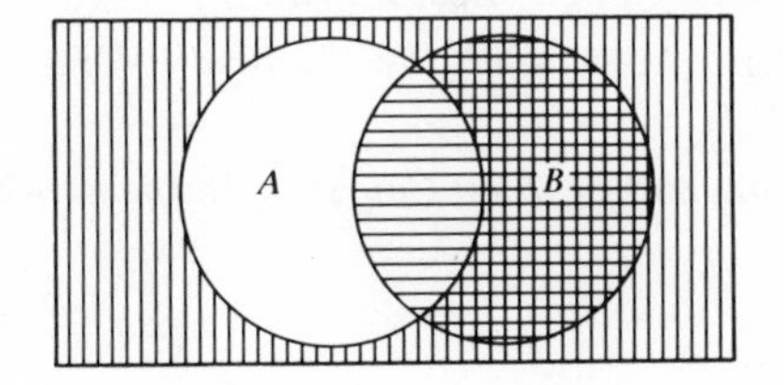

Figure 2.6

P	Q	$P \to Q$	
T	T	T*	S_X
T	F	F	
F	T	T*	
F	F	T*	

The truth set S_X, identified by the rows with asterisks in the truth table, corresponds with the lined area in the Venn diagram which represents $X_s = \sim A \cup B$. It should be noted that the unlined area is $A - B$. The complement of $A - B$, denoted $\overline{A - B}$, is $\sim A \cup B$ which as previously demonstrated is $\sim (A \cap \sim B)$.

Exercise 2.15. Determine the truth sets for each of the following statements. Determine which statements are equivalent. Determine which, if any, statements implies any other statements.

(a) $(P \vee \sim Q) \to R$

(b) $P \to P$

(c) $(\sim P \wedge Q) \vee R$

(d) $\sim (P \wedge R) \to \sim Q$

(e) $P \to (P \vee Q)$

(f) $(P \wedge R) \vee \sim Q$

REFERENCES

Christian, Robert. *Introduction to Logic and Sets.* Boston: Ginn, 1958.

Copi, Irving M. *Symbolic Logic.* New York: Macmillan, 1965.

Copi, Irving M. *Introduction to Logic.* New York: Macmillan, 1968, pp. 337-415.

Dixon, John R. *A Programmed Introduction to Probability.* New York: Wiley, 1964.

Lipschutz, Seymour. *Theory and Problems of Finite Mathematics, Schaums Outline Series.* New York: McGraw-Hill, 1966.

Quine, W. V. O. *Elementary Logic.* Boston: Ginn, 1941.

Stoll, Robert R. *Sets, Logic and the Axiomatic Method.* San Francisco: Freeman, 1961.

Suppes, Patrick, and Hill, Shirley. *First Course in Mathematical Logic.* New York: Blaisdell, 1964.

Chapter 3 / Plausible Reasoning

3.1 INTRODUCTION

In Chapter 2 the techniques stressed were those used to determine the truth value of compound statements and the validity of arguments when the statements involved were either true or false. What was not stressed is the diminished applicability of the techniques, truth tables, syllogisms, of propositional calculus in determining the credibility of compound statements and the plausibility of arguments when uncertainty exists in either the truth value of statements or in the validity of the implications (or argument).

This chapter describes additional techniques (under the name of *plausible reasoning*) which assist the substantive analyst in

1. Establishing the credibility (level of truth) of information.
2. Establishing the plausibility (level of validity) of arguments $A \rightarrow B$.
3. Analyzing the credibility relationships which exist between premise and consequence in *valid arguments* when the premise and/or consequence are credible but not specifically true or false.
4. Analyzing the credibility relationships which exist between premise and consequence in *other than valid* arguments when the premise and/or consequence are true, false, or credible.
5. Determining the credibility of statements on the basis of their substantive ties to other statements of known credibility levels.

Thus this chapter provides a transition from the formal structures and methods of logic to the less formal structures and methods of plausible reasoning. The chapter contains many examples which illustrate specific applications of the different forms of plausible reasoning to the methodological synthesis and analysis of substantive information.

3.2 REPORT CREDIBILITY

In practice information sources include a variety of physical sensors (photo, electronic, etc.) or human sensory systems. The collecting or reporting agent or agency, when possible, gives an evaluation of the reliability of their source and sensory system and the credibility of the information obtained from the source. It is standard for the reporting agent or agency to employ the credibility code in Figure 3.1 as a guide to evaluating the credibility of the information contained in the information report (IR) and of the information source. This credibility code is recommended as a standard for all substantive analysts and will be used as such in the remainder of this text.

Source	Information
A. Completely reliable	1. Confirmed by other
B. Usually reliable	2. Probably true
C. Fairly reliable	3. Possibly true
D. Not usually reliable	4. Doubtful
E. Unreliable	5. Highly unlikely
F. Do not know	6. Cannot be judged

Figure 3.1 Credibility Code

For example, if the originator of an information report evaluates his report B4, he means that the source (person, investigator, photo, defector, etc.) is usually relaible (B), but in this instance the information the source provided is doubtful (4). A typical information report evaluated B4 may say that an important diplomat has been seen in a specific city. The diplomat's mission and the people he visited may be listed in the collector's report. On receiving the report, the analyst must decide whether he will accept the evaluation of the originator of the report. He formulates the following questions: How good is the evaluator's evaluation? Was the diplomat's mission accurately assessed by the source at the 4 level? Was the source friendly or an enemy plant? How reliable have been this originator's previous reports? The analyst cannot take the collector's evaluations at face value. He must know his sources and their biases. Knowing these, he can modify their evaluations accordingly. Techniques used to evaluate the reliability of a source are discussed further in section 3.19. Reports judged as F6 require special consideration. The F6 report should not be considered worthless. The F6 credibility level allows the reporter an opportunity to state that he has no way of judging the source or the information supplied by the source. Hence, in some instances, the information contained in the F6 report is the only information on the subject and it is recognized that information was obtained from sources which escape evaluation. Rumors, cocktail talk, a quick viewing may result in the F6 report. It could be the precursor to some valuable intelligence. Hence, many F6 reports will warrant retention until additional information confirms or denies their contents.

3.3 INFORMATION ANALYSIS

The information analyst analyzes and evaluates the credibility of each piece of information he receives. He examines the plausibility of possible relationships among the items of information he processes. He attempts to synthesize items of information into an "incomplete" whole. How are the information items connected? Is one item a consequence of another? Are there any cause-effect relationships? Is the information item necessary to the conclusion of the analysis? Is it sufficient? Further, he attempts to determine whether the relationships (vis: cause-effect, stimulus-response, input-output of information) he hypothesizes are plausible and complete. If these relationships are lacking in plausibility or completeness, the analyst seeks more information through additional collection requirements in order that he may generate and test more hypotheses. It is in the testing of hypotheses, either objectively or subjectively, that the analyst exercises the statistical aspects of plausible reasoning to the greatest degree.

3.4 INFORMATION CREDIBILITY

Specifically, analysts deal with information items and with the relationships of how the information items are connected. Varying degrees of uncertainty exist in the information items and in their interrelationships in most practical information evaluation situations. Information has a continuous range of credibility. Each item of information can be evaluated as false or true or can range over a scale of credibility between true and false. One such scale of evaluation of information is suggested in a modified version of the Kent chart reproduced as Figure 3.2. Note that this scale applies to the credibility of information items which are simple statements. These items may be part of a complete information report. The credibility code generally is used to evaluate the reliability of an information source and the credibility of the information contained in its report. The information in the report can be a simple statement such as "Diplomat *X* has been assigned to embassy *Y*," or it can be a complete report covering diplomatic views on an important policy matter. Thus, the credibility code reflects a macro evaluation of the source of the report and information content, while the Kent chart reflects a micro evaluation of well-defined information items which may or may not be part of a report.

The first line (row) of the chart indicates the type of evaluation: numerical or nominal. The entries in the second row indicate whether the column contains

	Numerical Value				Nominal Values	
	Average		Range		Range	
	For	Against	For	Against	For	Against
	1.00	0	1.00	0	Certain	Impossible
Realm of possibility	0.93	0.07	0.86-0.99	0.01-0.14	Almost certain, highly likely	Almost impossible, highly unlikely
Realm of possibility	0.70	0.30	0.55-0.85	0.15-0.45	Probably likely, we believe	Doubtful, unlikely, improbable
Realm of possibility	0.50	0.50	0.46-0.54	0.46-0.54	About an even chance, about equally likely	About equally unlikely, about an even chance
Realm of possibility	0.30	0.70	0.15-0.45	0.55-0.85	Doubtful, unlikely, improbable	Probably likely, we believe
Realm of possibility	0.07	0.93	0.01-0.14	0.86-0.99	Almost impossible, highly unlikely	Almost certain, highly likely
	0	1.00	0	1.00	Impossible	Certain

Figure 3.2. Modified Kent Chart

From a privately circulated memo titled "Words of Estimated Probability" by Dr. Sherman Kent.

average numerical values or a range of values. From the third to the tenth row, the principal columns are each subdivided into two columns. The first column of the two is headed by the word "for" which signifies "the probability for." The second column is headed by the word "against" which signifies "the probability against." These entries can be interpreted as weights or as probabilities as will be noted later in this section.

Example 3.1 (Use of Kent Chart). To illustrate the use of the Kent chart we consider an analyst receiving covert information on the appearance of an important Russian diplomat in one of Russia's satellite countries. The analyst feels that the information is significant and highly likely. However, the analyst retains some doubt because collateral information does not confirm (or deny) the appearance of the diplomat. Using the Kent chart, the analyst asserts this doubt by evaluating his information as "likely" rather than "highly likely" or "certain." The analyst will retain the nominal value for the information on the diplomat until additional information causes the analyst to reassess the credibility of his information.

Analysts are often times asked to assess the possibility of an event occurring. For example, a typical question is "What is the possibility that Country X will have a bad wheat crop this year?" Typical answers are "it is possible," "it is clearly possible" or "there is a good possibility." Such usage and modifications of the word possible do not clarify what needs clarifying. Specifically, when one says that an event is possible, one is saying only that the event is not impossible. No statement is made relative to the level of possibility. What is needed is a means of subdividing the "realm of possibility" into meaningful subranges. The Kent chart does this. It accepts that every one agrees to the meaning of "certain," "almost certain," "impossible" and "almost impossible." Also, the Kent chart averts the confusion caused by modifying the word possible. The nominal values assigned to the subdivisions of the "realm of possibility" avoid the use of the word "possible" which carries with it the complete realm of possibility as identified in the Kent chart. It is recommended that analyst not modify the term "possible" to indicate various levels of probability. Modification of the terms "likely," probably and doubtful is preferred and should lead to more uniform usage and understanding.

In employing the Kent chart, the analyst must be careful in deciding whether his experience with the source and content of his information is sufficient to assign verbal or numeric values to the information item at hand. For example, the analyst may have received twenty highly reliable reports on the firing of a specific type of surface-to-air missile (SAM); sixteen of the twenty reports indicate successful firings–launch through target destruction. Based on this information the analyst will assign the probability of 0.8 to the statement

Credibility Code			Kent Chart					
			Numerical Value				Verbal values	
Source	Information		Average		Range		Range	
			For	Against	For	Against	For	Against
			1.00	0	1.00	0	Certain	Impossible
Completely reliable (A)	Confirmed by others (1)		0.93	0.07	0.86-0.99	0.01-0.14	Almost certain, highly likely	Almost impossible, highly unlikely
Usually reliable (B)	Probably true (2)		0.70	0.30	0.55-0.85	0.15-0.45	Probably likely, we believe	Doubtful, unlikely, improbable
Fairly reliable (C)	Possibly true (3)	Realm of possibility	0.50	0.50	0.46-0.54	0.46-0.54	About an even chance, about equally likely	About equally unlikely, about an even chance
Not usually reliable (D)	Doubtful (4)							
			0.30	0.70	0.15-0.45	0.55-0.85	Doubtful, unlikely, improbable	Probably likely, we believe
Unreliable (E)	Highly unlikely (5)							
			0.07	0.93	0.01-0.14	0.86-0.99	Almost impossible, highly unlikely	Almost certain, highly likely
			0	1.00	0	1.00	Impossible	Certain

Figure 2.3 Combined Charts

"SAM will destroy our bombers if the bombers fly within the range of the SAM." The 0.8 figure was obtained from the ratio of relative frequency of successful firings (16) to the total number of observed firings (20), and on this alone, assuming that no other information (such as the effects of ECCM, evasive action, etc.) was available to the analyst.

Continuing the example, suppose that the intelligence analyst receives only one report which states that the observer reported *many* (not twenty) SAM firings and that *most* of the firings were successful. In this case the analyst must evaluate the credibility of the observer based on his past performances. And he must assess what the observer means when he uses the terms "most" and "many." If the observer is highly credible, the interpretation of "most" could be "almost certain" or it could be "probably." Thus, the ranges of 0.86-0.99 or 0.55-0.85 can be selected, based on the analyst's interpretation of the reporter's usage of "most."

There are many types of statements whose credibility cannot be quantified through use of the Kent chart. Specifically, a general statement of the type "Within five years a major conflict between Communist China and Russia is inevitable," is outside of the realm of evaluation by the Kent chart. Rather, many of the premises from which this general statement was deducted might be evaluated in terms of the Kent chart and hence contribute to a more meaningful evaluation of the credibility of the general statement.

At this point it is informative to combine the charts given in Figures 3.1 and 3.2 into one chart and ascertain the interrelationships which exist between the credibility of new information reports as given by the credibility code and the probability or credibility of process information (facts, statements, events) as given by the modified Kent chart. The combination is given in Figure 3.3. However, keep in mind that one section applies to reports (macro) and the other to information items (micro). ■■

In this discussion credibility and probability will be used somewhat synonymously. The chart given in Figure 3.4 indicates the numeric ranges and the approximate interchangeability of the two terms. Note and compare the names attached to the various credibility levels. Note also that in the credibility code, the use of certain and impossible are avoided. This provides a hedge for human error and the unexpected.

The abbreviations SLC and SMC used in Figures 3.4 and 3.5 have the following meanings: In the middle of the credibility scale, the credibility level is designated as "even"; as the credibility of a statement is lowered beyond the 0.45 point at the lower end of the "even" level, the statement is called "less credible" or LC. If the credibility is lowered beyond the 0.15 end point of the LC region, the statement is called "somewhat less credible" or SLC. Comparable

Credibility	Credibility code	Probability	
False (F)	None	0	Impossible
Somewhat less credible (SLC)	Unreliable	0.01-0.14	Highly unlikely
Less credible (LC)	Not usually reliable	0.15-0.45	Improbable
Even (E)	Fairly reliable	0.46-0.54	Equally likely
More credible (MC)	Usually reliable	0.55-0.85	Probable
Somewhat more credible (SMC)	Completely reliable	0.86-0.99	Highly likely
True (T)	None	1.00	Certain

Figure 3.4. Interchangeability

meanings are applied to the regions above the even region. The key point to note here is that the terms LC, SLC, MC, and SMC are credibility assessments relative to midrange of statement credibility (even).

Few information reports contain information items which can readily be evaluated as absolutely true or false. Most information items are subject to error or deception and hence must be assessed as having only limited credibility if statements or as having a probability of occurring if events. Because of this uncertainty in the credibility of a statement or in the occurrence of an event, it is convenient to think in terms of a scale of credibility as indicated in Figure 3.5. In the figure the *A* scale refers to a statement or event designated by *A* and its levels of credibility or probability. Starting from the left, we note the letter F (signifying False) designating the left end point of the scale. At this point *A* has truth value (credibility) F or probability 0. Proceeding from the left to right, the credibility of *A* increases from F to higher levels until the right end point of the *A* scale is reached. At this point, the credibility of *A* is T. Between the

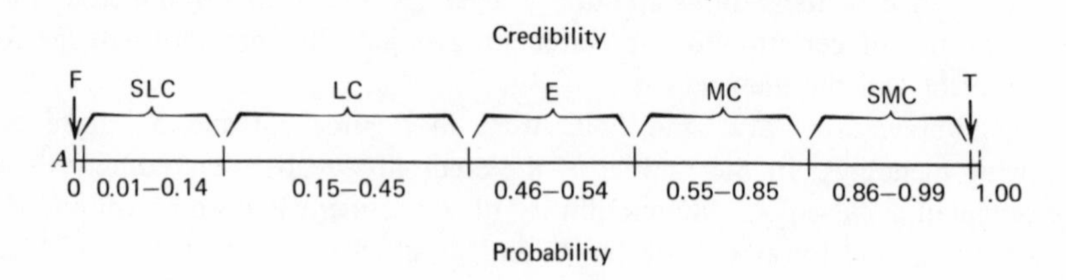

Figure 3.5 Credibility scale.

two end points, A assumes credibility levels ranging from SLC to SMC. The probability ranges for A change similarly from 0.01-0.14 to 0.86-0.99.

3.5 RELATIONAL EVALUATION

In this section we consider the credibility of two statements simultaneously. To do so we need to align two credibility scales, one for each statement, and make direct comparisons as indicated in Figure 3.6. In Figure 3.6 the arrows over the A scale and under the B scale indicate the level of credibility assigned to statements A and B. On the scales shown, A is assessed in the LC region and B in the MC region.

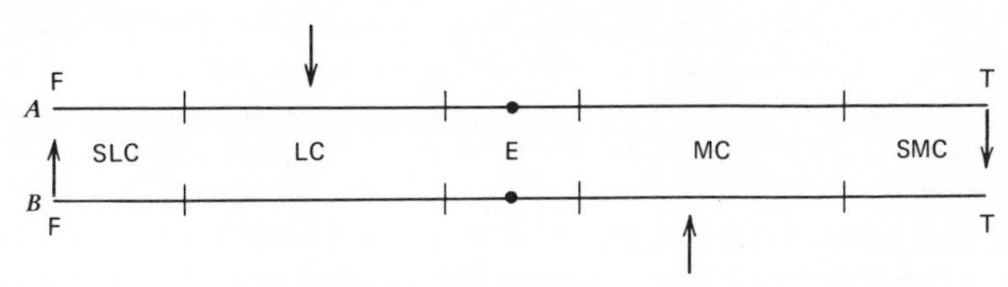

Figure 3.6 Paired credibility scales.

Through the use of the paired credibility scales, it is possible to assess the credibility of statements which are related through direct implication. Returning to the "formal" logic of statements, we recall the following two patterns of the deductive argument:

Pattern 1:		Pattern 2:	
	$A \to B$		$A \to B$
	B is false		A is true
	$\therefore A$ is false		$\therefore B$ is true

Pattern 1 of the deductive argument states that if the consequence B of an implication is false, then the premise A of the implication *must be* false. Referring to the paired credibility scales, pattern 1 places the truth value of B at the F position of the B credibility scale and assigns the credibility value F to statement A in the implication $A \to B$. We denote this by placing an arrow (implication) between the two scales at the F level pointing from the B scale to the A scale.

Pattern 2 of the deductive argument states that if the premise A of an implication is true, then the consequence B of the implication *must be* true. Referring to the paired credibility scales, pattern 2 places the truth value of A at the T position of the A credibility scale and assigns the credibility value T to statement B in the implication $A \to B$. We denote this by placing an arrow (implication) between the two scales at the T level pointing from the A scale to the B scale. Thus, in dealing with credibility in the evaluation of the truth value of statements, the end points of the paired credibility scales yield the credibility relationships between the premise and consequence in a deductive argument whenever the extreme values of T or F can be assigned to the statements A or B, respectively.

When the credibility of statements A or B is other than a definitive T or F in the deductive argument patterns 1 or 2, the paired credibility scales can still be employed gainfully in understanding the credibility relationships which exist between the premise and consequence.

Starting with the deductive argument pattern 1, let us assume that the credibility of statement B is E (even). Now let us observe how the credibility of statement A might vary as the credibility of statement B varies from E to F.

From Figure 3.6 it is noted that as the truth value of B tends towards F on the B scale but does not achieve the value of F, the truth value of A is not restricted to any credibility range on the A scale. However, it is reasonable (credible) to accept the truth value of A to be less than the truth value of B in the pattern 1 argument whenever we accept B as having a truth value other than false. This assumption is reasonably justified as follows: As the truth value of statement B tends to F from its value E, eventually the truth value of A must tend toward F. Thus, when the truth value of B is close to F, we can reasonably expect the truth value of A to be at least as close to F as B. A safe but pessimistic assumption is that the truth value of A is closer to F than the truth value of B wherever B is on the F side (left of E) of the truth value scale. Figure 3.7 illustrates the relationship between the truth values of A and B when B is on the F side of E on the truth scale.

Starting with truth value B_0 and progressing to the left through truth values B_1, B_2, B_3 and F for B, we can plausibly state that A will have approximately the corresponding truth values A_0, A_1, A_2, A_3 and F, respectively. Note that *we are not forced* to place the truth values of A at the intermediate positions

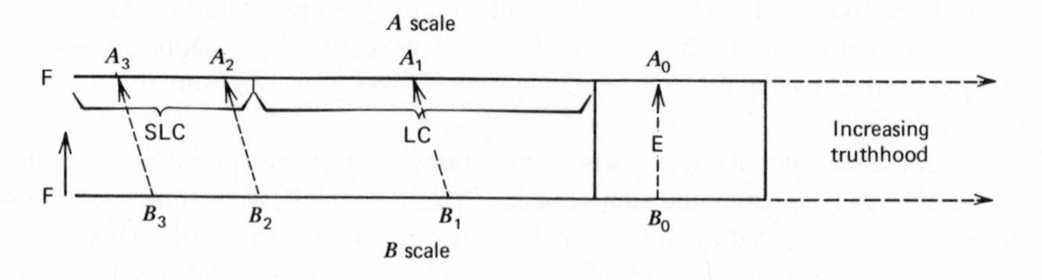

Figure 3.7 Credibility relationships.

(A_0, A_1, A_2, A_3) shown. We are only forced to value A at position F whenever we value B at position F. We only state that as the truth value of B tends toward the value F, the truth value of A *must eventually* tend toward F. The closer the truth value of B is to F, the more credible is the statement that the truth value of A is as least as close as the truth value of B to F.

Example 3.2 (Credibility Patterns). Consider the two statements:

A: Country X is recalling its ambassador from country Y.

B: Country X is initiating a trade embargo against country Y.

First, assume that every time country X recalls its ambassador from a country, it subsequently initiates a trade embargo against that country. On the basis of this assumption, we can state that $A \rightarrow B$ is a valid implication.

Now suppose that it is difficult to obtain information on the intentions of country X relative to recalling its ambassador. Yet country Y can, through trade sources, obtain some information on whether country X intends to initiate a trade embargo, i.e., it is difficult for country X to cover the vast preparations required to initiate a trade embargo. Hence country Y assigns its resources to the collection of intelligence on a possible trade embargo in order to anticipate the recall of the ambassador of country X. When intelligence indicates that a trade embargo is highly likely, then country Y can anticipate the recall of country X's ambassador. Yet there is some reluctance on country Y's part to say that country X's recall of its ambassador is highly likely. Rather, the tendency is to have the likelihood of A occurring lag the likelihood of B occurring. If evidence continues to accumulate and changes the likelihood of a trade embargo to be highly unlikely, then there is a strong tendency to evaluate the possibility of country X recalling its ambassador at a still lower level.

At the other end of the truth scale we have the deductive argument pattern 2 repeated as:

$$\begin{array}{c} A \rightarrow B \\ A \text{ is true} \\ \hline B \text{ is true} \end{array}$$

In this deductive pattern, the truth of the premise A guarantees the truth of the consequence B. Now let us investigate what can be said of the truth value of B when the truth value of A is less than T. Starting with the truth value E for statement A and allowing the truth value of A to progress from E to T on the A scale, we are forced to accept that the truth value of B on the B scale will be T when the truth value of A arrives at T on the A scale. At intermediate truth value positions E, A_1, A_2 and A_3 for A on the A scale, it is plausible to associate values E, B_1, B_2 and B_3 for B on the B scale, as indicated in

Figure 3.8. Note that we have placed the truth values for B slightly higher than the corresponding values for A. We are *not forced* to do this. We only state that it is plausible that we do so to guarantee that B will be true when A becomes true. It is quite possible for B to be true even if A is completely false, and thus reasonable, in cases of uncertainty, to assign B a slightly higher truth value than A. ■■

Example 3.3 (Credibility Relationships). Consider the two statements A and B listed in Example 3.2. Again, suppose that each time country X recalls its ambassador from a country, it subsequently initiates a trade embargo against that country. Hence, we can state that $A \to B$ is a valid implication.

Now suppose that through an outstanding information collection effort country Y is able to obtain information on the intentions of country X relative to recalling its ambassador from country Y. As evidence accumulates supporting the recall intention country Y feels at least equally assured that country X will initiate a trade embargo against them. If the evidence accumulates and indicates a trend toward country X continuing diplomatic relations, then country Y can be at least equally assured that a trade embargo will not be initiated.

At this point, with the assistance of Figures 3.7 and 3.8, it is possible to summarize some intuitively acceptable practices in interpreting *deductive* patterns when the statements therein have credibility other than T or F.

Interpretation 1: When the credibility of A or B is in the E range, i.e., the truth or falsity of A or B is about even, the following patterns evolve:

$$\begin{array}{c} A \to B \\ B \text{ is E} \\ \hline A \text{ is E} \end{array} \qquad \qquad \begin{array}{c} A \to B \\ A \text{ is E} \\ \hline B \text{ is E} \end{array}$$

The two patterns indicate that when two statements are linked together through a direct implication, and either statement has a truth value of approximately E,

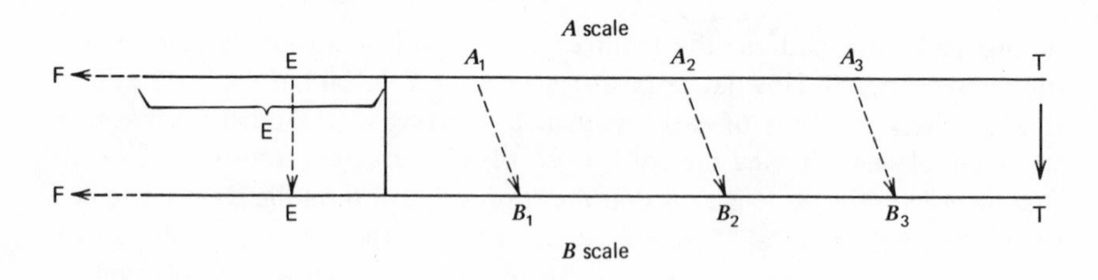

Figure 3.8 Credibility relationships.

it is plausible to accept the truth value of the remaining statement at approximately the E level. If one interprets the E region as the region of uncertainty, then the two patterns state that uncertainty in premise or conclusion begets uncertainty in conclusion or premise, respectively.

Interpretation 2: When the truth value of B tends toward F, it is plausible to conclude that the truth value of A tends slightly more rapidly than the truth value of B toward F, i.e., the credibility of A remains less than that of B as B tends toward F and assumes truth values close to F.

$$\begin{array}{l} A \rightarrow B \\ B \text{ is LC} \\ \hline A \text{ is SLC} \end{array}$$

Interpretation 3: When the truth value of A tends toward T, it is plausible to conclude that the truth value of B tends slightly more rapidly than the truth value of A toward T, i.e., B remains more credible than A as A tends toward T and assumes truth values close to T.

$$\begin{array}{l} A \rightarrow B \\ A \text{ is MC} \\ \hline B \text{ is SMC} \end{array}$$

Interpretation 4: It is acceptable to have A false in the implication $A \rightarrow B$ and still have B true. As A becomes more true, i.e., more evidence is gathered to support A, then based on the credibility of A, the credibility of B exceeds the credibility of A and tends more rapidly than A toward higher levels of credibility.

Interpretation 5: It is acceptable to have B true in the implication $A \rightarrow B$ and still have A false. As B loses its credibility, i.e. tends toward falsehood, then based on the loss of credibility in B, the credibility of A lags B and tends more rapidly toward lower levels of credibility.

The credibility relationships obtained between A and B, as the truth values of A and B range between T and F in the implication $A \rightarrow B$, can be summarized

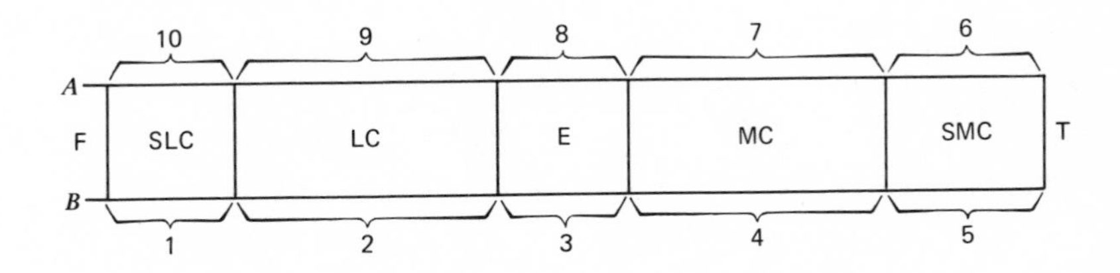

Figure 3.9 Credibility regions.

through the use of the credibility scale diagram shown in Figure 3.9 and the ten deductive logic patterns shown in Figure 3.10.

Again we assume that the credibility (level of truth) value of *A* or *B* is continuously variable and can be measured between F and T through the intermediate ranges SLC, LC, E, MC and SMC, as indicated in Figure 3.9. The argument pattern numbers of Figure 3.10 and the numbers in Figure 3.9 are related as follows: The second statement in the premise involves statement *B* in patterns 1-5, and statement *A* in patterns 6-10. In each pattern the pattern number corresponds to the credibility region in which the credibility value of *A* or *B* lie. For example, in pattern 3, *B* assumes a credibility value of E which lies in region 3. In pattern 7, the credibility value of *A* lies in region 7.

Starting with the deductive pattern 1 in Figure 3.10 we note that this pattern restates the previous finding that *A* must be false if *B* is false. Parenthetically, pattern 1 indicates that if the credibility value of *B* ranges in the SLC region, we expect the credibility value of *A* to range in the same region and be slightly lower than *B*.

When the credibility value of *B* ranges in the LC region, as indicated by pattern 2 in Figure 3.10, and as located in the credibility scale of Figure 3.9, it is more difficult to state definitively what the corresponding credibility value of *A* should be. As long as we have assessed the credibility value of *B* to be less than even, the tendency is to assign a credibility value to *A* at least as low as *B*.

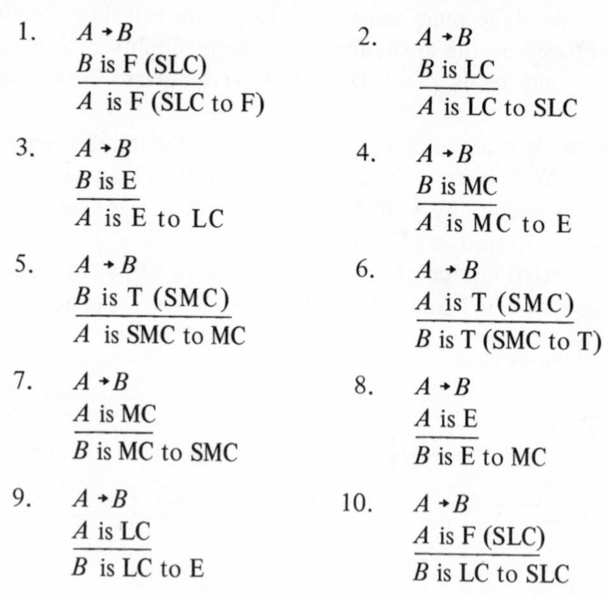

Figure 3.10 Argument Patterns

Thus, pattern 2 indicates that when B is LC, A can plausibly be assigned credibility values from LC down through SLC.

Patterns 3, 4, and 5 all exhibit a lower assessment of the credibility value of the premise (A) relative to the assessment of the conclusion (B), regardless of the assessed credibility value of the conclusion. This relationship persists in the implication $A \rightarrow B$ because B is the necessary condition for A. It supports the credibility of A but it is not sufficient for (does not guarantee) the truth or existence of A. One can note in pattern 5 that it is not permissible to assign A the truth value T while B is assigned a credibility value other than T. Hence, the tendency is to have the credibility value of A lag the credibility value of B as B tends toward a truth value T.

One might call the above observation the "principal of over compensation," i.e., not having faith in the credibility of the conclusion begets less faith in the credibility of the premise from which the conclusion was derived.

Patterns 6-10 indicate a more optimistic evaluation of the credibility of the conclusions when viewed through the credibility value assigned to the premises. For example, pattern 6, which has previously been discussed, indicates that when A is assigned the truth value T, B must be assigned the truth value T. If A is SMC, then B will range in SMC to T.

The optimism persists in patterns 6-10 because the premise A is the sufficient condition for B; A guarantees the credibility, truth, or existence of B. One can note in pattern 10 that it is not permissible to assign a truth value of F to B while A is assigned a credibility value other than F. Hence, the tendency is to have the credibility value for B higher than the credibility value of A as A tends toward the truth value F.

In summary, the ten patterns indicate the following two general interpretations:

1. If one does not accept your conclusions, then one will be less inclined to accept the premises from which the conclusions were validly drawn.
2. If one accepts your premises, then one will be more inclined to accept the conclusions which were validly drawn from your premises. ■■

Example 3.4 (Argument Synthesis). There are many situations where analysts are asked to assist in determining whether some conjecture, hypothesis or statement is true or credible. For example, some user of intelligence may make the statement *P: The Russians are building a new airfield in Siberia!* Analysts from many functional areas of intelligence would be asked if they could supply information which would help establish the truth or falsity of some aspect of the statement *P*. The following is an example of how intelligence specialists could answer this question and supply inputs to other intelligence analysis or users.

The biographics specialist notes that if statement P is true, then several airfield construction specialists should be located at the airfield construction site. Going through his files, he checks on the whereabouts of all key military and civilian airfield constructions specialists. His search results in determining that two military and three civilian airfield construction specialists have been assigned to the construction site within the previous month. He reports back to the intelligence user the statement *Q: The two military and three civilian airfield construction experts are in the area of the alleged airfield construction site.*

At this point the intelligence user, who originally made the statement P, has the following information at hand:

P: Airfield in Siberia under construction (hypothesis).

Q: Airfield construction experts transferred to alleged construction site.

Additionally, he notes the implications:

If P, then Q, and if Q, then P

with the biographics specialist's judgement that observation Q has high credibility.

Independently, the airfield specialist intelligence analyst offers the user the following observation, *R: The geophysical and environmental attributes of the alleged aircraft site meet both military and commercial airfield site specifications.* The airfield specialist also notes that if an airfield is being constructed at the Siberian site, one should expect the presence of quite a few pieces of heavy construction equipment and shelters for construction personnel. The airfield analyst initiates a special intelligence collection request (SICR) to ascertain the presence of equipment or shelters. The analyst receives the report *S: Both heavy construction equipment and personnel shelters have been observed in the area of the airfield site.* Previous sightings of the area did not reveal either equipment or shelters in the area. The analyst reports his new findings to the user.

The user, who is also an intelligence specialist, assesses his intelligence information as follows:

P: Airfield in Siberia under construction (hypothesis).

Q: Airfield construction experts operating at the alleged site.

R: Geophysical and environmental attributes of site meet airfield construction specifications.

S: Heavy construction equipment and personnel shelters are at the site.

The user and his supporting analysts, noting that they cannot verify directly and positively the truth or falsehood of the statement P, address their attention to the verification of the credibilities of statements Q, R and S and their substantive connections to P. The reasoning pattern employed is: Whenever the credibility of a statement P, for some reason, cannot be verified directly, attention should be shifted to the collection and assessment of other intelligence information which could shed light on the credibility of the statement P. Both the credibility of the new information (statements) Q, R and S and their logical (plausible) connections to the statement P can be employed to estimate the credibility of P.

3.6 SUMMARY OF DEDUCTIVE ARGUMENT SYNTHESIS AND EVALUATION

To evaluate the credibility of a statement P of unknown credibility from statements Q_1, Q_2, ..., Q_n of known credibility, the following argument synthesis and evaluation procedure is suggested.

1. Statement P's credibility is questioned.
2. Credibility of P not directly verifiable (assumption).
3. Collect evidence Q_1, Q_2, ..., Q_n with logical (substantive) ties to P.
4. Establish implication relationships between P and Q_1, ..., Q_n, where $P \rightarrow Q_1, Q_1 \rightarrow P, P \rightarrow Q_2, Q_2 \rightarrow P, ..., P \rightarrow Q_n$ and $Q_n \rightarrow P$.
5. Assess the credibility of the statements Q_1, Q_2, ..., Q_n and the validity of the arguments, $P \rightarrow Q_i$, $Q_i \rightarrow P$, $i = 1, 2, ..., n$.
6. Using the deductive patterns of Figure 3.10, evaluate the credibility of the statement P based on the credibility of the statements Q_1, Q_2, ..., Q_n and those arguments $P \rightarrow Q_i$, $Q_i \rightarrow P$ which are found to be valid or strongly so.

In following the suggested synthesis and analysis procedure, it should be kept in mind that the argument $P \rightarrow Q$ is considered valid whenever one must accept the conclusion Q as true if one accepts the premise P as true. Another way of saying that the argument $P \rightarrow Q$ is valid is: If a conclusion validly derived from the premise is found to be false, the premise must be rejected. However, in a valid argument, if a large number of conclusions are found to be true, the truth of the premise is not guaranteed. In Example 3.4 the truth of Q, R and S render the airfield hypothesis quite probable, but it is conceivable, though unlikely, that the construction experts have gathered simply for a party, bringing equipment to construct a skating rink.

Stated symbolically, the point is that in Example 3.4 $(P \wedge Q \wedge R) \rightarrow P$ is not a totally valid implication. It is, however, the less than valid sort of implication with which the information analyst must deal. This brings us to the next section which discusses the specific problem.

3.7 RELATIONAL EVALUATION (Inductive)

Up to this point we have discussed the transfer of credibility value from A to B or B to A under the condition that the implication $A \rightarrow B$ is valid, i.e., A true yields that B is true and, B false yields that A is false. At this point we evaluate the transfer of credibility from A to B or B to A through other than totally valid implications. We will illustrate how to assess the plausibility of arguments based on substantive connections between premise and conclusion and how this affects the credibility relationships between premise and conclusion. We introduce a new term which will be used quite widely in the sequel. The term is *belief* and it is used to express subjective or objective evaluations of statements or arguments and hence combines the terms and notation of credibility, plausibility, and probability.

3.8 BELIEFS

Before we proceed we need clarify what is meant by the term "belief." It will be shown that we can use "beliefs" in a manner analogous to credibility, plausibility, and probability in the field of substantive analysis.

In the context of intelligence, the analyst is seldom certain that the observations, collections, and reports made available to him reflect the true state of the event being reported on. The information provided the analyst reflects the reliability* of the collection system (optical, human, etc.) supporting him. If the information source is human, then the reliability (and hence credibility) assigned to the information will in most part reflect the capabilities of the human collector. It is entirely possible for an analyst or a collector to believe an erroneous or false report and reject a correct or truthful report. The uncertainties associated with information collection and assessment do not allow reporters or analysts to exercise extreme judgments (i.e., the analyst's evaluation of the credibility of the information is often hedged). However, a good substantive analyst will always admit to specific degrees of belief (i.e., the analyst reports the information as "very probable," "quite reliable," etc.). *Thus we shall view a belief as a measure of the confidence one places in his observations or information actually reflecting the true event or state of nature.* Note that this description of belief is a personal one, i.e., we do not expect identical beliefs from two different individuals viewing the same event or state of nature. One need only serve on a jury and listen to the testimony provided by witnesses to become a believer in the inability of observers to report accurately and objectively, even when they are first hand observers. This is not to say that

* A more detailed discussion on reliability of a source is discussed in Section 3.19. The assumption is that the credibility and reliability of a source are essentially equal provided one evaluates reliability as suggested in Section 3.19.

all beliefs are subjective and replete with inaccuracies. To the contrary. Scientists form many beliefs and they voice these beliefs as "theories," "hypotheses," etc. The difference with scientists is that in forming their "belief" they have objectively evaluated the uncertainty, if any, associated with their belief. The scientist will admit beliefs only if they are objectively quantified.

The substantive analyst should also prefer and strive to obtain objectively derived information or beliefs. It is realized, of course, that in some situations supporting information is not available. Yet the substantive analyst must make decisions or state opinions on the information available. Knowing whether the information (beliefs) were subjectively or objectively derived would be of great assistance to the analyst in his decision-making process. For he too, like the scientist, must be aware of the uncertainties associated with the information on which his decisions are based.

Under some circumstances the analyst will be able to observe a number of events and compiled statistics on the occurrence of an event. From the statistics the analyst can form a belief on the occurrence of the event in some future time. For example, the analyst might have reports of 15 firings of a test missile. The reports indicate that 12 hit their targets. The analyst might be asked his opinion on the effectiveness of the missile. Normally, the answer will be that the missile is 80% effective. But some analysts are not happy with a sample of 15. If they knew of 30 or more firings with successes at the 0.8 level of achievement, they would be more confident in providing the answer of 0.8 to the requestor. Many would hedge and lower their estimate to 0.75 or 0.7 for two reasons. First, the sample size may not be large enough and second, the estimate contains an element of risk in that it predicts that the success ratio will be maintained in future firings of the missile.

Under other circumstances, experience and expert judgment based on analogy or observed causal relations may provide an acceptable rationale behind the formation of a belief. Such beliefs can be expressed qualitatively and hence are mainly subjective in nature. For example, a belief that war between two nations is inevitable is highly subjective. For our knowledge on the causes of war is largely based on experience with previous wars and their causal factors.

3.9 PROPERTIES OF BELIEFS

We introduce the following notation. Events, statements, and observations will be denoted A, B, C, and degrees of belief in the events occurring, statements being true, and observations representing the state of nature will be denoted $b(A), b(B), b(C), \ldots$.

First we assert that any belief can be expressed as a number whose value lies between 0 and 1. For a statement A, if one believes that A is true, then

$b(A) = 1$. If one believes A is false, then $b(A) = 0$. Otherwise, the belief can be converted to a number between 0 and 1, $0 < b(A) < 1$.

If a belief on the credibility of a statement A is arrived at subjectively, then the Kent chart or credibility code can be employed to convert that belief to a numerical range or value between 0 and 1. If a belief is arrived at objectively through statistics, probability, or purely logical processes, then the numbers so obtained can be converted to the 0-1 scale as recommended in the following sections.

Example 3.5 (Belief Evaluation). This example illustrates how an analyst can evaluate his beliefs and convert his independent beliefs into relative beliefs. Suppose that an analyst decides, on the basis of information supplied him, that four and only four courses of action (events) are open to country A. He might list the events as below:

U: Country A will withdraw its ambassador from country B.
V: Country A will amass troops on the border of country B.
W: Country A will sign a nonaggression pact with country C.
X: Country A will stop trade with country B.

The analyst can assess his belief of each event occurring independent of any of the others occurring.

The analyst subjectively may feel that events U and V are "probably likely," that event W is "doubtful," and that event X is "almost certain." Consulting the Kent chart, these subjective assessments of the "independent events" can be converted to average values of beliefs as listed in the following table:

Event	Independent Belief
U	0.70
V	0.70
W	0.30
X	0.93

If the analyst intuitively feels that country A will initiate one and only one action against country B, then the analyst should consider all his beliefs simultaneously to determine what action is preferred by country A. His initial independent evaluations indicated that he believed that country A most favors initiating event X. The analyst can now express this belief as a probability of event X occurring relative to the probability of occurrence of the three other events. To do this he performs the following simple computation to "normalize" his independent beliefs:

$$b(U) = \frac{\text{Independent belief in } U}{\text{Sum of independent beliefs of all events}}$$

$$= \frac{0.70}{0.70 + 0.70 + 0.30 + 0.93}$$

$$= 0.26$$

Going through the same form of computations for V, W, and X, we obtain

$$b(V) = 0.26 \qquad b(W) = 0.11 \qquad b(X) = 0.37$$

The analyst can now state a relative subjective belief that event X has a probability of 0.37 of occurring, events U and V each have a probability of 0.26 of occurring and event W a probability of 0.11 on his assumption that country A will take only one action. ■■

Beliefs in statements can be thought of as the assessed levels of credibility of the statements. Beliefs in arguments can be considered as equivalent to the argument's plausibility. Thus, a belief is quite similar to a subjective (or objective) measure of the probability of an event, the plausibility of an argument, or the credibility or level of truth of a statement. The relationship between beliefs, credibility, and plausibility can be illustrated on paired credibility scales. Thus, on a paired credibility scale, as given in Figure 3.11, statements A and B are given credibility or belief values $b(A)$ and $b(B)$ and are so marked on the appropriate scale. The implication $A \rightarrow B$ is also assigned a plausibility or belief value $b(A \rightarrow B)$ which is denoted by the dashed line between the two scales. How the credibility of statements and plausibility of arguments are related to each other will be described in Section 3.19. Before we discuss these relationships it is necessary to describe some fundamental properties of beliefs.

3.10 MUTUALLY EXCLUSIVE SETS OF BELIEFS

In Example 3.5, it was assumed that the events U, V, W, and X were mutually exclusive, i.e., country A would not elect to follow two or more courses of

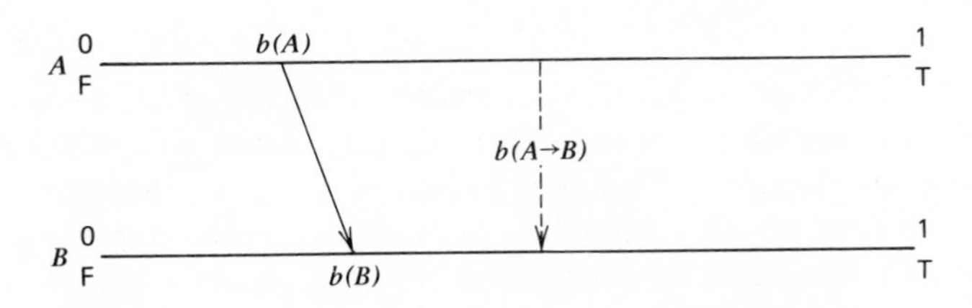

Figure 3.11 Paired belief scales.

action (events) simultaneously. This is equivalent to saying that the analyst's relative belief that event U or event V might occur is equal to the sum of his relative beliefs that each event U and V will occur. The following formula is an equivalent statement for events U and V being mutually exclusive:

$$b(U \cup V) = b(U) + b(V)$$

i.e., the belief that event U or V will occur is equal to the sum of the independent beliefs that event U and event V will occur.

Further, the relative beliefs $b(U) = b(V) = 0.26$, $b(W) = 0.11$, and $b(X) = 0.37$, obtained in Example 3.5 can be positioned on a credibility scale as illustrated:

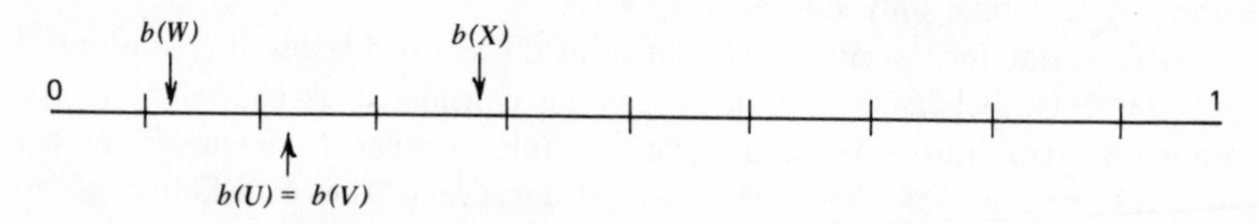

In the sense of relative beliefs described in Example 3.5, we can now state another property of beliefs. If A, B, ..., N are mutually exclusive events, in that no two of them can occur (or be true) simultaneously, then the evaluation of the belief of the occurrence of at least one of the events can be expressed in terms of the beliefs in the occurrence of the individual events, i.e.,

$$b(A \cup B \cup \cdots \cup N) = b(A) + b(B) + \cdots + b(N)$$

Note that no statement is made about whether the events exhaust the possibilities. When the events are exhaustive, the formula is further simplified as indicated in the next section.

3.11 EXHAUSTIVE SETS OF BELIEFS

If the events A, B, ..., N exhaust the possibilities, i.e., no other events are possible, and if the events are mutually exclusive, then we can state

$$b(A) + b(B) + \cdots + b(N) = 1$$

Example 3.6 (Exhaustive Beliefs). Consider a simple event designated A: Submarine was sighted at location X at time T. Based on the information supplied on the sighting, the substantive analyst forms a belief $b(A)$, i.e., he states an opinion on the credibility of the facts supporting the sighting. Simultaneously, he states an opinion on the lack of credibility of the facts supporting the sighting. This belief is denoted $b(\bar{A})$ when $\bar{A}$ designates the denial of the sighting. Since A and $\bar{A}$ are mutually exclusive and exhaustive,

$$b(A) + b(\bar{A}) = 1$$

In another instance, the question of whether a submarine was sighted may have been answered in the affirmative. The important question then transfers to What type submarine was sighted? The events can then be restricted to the sighting of specific classes or types of submarines. Mutually exclusive and exhaustive events such as,

A: Class G submarine sighted.

B: Class Z submarine sighted.

C: Class W submarine sighted.

can be hypothesized, where it is known that these are the only three types of submarines operating in the area. In this case we have

$$b(A) + b(B) + b(C) = 1$$

wherein we have accepted that a submarine (but only one) was actually sighted and our only concern is to determine which class submarine was sighted.■■

In practice it may be very difficult to ensure that one's alternatives are in fact completely exhaustive. Overlooked possibilities can easily destroy the value of an evaluation.

3.12 IMPLICATION BELIEF

Our next concern is to assess our belief in the plausibility of the implication $A \to B$. As discussed in Section 3.13, our value judgment on the plausibility of the implication is dependent on our previous experience with, and knowledge of, analogy and causal relationships existing between A and B. If our judgment indicates that B *must be* true whenever A is true, we express our belief as

$$b(A \to B) = 1$$

If we believe that A *must* be false whenever B is false, we similarly express our belief as

$$b(A \to B) = 1$$

When the truth of B is not assured by the truth of A, then our previous experience might give us a clue as to the relative frequency with which B is true when A is true or A is false when B is false. In either case, the relative frequency can serve as an approximation for $b(A \to B)$.

When dealing with "one time" events, like a declaration of war, an attack, or a landing on Mars by 1980, we are forced to use value judgments to evaluate the plausibility of any implication $A \to B$. For example, suppose

A: We will spend $3 billion a year until 1980 on space programs leading to a manned space vehicle landing on Mars.

B: A U.S. citizen will land on Mars by 1980.

It is obvious that the validity of the implication $A \to B$ is dependent on value judgments of the success of previous space programs and our ability to identify and solve the technological problems confronting such an ambitious space program. If we feel highly confident that the level of expenditure described in A would indeed lead to a landing by 1980, then we will assign a high plausibility value to $b(A \to B)$; if we think the problems could almost surely not be solved by 1980, even with twice the expenditure or $6 billion a year, then we would assign a very low value to $b(A \to B)$. At this point let us summarize our statements and notation.

1. By $b(A)$ we denote our belief in the truth of A, or in the occurence of A as an event.
2. Our beliefs will be assigned numbers between 0 and 1,
$$0 \leqslant b(A) \leqslant 1$$
3. If A is true, $b(A) = 1$. If A is false, $b(A) = 0$.
4. If the events $A, B, C, \ldots, N$ are mutually exclusive, then
$$b(A \cup B \cup C \cup \cdots \cup N) = b(A) + \cdots + b(N)$$
5. If the events $A, B, C, \ldots, N$ are mutually exclusive and exhaustive, then
$$b(A) + b(B) + \cdots + b(N) = 1$$
6. Our belief $b(A \to B)$ in the plausibility of the implication $A \to B$ is a number between zero and one,
$$0 \leqslant b(A \to B) \leqslant 1$$

3.13 PLAUSIBILITY OF ARGUMENTS

Up to this point, we have alluded to the evaluation of the plausibility of an argument. Now we describe some specific techniques for estimating the plausibility of arguments of various types. Arguments, in their simplest form, can be reduced to an implication $P \to Q$. The premise P and the consequence Q can consist of the logical combination of many simple statements. For example, we can have $(P_1 \wedge P_2) \to Q$, $[P_1 \vee (P_2 \wedge P_3)] \to Q$, $P \to (Q_1 \wedge Q_2)$, or $P \to (Q_1 \vee Q_2 \vee Q_3)$. It will be shown that the strength of the argument is dependent on the logical structures of, and the substantive connections between, the premise and the consequence.

To evaluate an argument, we must, in essence, answer the questions: To what degree are we forced to accept Q whenever we accept P? If we do not accept Q, to what degree are we forced not to accept P? The questions can be partially answered using and combining the following methods of evaluation.

Cause-Effect

The objective of this section is to gain some knowledge on how to evaluate causal connections between premise and conclusion. For example, the presence of oxygen is a cause for burning. With oxygen present, substances will burn if their kindling temperature is reached. Substances will not burn in the absence of oxygen. Nor will substances burn if they fail to reach their kindling temperature. Both temperature and the presence of oxygen are causes for burning in the sense that if one is eliminated, substances will not burn. We normally consider reaching the kindling temperature as the only cause for burning. We take for granted that oxygen is always present. But it should be noted that each is only a necessary condition for substances to burn. Together they constitute a sufficient condition (they guarantee burning).

In medical diagnosis the emphasis is on finding the causes for the illness to be treated or the symptoms under observation. When a complete set of causes are determined, then a prescribed program of treatment is directed at eliminating each of the causes. For example, certain strains of viruses cause certain forms of influenza. When the form of flu is identified, the drug which is most successful in eliminating the virus is normally prescribed. Thus a virus is a cause for the flu and in a sense viruses are a sufficient condition for the flu when present in sufficient quantities. When the cause (the virus) is eliminated, the symptoms disappear.

As indicated above, there are instances where many causes must occur simultaneously before the effect is noticed. For example, the cause-effect argument "If oxygen is present, wood will burn" is a weak one. For oxygen exists in the presence of wood and wood does not burn unless other conditions are satisfied. The cause-effect argument "If oxygen is present and the temperature exceeds the kindling temperature of wood, then wood will burn" is a physical fact and hence a valid argument. Most people would consider the cause-effect argument "If the temperature exceeds the kindling temperature of wood, then wood will burn" a highly plausible argument. In the latter however, it is understood that oxygen is present.

The fields of political science, criminal investigation, law, and psychology are filled with cause-effect type of arguments which in most instances are harder to evaluate than the more objective types of cause-effect arguments found in the sciences. For example, consider the following:

If Czechoslovakia fails to accept the party line, Russia will take military action against it.

If a dope addict cannot maintain his daily requirements, he will turn to crime to achieve them.

If more border disputes break out into armed conflict, the United Nations will develop a stronger international armed force to resolve them.

In the above cause-effect type of arguments, the evaluation of the plausibility of the arguments lies in previous experiences of the same type (analogy), accepted practices, ideologies involved, and normative judgments. Regardless of whether we have the wherewithal to evaluate arguments of the above type, these arguments are made and will continue to be made by people in responsible and authoritative positions. Hence it is a necessity that some attempts at analysis be undertaken. The following illustrates how the plausibility of arguments based on cause-effect can be evaluated.

Causes can be interpreted in several senses. First, a cause can be interpreted as a necessary condition. For example, a temperature of only X degrees is necessary to boil water at a high altitude. If the altitude is reduced, the temperature of X will no longer cause water to boil. In this sense the temperature of X is a necessary condition because there are conditions under which X degrees does not guarantee that water will boil. Second, a cause can be interpreted as a sufficient condition. For example, if your heart stops beating for a preset period of time, death will occur. If carbon or tungsten is mixed in proper proportions with iron, a hardened steel will result. These causes guarantee the realization of the effects, death in the first example, hardened steel in the second. Third, a cause can be interpreted as that event which precipitated the effect with the knowledge that other normally prevailing conditions existed. In this sense the cause need not be a necessary or sufficient condition. For example, if a match is struck and applied for a short period of time to a large piece of wood, normally the wood will not burn. However, if the piece of wood is saturated with gasoline and then the match applied to it, the wood will normally burn. The cause of the burning is the presence of gasoline. And the presence of gasoline in neither necessary nor sufficient for wood to burn. However, in this case, in the face of normally prevailing conditions, the presence of gasoline caused the wood to burn. Anyone who has started a fire in a fireplace or open campsite is well aware of this fact and has used it to his advantage.

In this third interpretation of cause, the cause may be classified as remote or proximate. A proximate cause is one that is clearly identified as the essential difference between occurrence or nonoccurrence. The gasoline was the proximate cause in the case of igniting a large piece of wood. However, some causes may be remote in the sense that A caused B, B caused C, and C caused D, which in turn caused the lights to go out. In this case, A might be an excessive amount of rainfall which caused B, the flooding of the basement, which caused C, an electrical short to occur in the wiring, which caused D, the circuit breakers to open, which in turn caused the lights to go out.

While there is no standard way of interpreting the sense of cause, there is a preferred way and we suggest it here as a guide. We shall consider causes as necessary conditions for their associated effects. We will work only with the most proximate causes. We will also seek the set of necessary causes which in conjunction with each other form a sufficient condition for the effect. Thus when we achieve sufficient conditions as causes, then the cause-effect argument will have plausibilities very close to valid. The evaluation of cause-effect arguments when the causes are necessary conditions will require the determination of the probability (or plausibility) of achieving the effect with only a limited number of necessary (but not sufficient) conditions satisfied. This technique is illustrated in Section 3.17 and will not be repeated here.

Analogy

An analogical argument is one wherein the plausibility of the argument is based on its structural and substantive similarities with previously accepted arguments. For example, if we want to evaluate the plausibility of an argument $P \rightarrow Q$, we can evaluate the logical or causal relationships between P and Q. If we cannot do this directly, we can turn to another argument $A \rightarrow B$ of known plausibility and establish that the premises P and A and the consequences Q and B are analogous to each other. The parts of the arguments are diagrammed below.

$$\begin{array}{ccc} & A \rightarrow B & \\ ? & \wr\ ?\ \wr & ? \\ & P \rightarrow Q & \end{array}$$

We can assert that since B is a plausible consequence of A and P is analogous ($\approx$) to A, then Q will be a plausible consequence of P. Note that question marks are placed over the implication symbol ($P \rightarrow Q$) and the analogous symbols ($Q \approx B$) and ($P \approx A$). The reasons for these question marks are as follows:

1. In most arguments, by analogy Q and B will be equivalent. However, in some arguments the condition Q will be weaker (or stronger) than B. This will affect the strength of the analogical argument $P \rightarrow Q$.
2. In most analogical arguments A will consist of a set of instances $A_1, A_2, \ldots, A_n$ which in some sequence or logical combination yields B as a plausible consequence; P can be a subset or an extension of the set A and hence only similar to A.

3. The question mark over $P \to Q$ indicates that the argument's plausibility is to be determined from the plausibility of the argument $A \to B$ and the analogical ties between A and P and B and Q.

Analogical arguments are nondeductive or inductive argument types. The conclusions do not necessarily follow from the premise. The analogical argument is used to show that the conclusions are reasonable or plausible, or probable. One is not to consider an analogical argument as conclusive or deductive unless it passes the formal logical tests required of it.

Example 3.7 (Analogical Argument). It is known that when country *A* did not follow the party line of country *B* in its relations with other countries, country *B* used military force to enforce the party line. Country *C* is in the same coalition as *A* wherein *B* is the coalition's dominant power. The argument that requires evaluation is the following: If country *C* does not follow country *B*'s party line, then country *B* will use military force to enforce the party line. The argument can be abbreviated as

A uncooperative → *B* uses military force
C uncooperative → *B* uses military force

How good the second argument proves to be depends on the strength of the analogy between *A* and *C*–i.e., will *B*'s reaction to *C*'s actions be the same as its reaction to *A*'s actions, when *A*'s and *C*'s actions are similar. The second argument can be made more plausible if its conclusion is weakened. For example, if we rewrite the arguments as follows:

A uncooperative → *B* uses military force
C uncooperative → *B* sends diplomatic protest

The second argument in this example is more plausible or probable than the second argument in the first example.

The analogical argument in the first set of arguments can be strengthened if we strengthen the premise while leaving the conclusion unchanged. For example, we change the argument as follows:

A uncooperative	→	*B* uses military force
C uncooperative and *C* expels diplomat of *B*	→	*B* uses military force

■■

There are several ways to strengthen or weaken an analogical argument. It is necessary that an analyst know how analogical arguments can be strengthened or weakened. Knowing the techniques will help in evaluating analogical arguments. The conditions which establish, strengthen, or weaken analogical arguments will now be described. But before we proceed, we define the principal parts of an analogical argument.

An analogical argument normally consists of a number of *cases* or *examples* of previous events or evaluated causal relations of the form:

When A does u, v, w, x, Y does Z.

When B does u, v, w, x, Y does Z.

When N does u, v, w, x, Y does Z.

These are called *supporting cases* and each case consists of premises followed by conclusions. Some writers will call each case a premise but we will not follow that convention here. The symbols u, v, w, and x are called *respects, actions* or *events*. They may all appear in each supporting case, or different subsets of them may appear in any of the supporting cases. The supporting cases are normally followed by a statement of the form

If R does u, v, w, x, then Y will do Z.

and this statement is called the *concluding statement*. It consists of a premise containing the same or modified respects that the premises of the supporting cases contain, and a conclusion which will be similar to, but could be stronger or weaker than, the conclusion in the supporting cases. When the concluding statement is appended to the supporting cases, the result is an analogical argument of the following form:

When A did u, v, w, x, then Y did Z.

When B did u, v, w, x, then Y did Z.

⋮

When N did u, v, w, x, then Y did Z.

Therefore,

If R does u, v, w, x, then Y will do Z.

The above form for an analogical argument is the ideal form. Many variants are to be expected and evaluated. The respects need not all be similar in all premises, and the conclusion Z may be varied among the supporting cases and in the concluding statement.

3.14 EVALUATING ANALOGICAL ARGUMENTS

The objective of this section is to demonstrate the existence of criteria through which one can evaluate the plausibility of an analogical argument. It is repeated that analogical arguments are never valid. They only achieve a level of plausibility (or probability) in the sense that their conclusions probably follow from the premises or are caused by the events of the premise through a plausible causal relation.

Since we have no specific numerical techniques with which to appraise analogical arguments, we will be using relative terms such as stronger, weaker, more cogent, and more probable, in assessing arguments. Further, we will illustrate the evaluation criteria through a series of examples. We will attempt to show how changes in the premises or conclusions affect the strength or cohesion of an analogical argument. And finally, we will discuss the subjects of relevance and causal relations as two important criteria to be satisfied for the achievement of good analogical arguments.

Frequency of Similar Simple Arguments

Analogical arguments are evaluated on the basis of the number of previously established arguments one can identify which contain simple premise-conclusion relationships similar to the argument one wishes to evaluate. For example, if we wish to establish analogically that country *Y* will use military force if country *A* does not grant country *Y* territorial concessions, we will trace through recent history and note that when country *B*, country *C*, and country *D* did not grant territorial concessions to country *Y*, country *Y* used military force to gain them. The argument form is as follows:

> Country *B* took action *U*; country *Y* responded with *Z*.
> Country *C* took action *U*; country *Y* responded with *Z*.
> Country *D* took action *U*; country *Y* responded with *Z*.

Therefore,

> Country *A* takes action *U*, country *Y* will respond with *Z*.

If it can be established that country *Y* views countries *A*, *B*, *C*, and *D* analogously with respect to territorial claims, then the argument drawn will be quite plausible. Note that one may have been able to find but one analogous situation in history, say only country *B*. If this were the case, then the argument would lose some of its plausibility. This is not to say that the plausibility of an argument is directly related to the number of supporting cases that can be uncovered, i.e., the plausibility of the argument based on three cases involving countries *B*, *C*, and *D* is not three times the plausibility of the argument based on one case involving only country *B*.

The supporting cases referred to above were very simple in form. The arguments involved only one action and one response. The next type of analogical argument to be evaluated involves several actions, and a single response.

Frequency of Similar Complex Arguments

The previous discussion established the number of supporting simple cases as a criteria for evaluating the plausibility of an analogical argument. This section illustrates that as the analogical argument becomes more complex, i.e., supporting cases agree with each other in more than one action or respect, the analogy increases in plausibility. As an example, consider the previous supporting cases modified to agree with each other in more than one action. This change is illustrated below.

We still wish to establish that country Y will use military force against country A. But this time we note that country A does not care to grant country Y territorial concessions (u), trade concessions (v), special military bases (w), and sign a nonagression pact (x). Historically we noted that when country B refused to grant country Y territorial concessions and sign a nonagression pact, country Y initiated military action (Z). When country C refused to grant trade concessions and special military bases to country Y, country Y used military force to gain them. When country D refused to grant territorial concessions, country Y initiated military action against country D. These cases can be abbreviated and diagrammed as follows:

B refused u and x, Y responded with Z.
C refused v and w, Y responded with Z.
D refused u, Y responded with Z.

Therefore,

If A refuses u, v, w, and x, then Y will respond with Z.

The complex cases could have all been stated as B, C, and D refusing to do u, v, w, and x. This would have weakened the argument since it would not have established that one or two actions were sufficient to have Y respond with action Z. In other words, if u, v, w, or x each alone caused Y to react with Z, then actions which are combinations of u, v, w, and x would have a higher probability of having Y respond with Z.

Again the plausibility of this analogical argument is not four times that of the simple argument described previously. Using four actions or respects u, v, w, and x as against one action or respect u, serves to increase the plausibility of the analogical argument but not in any preordained ratio. Also important is the causal relationships and relevance that the actions u, v, w, and x have with the action Z which appears in each supporting case and the concluding statement. More will be said on the importance of the respects or actions being relevant to the conclusion in the sequel.

Strength of Conclusions Relative to Premises

In the two previous evaluation criteria, the conclusions were not varied among the cases and the concluding statement. The conclusions in the concluding statement of an analogical argument need not be the same as the conclusions stated in the supporting cases or instances. For example, consider the simple arguments discussed previously.

B took action *u*, *Y* responded with *Z*.

Therefore

If *A* takes action *u*, *Y* will respond with *Z*.

The response *Z* means that *Y* will use military force. Now suppose that instead of *Z* we conclude that *Y* will respond with *Z′*, where *Z′* means some less extreme action such as recalling a diplomat, severing diplomatic relations, or involving a trade embargo. The new analogical argument will be more plausible than the original one because *Y* is more likely to invoke a lesser action *Z′* than an extreme action *Z*. This is not to say that the original analogical argument is not plausible. We merely state that if the concluding statement in the alternative analogical argument is weakened relative to the conclusion in the supporting cases, then the plausibility of the alternative argument exceeds that of the original argument. Another example of weakening the conclusions follows.

On six previous bombing missions the bombers destroyed 80% of the areas targeted. The squadron leader can report that his squadron will destroy 80% of any targeted areas in its future missions and have a plausible argument. He can report instead that his squadron will destroy 60% of any targeted areas in it future missions. This argument will be more plausible than the previous one Note that the conclusion was weakened relative to the premise and hence it realization was more probable, making the argument a more plausible one. the squadron leader reports that his squadron will achieve a destruction level o 90% on its next mission, all will agree that the plausibility of his argument i reduced considerably. In this example, it was understood that all the respects o actions which lead to the squadron's performance would remain in effect o the next bombing mission. Some of these respects would be the pilots, the type of planes employed, the weather, the antiaircraft activity over the target area, and the squadron's leadership.

Strength of Premises Relative to Conclusion

We have seen that when the conclusions in the concluding statement are weakened relative to their statement in the supporting cases, the plausibility of the argument is strengthened. We will now show that when the premises in the concluding statement are weakened relative to their statement in the supporting cases, the argument's plausibility is weakened. Consider the following example.

In observing the construction of airfields in four previously recorded cases, four conditions were noted prior to and during the actual construction of the airfield; these were:

U: Airfield site meets geophysical requirements.

V: Heavy construction equipment was at site.

W: Personnel and equipment shelters were erected early in the construction cycle.

X: Airfield construction experts were on site prior to start of actual construction.

Thus for four cases, *U, V, W,* and *X* we can state the following:

When conditions *U, V, W,* and *X* were satisfied at site *Y*, an airfield was constructed (*Z*).

Now suppose we are observing some new construction activity at another site. If we note that conditions *U, V, W,* and *X* are all satisfied, we can state quite plausibly that an airfield is being constructed at the new site. However, if we note that *U* is only partially satisfied–i.e., not all geophysical requirements for building an airfield are satisfied at the site–and that *X* is only partially satisfied–i.e., fewer airfield construction experts have been noted at the site–then the argument supporting the conclusion that an airfield is under construction is not as plausible. The several points of differences (*U* and *W*) casts some doubt on the plausibility of the conclusion. Again, we can not equate the degree of weakening of the argument to the number of points of difference. We can only say that when points of difference occur in the premises of the concluding statement these points of difference tend to weaken the argument. It should be kept in mind that points of difference can strengthen the argument if their statement in the concluding argument is stronger than their statements in the supporting cases.

Dissimilarity of Premises Relative to Conclusion

When cases which support the plausibility of the argument under consideration are being assembled, the tendency is to attempt to list as many similar cases as possible. The objective is to reduce the probability that the cases will contain instances which are disanalogous to the premises of the concluding argument. For example, in the argument

B took action *U, Y* responded with *Z*.

Therefore,

If A takes action U, Y will respond with Z.

If B and A both border on country Y's boundaries, have sided with each other in disputes with Y, and have differed with Y ideologically, then the argument is considered to be quite plausible.

Now suppose that the two cases become history and we now are analyzing another territorial dispute. This time it is with country C. If C is very much like A and B, and if C refuses territorial claims from Y, then the conclusion that Y will respond with Z is plausible. However, if C does not border on Y, has not had any previous disputes with Y, and agrees with the ideology of Y, then the conclusion that Y will respond with Z loses some of its plausibility.

Now consider the situation where countries A, B and C took action. Previous to taking action, A sided with Y, B opposed Y in all its actions and C sometimes sided with and sometimes opposed Y. Under these dissimilarities Y still responds with Z—military action against A, B and C. If a new case of territorial dispute comes up with country Y against country D, then we could assert with confidence that country Y will use military force if country D does not agree to its territorial claims. The reason for this confidence is that three previous cases with countries A, B and C will reduce the probability that country D will introduce respects which are disanalogous to respects in the preceding supporting cases wherein country Y reacted with military force.

Evaluation Criteria Summary

From the discussions on the five evaluation criteria for analogical arguments, we can state what constitutes a good analogical argument. First, if we can establish a number of cases wherein the premises support conclusions like those in the concluding arguments, then the concluding argument is established with greater probability than if only a single case was established. Second, if the premises of the supporting cases contain more and diverse respects which favor the conclusion, and if the premise of the concluding statement contains these respects, then the concluding argument will be more plausible. Third, if we weaken the conclusions relative to the conclusions in the supporting statements, the argument becomes more plausible. Fourth, if we strengthen the premises in the supporting cases, or make them disanalogous relative to their statement in the concluding statement, then the argument is weakened. We could leave the premises in the supporting cases unchanged and just weaken the premise in the concluding statement and obtain the same weakening effect in the plausibility of the argument. Fifth, we can strengthen the argument by selecting the supporting cases to be dissimilar so as to avoid possible disanalogies with the concluding statements. Understanding these techniques results in the intelligent structuring and analyzing of analogical arguments. However, there is an important addition to these techniques which will result in an even greater plausibility for the arguments so structured. Briefly, the analogical arguments will be more plausible if, on structuring the arguments, one concentrates on finding supporting cases which establish a high degree of relevance between the premise and conclusion of each supporting case. Specifically, premises should be well-recognized causes for the conclusion in each supporting case. The premise-conclusion relationship should have a strong cause-effect flavor. As an example, consider the argument that concluded that *Y* would respond with military force if *B* did not grant *Y* some territorial concessions. The argument was made plausible by assembling a set of supporting cases. Suppose that it was also established that *Y* lost the territory it claims to *B* in a previous conflict, that *Y* needs the territory to gain access to the seaways, that the ethnic structure of the territory is more akin to *Y*'s than to *B*'s, and that *Y* claims that it needs the territory for its strategic resources. These additional respects are all relevant to the type of action *Y* will take to obtain the territorial concessions. The previous supporting cases established the tendency for country *Y* to react with military power when its territorial claims were denied. The additional respects demonstrates the causes for country *Y* to act as it does. Each respect is a cause for some action. Together they provide the cause-effect rationale behind predicting that *Y* will use military force.

Many other respects such as trade relations, diplomatic relations, mutual investments, religion, population, GNP, travel, and communications could have been considered in attempting to determine *Y*'s reactions to *B*'s refusal. This set could have shed some light on the strength of *Y*'s reaction, but they lack the high relevance or cause-effect relationship of the previous listed respects. We could go further and say that *B* and *Y* have beautiful scenery, clean cities, excellent farms, high technologies, and good educational levels. The similarities in these respects would have little relevance to *Y*'s reaction to *B*'s refusal. And, if used alone to support the concluding statement, they would present an argument of very low plausibility.

Thus the sixth criterion for the evaluation of analogical arguments is the relevance of the premises to the conclusions in the supporting cases and in the concluding statement. Strong cause-effect relationships between the premises and the conclusions will materially support higher plausibilities for analogical arguments. If high relevance is achieved in the premise-conclusion relationship, then the need for a large number of supporting cases containing many dissimilar respects is reduced but not obviated. A number of good supporting cases are always an asset, but if relevance is not present in the supporting cases, the argument's plausibility will be weakened.

Relative Frequency

In some implications or arguments it is possible to measure the degree of association between premise and consequence. For example, let P and Q be the statements:

P: I aim the gun properly.

Q: I hit the target.

Then $P \to Q$ is a plausible argument. One can set up an experiment and, under carefully controlled conditions, aim and fire the gun 100 times. If 95 hits occur, then the plausibility of the argument can be assessed at 0.95. Note that the plausibility of this argument is dependent on the condition of the gun and ammunition employed and on the control of the conditions under which the gun is fired. To obtain a more accurate measure of the plausibility of the argument, the experiment should be repeated many times. The results should be averaged to obtain a "mean" plausibility and the standard deviation should be computed to give a measure of the dispersion of sample averages about the mean. These statistical techniques will be discussed in Chapter 4.

Not all relative frequency measurements are amenable to statistical calculations. For example, an analyst may be interested in the number of times country X voted for, against, or abstained from voting on issues like admitting Red China into the United Nations. One cannot conduct repeated controlled experiments to obtain this data. It is historical fact. And from historical facts the analyst can obtain some information which could help predict the outcome of the next vote prior to its taking place. In analysis of this type a great deal of emphasis is placed on the relative frequency with which each country has voted for, against, or abstained from voting, on issues under consideration.

3.15 BELIEFS OF ARGUMENTS CONTAINING COMPOUND STATEMENTS

In the sections which follow we will be evaluating arguments of the form $A \to B$ and $\bar{A} \to B$. To evaluate the plausibility of these arguments we need to know the logical structure of A and $\bar{A}$, when A is other than a simple statement. We need to consider three cases wherein A is the conjunction (product) of simple statements, A is the union (sum) of simple statements, and A contains both the product and sum of simple statements. We shall illustrate the three cases wherein A consists of only three statements. The extension to cases wherein A contains more than three statements will be obvious from the context.

Case 1. Let A be the conjunction (product) of three simple statements A_1, A_2, and A_3.

In this case the negation or complement of A can be expressed by either of two equivalent forms. The first form is the union of the negation of the three simple statements:

$$\overline{A_1 \cdot A_2 \cdot A_3} = \bar{A} = \bar{A}_1 + \bar{A}_2 + \bar{A}_3$$

The second form is the union of seven products of the three simple statements:

$$\overline{A_1 \cdot A_2 \cdot A_3} = \bar{A}_1\bar{A}_2\bar{A}_3 + \bar{A}_1\bar{A}_2A_3 + \bar{A}_1A_2\bar{A}_3 + A_1\bar{A}_2\bar{A}_3 + \bar{A}_1A_2A_3 + A_1\bar{A}_2A_3 + A_1A_2\bar{A}_3$$

The second form is normally the preferred form because the seven terms in the union (sum) are mutually exclusive. An example of how to handle this second form in a simple situation follows. The situation is simple in that it is easy to evaluate the credibilities of the negations of the three statements which make up the compound statement. ■■

Example 3.8 (Compound Statement Argument) Consider the following argument where the premise P consists of the conjunction (product) of three statements.

P_1: Oxygen is present.

P_2: Kindling temperature of wood is reached.

P_3: Wood is saturated with gasoline.

The conclusion Q is the simple statement,

Q: Wood will ignite and burn.

The premise P is written as $P = P_1 \wedge P_2 \wedge P_3$. The premise is a sufficient condition for Q, i.e., whenever the first two conditions of P are satisfied, wood will ignite and burn. Hence we have satisfied the conditions that

$$b(P \rightarrow Q) = 1$$

The statement P_3 is superfluous and its satisfaction is not required, i.e., wood will burn without gasoline. The negation of P is the statement

$$\overline{P_1 \wedge P_2 \wedge P_3} = \bar{P}_1 \vee \bar{P}_2 \vee \bar{P}_3 = \bar{P}$$

It can be rewritten as the sum of the following seven mutually exclusive events:

$$\left.\begin{array}{l}\bar{P}_1 \wedge \bar{P}_2 \wedge \bar{P}_3 \\ \bar{P}_1 \wedge \bar{P}_2 \wedge P_3 \\ \bar{P}_1 \wedge P_2 \wedge \bar{P}_3 \\ \bar{P}_1 \wedge P_2 \wedge P_3 \\ P_1 \wedge \bar{P}_2 \wedge \bar{P}_3 \\ P_1 \wedge \bar{P}_2 \wedge P_3\end{array}\right\} \rightarrow \text{Wood will not burn.}$$

$$P_1 \wedge P_2 \wedge \bar{P}_3 \quad \rightarrow \text{Wood will burn.}$$

As indicated above, for six event conditions of P, wood will not burn. For the seventh event condition $P_1 \wedge P_2 \wedge \bar{P}_3$, wood will burn because the necessary conditions P_1 and P_2 are not negated. Thus, $\bar{P}$ contains an event which is sufficient for the occurrence of Q. Hence we evaluate

$$b(\bar{P} \rightarrow Q) = 1$$

Example 3.8 suggests that the premise of an argument can be synthesized to guarantee that the argument $P \rightarrow Q$ satisfies

$$b(P \rightarrow Q) = 1 \qquad \text{and} \qquad b(\bar{P} \rightarrow Q) = 0$$

Briefly, P should be formed as the conjunction of necessary conditions $P_1, P_2, \ldots, P_n$ for Q, i.e.,

$$P = P_1 \wedge P_2 \wedge P_3 \wedge \cdots \wedge P_n$$

wherein for each necessary condition P_i,

$$b(\bar{P}_i \rightarrow Q) = 0 \qquad\qquad i = 1, 2, \ldots, n$$

The conjunction of the necessary conditions P_i must form a sufficient condition for Q, i.e., $b(P \rightarrow Q) = 1$. It should be kept in mind that a condition P_i is a necessary condition for Q if Q cannot occur when P_i is absent or false. A condition P_i is a sufficient condition for Q if Q must occur or be true whenever P_i occurs or is true.

In Example 3.8 the condition "P_3: Wood is saturated with gasoline" is not a necessary condition for wood to burn. Hence the term $P_1 \wedge P_2 \wedge \bar{P}_3$, which contains $\bar{P}_3$ does not deny the possibility that wood will burn. The condition $P_1 \wedge P_2$, being a sufficient condition for wood to burn, guarantees the occurrence of the event.

Case 2. Let P be the disjunction (union or sum) of three simple statements P_1, P_2, and P_3. In this case the negation or complement of P can be expressed as the product of the negations of the three simple statements

$$\overline{P_1 + P_2 + P_3} = \bar{P} = \bar{P}_1 \cdot \bar{P}_2 \cdot \bar{P}_3$$

Using Example 3.8, $b(\bar{P} \to Q) = 0$, since $\bar{P} = \bar{P}_1 \cdot \bar{P}_2 \cdot \bar{P}_3$ negates the necessary conditions P_1 and P_2. Hence, wood will not burn. ■■

Case 3. Let P be a combination of disjunctions and conjunctions of the three simple statements. In this case the negation or complement of P can be expressed in a multitude of equivalent forms. Depending on the logical connectives used in P, there may or may not be a preferred form for evaluating $\bar{P} \to Q$. The possible forms for $\bar{P}$ are illustrated by several examples.

$$P = P_1P_2 + P_2P_3$$
$$\bar{P} = \overline{P_1P_2 + P_2P_3} = \overline{P_1P_2} \cdot \overline{P_2P_3} = (\bar{P}_1 + \bar{P}_2) \cdot (\bar{P}_2 + \bar{P}_3)$$
$$P = P_1 + P_2P_3$$
$$\bar{P} = \overline{P_1 + P_2P_3} = (\bar{P}_1) \cdot (\overline{P_2P_3})$$
$$\bar{P} = \bar{P}_1 \cdot (\bar{P}_2 + \bar{P}_3)$$
$$P = P_1 + P_2 + P_1P_3$$
$$\bar{P} = \overline{P_1 + P_2 + P_1P_3} = (\bar{P}_1) \cdot (\bar{P}_2) \cdot (\overline{P_1P_3}) = (\bar{P}_1) \cdot (\bar{P}_2) \cdot (\bar{P}_1 + \bar{P}_3)$$

When a statement P is a compound statement composed of several simple statements (or events), the belief we have in P is dependent on the beliefs we have in simple statements (or events) and how they are conjoined logically to form the statement P. The value of $b(P)$ is also dependent on whether the simple statements (or events) are independent, mutually exclusive, or exhaustive sets of statements (or events). Section 3.8 discussed the calculation of beliefs for independent, mutually exclusive, or exhaustive sets of statements (or events). We will now provide formulas for the calculation of beliefs of compound statements consisting of the union or product of simple statements which may not be independent, mutually exclusive, or exhaustive. We simply state the formulas here and will give them more analytic coverage in the chapter on probabilities.

When the statement P consists of the sum of three simple statements P_1, P_2, and P_3 and no conditions are placed on the Pi, $i = 1, 2, 3$, then we express our belief in P as follows:

$$P = P_1 + P_2 + P_3$$
$$b(P) = b(P_1) + b(P_2) + b(P_3) - b(P_1P_2) - b(P_1P_3) - b(P_2P_3) + b(P_1P_2P_3)$$

It should be noted that this formula is identical in form as the counting formula in Chapter 1. The belief $b(P_1P_2)$ is the joint belief that statements P_1 and P_2 are simultaneously true. If the three statements are mutually exclusive, then

$$b(P_1P_2) = b(P_1P_3) = b(P_2P_3) = 0$$

and $$b(P) = b(P_1) + b(P_2) + b(P_3)$$

If P_1, P_2, and P_3 are also exhaustive, then

$$b(P) = b(P_1) + b(P_2) + b(P_3) = 1$$

When the statement P consists of the product of three simple statements P_1, P_2, and P_3 which are independent, then the following formula holds

$$P = P_1 \cdot P_2 \cdot P_3$$
$$b(P) = b(P_1) \cdot b(P_2) \cdot b(P_3)$$

If the simple statements (or events) are not independent, then

$$b(P) = b(P_1/P_2P_3)\, b(P_2/P_3)\, b(P_3)$$

where the notation $b(P_2/P_3)$ is used for the conditional belief of P_2 under the assumption that P_3 is true.

3.16 ON CONSTRUCTING DEDUCTIVE OR INDUCTIVE IMPLICATIONS

In formal logic no conditions on relevance were placed on the substantive content of the statements A, B in the implication $A \to B$; A and B could be totally unrelated statements. For example, $A \to B$ and $B \to C$ yields $A \to C$ independent of the substantive connections between A and C. However, in information analysis, it is the substantive content of A and B in the implication $A \to B$ which guides the analyst in stating whether $A \to B$, $B \to A$, or neither. The analyst's interests lie in determining the plausibility of the substantive connection, if any, in the implication $A \to B$.

In this section we will illustrate the procedure good information analysts follow in constructing a plausible implication, $A \to B$, where A and B are only two seemingly relevant statements. Let us consider the statement B: *Heavy construction equipment is located at site X.* An analyst raises the question, what could cause, or be caused by statement B? Systematically, the analyst can list some causes or effects of statement B as statements A_i: $i = 1, 2, 3$.

A_1: Building an airfield at site X.

A_2: Building a military supply base at site X.

A_3: Site X is a heavy construction equipment storage site.

The analyst can now write three implications and proceed to check their plausibility.

$$A_1 \to B \qquad A_2 \to B \qquad A_3 \to B$$

As a *deductive* argument $A_1 \rightarrow B$ says two things: (a) if A_1 is accepted as true, then B must be accepted as true; (b) if B is accepted as false, A_1 must be accepted as false. Looking closer at the meaning of the implication, the analyst observes that airfields have been built without the use of heavy construction equipment. He states that although B may be false, he is not forced to say that A_1 will be false. Hence, he must admit that the implication $A_1 \rightarrow B$ is not a deduction. He believes that it is plausible but may not be 100% realizable. The level of plausibility he assigns to the implication $A_1 \rightarrow B$ depends on his evaluation of airfields being built without the use of heavy construction equipment. In essence, his belief in the implication $A_1 \rightarrow B$ rests on his assessment of the necessity of heavy construction equipment in building an airfield. This is truly a substantive judgment, not a strict logical conclusion.

If the analyst writes $A_3 \rightarrow B$, he notes that it is possible (but highly improbable) for no construction equipment to be on site X at the observation time, yet the site X can still be a storage site for heavy construction equipment. Based on his previous experience, the analyst may feel that $A_3 \rightarrow B$ is a stronger implication than $A_1 \rightarrow B$.

The analyst also questions whether $B \rightarrow A_1$ or $B \rightarrow A_2$, or $B \rightarrow A_3$. He notes that negating A_1 and A_2, i.e., admitting to no construction on site, still leaves a universe of possibilities open as to why heavy construction equipment would be located at site X. Hence he notes that both are weak implications. The implication $B \rightarrow A_3$ appears stronger because if he negates A_3–i.e., agrees that site X is not a heavy construction equipment storage site–he can readily accept (but still not 100%) the cause-effect rationale behind heavy construction equipment not being located at site X.

When dealing with a set of seemingly related statements, A, B, C, ..., M, the analyst can evaluate the premise-consequence relationship between each pair of statements. However, a little judgement on the analyst's part can quickly remove many pairings from consideration without extensive study. Those pairs which remain can be analyzed as follows, where A and B can be any two statements.

Step 1. Consider pairwise implications and their converses. For example, for statements A and B is (a) or (b) preferable?

(a) $A \rightarrow B$

(b) $B \rightarrow A$

Step 2. Question your opinion of the validity of (a) by deciding whether A must be false when B is false–i.e., how valid is $\sim B \rightarrow \sim A$? Note that $\sim B \rightarrow \sim A$ is equivalent to $A \rightarrow B$.

Step 3. If $\sim B \to \sim A$ is not valid, based on experience, cause-effect, analogy, and frequency counts, estimate the plausibility of $\sim B \to \sim A$, e.g., for what percentage of occurrences will $\sim B$ yield $\sim A$? This percentage, if it can be estimated, can serve as your belief in the plausibility of the implication $A \to B$.

Step 4. Question your opinion of the validity of (b) by deciding whether B must be false when A is false, i.e., how valid is $\sim A \to \sim B$?

Step 5. If $\sim A \to \sim B$ is not valid based on experience, cause-effect, analogy, or frequency counts, estimate the plausibility of $\sim A \to \sim B$, e.g., for what percentage of occurrences will $\sim A$ yield $\sim B$?

Step 6. Answer whether $A \to B$ is more plausible than $B \to A$. If both are valid or highly plausible, then A and B may be equivalent, i.e., $A \leftrightarrow B$. Otherwise evaluate $A \to B$ and $B \to A$ and record accordingly.

Up to this point we have illustrated how to evaluate the credibility of a statement A and translate this subjective or objective evaluation to a numerical scale. We call this credibility our belief in the statement A. For any two statements A and B, we have indicated how arguments $A \to B$, $B \to A$ can be synthesized and their plausibilities estimated using the techniques of relative frequency, cause-effect, analogy, and facts from past experiences. We are now in the position to estimate the credibility of one of the statements A or B, provided we know the credibility of the remaining statement and the plausibility of either of the arguments $A \to B$ or $B \to A$. The next section introduces some elements of belief evaluation and provides the rationale behind their applicability to the substantive analysis of information.

3.17 BELIEF FORMULA AND ITS APPLICATIONS

Now we can answer the question: Given $b(A)$, $b(\bar{A})$, $b(A \to B)$, and $b(\bar{A} \to B)$, how are these quantities related to $b(B)$? First, we write the relation and then we give the reasons why the relation conforms with the acceptable notions of formal logic, plausible reasoning, and probability. The relation is given by the following equation:

$$b(B) = b(A) \cdot b(A \to B) + b(\bar{A}) \cdot b(\bar{A} \to B)$$

The above equation states that our belief $b(B)$ in a consequence (B) of an argument $(A \to B)$ is equal to our belief $b(A)$ in the premise (A) weighted by our belief $b(A \to B)$ in the plausibility of the argument $(A \to B)$ to yield $b(A) \cdot b(A \to B)$. To this quantity is added the quantity $b(\bar{A}) \cdot b(\bar{A} \to B)$ where $b(\bar{A})$ is our belief that A is not true or will not occur. It is weighted by our belief $b(\bar{A} \to B)$ in the argument $\bar{A} \to B$ to contribute the quantity $b(A) \cdot b(\bar{A} \to B)$

to our belief $b(B)$ in B. This latter component in the evaluation of $b(B)$ has an interesting interpretation. It states that our knowledge of the credibility of a statement A and the plausibility of the argument $A \rightarrow B$ are not the only contributors to our knowledge of the credibility of B. Our knowledge of the credibility of $\bar{A}$ and the plausibility of the implication $\bar{A} \rightarrow B$ also contribute to our assessment of the credibility of B. In fact, the formula states that its contribution must be weighed equally with the contribution from the former product.

One should note that $b(A) + b(\bar{A}) = 1$. However, it is generally not true that $b(A \rightarrow B) + b(\bar{A} \rightarrow B) = 1$. For the premises A and $\bar{A}$ may have little or no association with the consequence. One can only state that,

$$0 \leqq b(A \rightarrow B) \leqq 1 \qquad 0 \leqq b(\bar{A} \rightarrow B) \leqq 1$$

The next example illustrates that the sum $b(A \rightarrow B) + b(\bar{A} \rightarrow B)$ can exceed the value 1.

Example 3.9 (Belief Evaluation). Let A and B be the statements

A: Wood is burning.

B: Oxygen is present.

The argument $A \rightarrow B$ is a highly plausible one. Hence we evaluate

$$b(A \rightarrow B) \cong 1$$

The argument $\bar{A} \rightarrow B$ is also a highly plausible one. For when wood is not burning, oxygen remains present. Hence we evaluate

$$b(\bar{A} \rightarrow B) \cong 1$$

The sum is

$$b(A \rightarrow B) + b(\bar{A} \rightarrow B) \cong 2$$

Now let us see what happens when we change to another statement for B. Let A and B be the statements

A: Wood is burning.

B: Kindling temperature of wood is reached.

The argument $A \rightarrow B$ is a highly plausible one. Hence we evaluate $b(A \rightarrow B) \cong 1$. The argument $\bar{A} \rightarrow B$ is not a plausible one. For if wood is not burning (in the assumed presence of oxygen), one does not expect the wood to have reached its kindling temperature. Hence we evaluate

$$b(\bar{A} \rightarrow B) \cong 0$$

The sum is

$$b(A \to B) + b(\bar{A} \to B) \cong 1$$

■■

Examples 3.10, 3.11, and 3.12 which follow show how the formula for $b(B)$ yields values which conform with our intuition and experience.

Example 3.10 (Uncertain Beliefs). Suppose there is insufficient reason to believe A over $\bar{A}$—i.e., we state that

$$b(A) = \frac{1}{2} \qquad b(\bar{A}) = \frac{1}{2}$$

Suppose also that we are equally undecided on whether $A \to B$ or $\bar{A} \to B$—i.e., we state that

$$b(A \to B) = \frac{1}{2} \qquad b(\bar{A} \to B) = \frac{1}{2}$$

We would expect the answer for our belief in B to be equally undecided, i.e., $b(B) = 1/2$. Substituting our values into the formula

$$\begin{aligned} b(B) &= b(A) \cdot b(A \to B) + b(\bar{A}) \cdot b(\bar{A} \to B) \\ &= \frac{1}{2} \cdot \frac{1}{2} + \frac{1}{2} \cdot \frac{1}{2} \\ &= \frac{1}{2} \end{aligned}$$

we obtain the expected answer. ■■

In the above example let us assume that the statements A and B are

A: Russians are amassing approximately 50,000 troops on border.

B: Russians will cross border in 30 days.

Let us assess the truthhood of A as 1/2, i.e., $b(A) = 1/2$. If we believe that the concentration of approximately 50,000 troops indicates a 50% chance of causing a border crossing in 30 days, then we would evaluate $b(A \to B) = 1/2$. At this point we must note that we are not forced to say that $b(\bar{A} \to B) = 1/2$. For if the Russians were not amassing 50,000 troops on the border ($\bar{A}$), they could be taking other actions like amassing many more or fewer troops which should cause us to agree to a higher or lower value of belief than 1/2 that the Russians will cross the border in 30 days. If $\bar{A}$ means that the Russians are amassing many more than 50,000 troops, then a border crossing is more likely—i.e., $b(\bar{A} \to B) > 1/2$. If $\bar{A}$ means that the Russians are amassing fewer than 50,000 troops, then a border crossing is less likely—i.e., $b(\bar{A} \to B) < 1/2$. If we accept $b(\bar{A} \to B) = 1/2$, then our belief formula indicates that we should accept $b(B) = 1/2$.

Example 3.11 (Certain Beliefs). Suppose that we accept A as being true, $b(A) = 1$, and suppose that our beliefs in $A \to B$ or $\bar{A} \to B$ are also known. We know that $b(\bar{A}) = 0$, since $b(A) = 1 - b(\bar{A})$. Our relation reduces to

$$\begin{aligned} b(B) &= b(A) \cdot b(A \to B) + b(\bar{A}) \cdot b(\bar{A} \to B) \\ &= 1 \cdot b(A \to B) + 0 \cdot b(\bar{A} \to B) \\ &= b(A \to B) \end{aligned}$$

This result conforms with our intuition because our belief in the validity of the implication $b(A \to B)$ measures our belief in the conclusion B when A is true. If we have agreed that A is true, and accept the validity of the implication, we would have

$$b(B) = b(A \to B) = 1$$

Now suppose that A is false–i.e., $b(A) = 0$ so that $b(\bar{A}) = 1$. In this case

$$b(B) = b(\bar{A} \to B)$$

and $b(\bar{A} \to B)$ measures our belief in B when A is false.■■

Example 3.12 (Range of Beliefs). When A is neither true nor false, when $0 < b(A) < 1$, then $b(B)$ should reflect the possibility of B being true even if A is not true. Intuitively we expect the degrees of belief in both $A \to B$ and $\bar{A} \to B$ to affect our degree of belief in B. When the implication is valid, $b(A \to B) = 1$, deductive pattern 6 of Figure 3.10 indicates that our belief in B should be at least as large as our belief in A. To illustrate that this is the case, we set $b(A \to B) = 1$. Evaluating the belief relation, we obtain

$$\begin{aligned} b(B) &= b(A) \cdot b(A \to B) + b(\bar{A}) \cdot b(\bar{A} \to B) \\ &= b(A) \cdot 1 + b(\bar{A}) \cdot b(\bar{A} \to B) \end{aligned}$$

Since $b(\bar{A} \to B) \geqq 0$ and $b(\bar{A}) \geqq 0$, we have $b(\bar{A}) \cdot b(\bar{A} \to B) \geqq 0$. Thus $b(B) \geqq b(A)$, in agreement with our intuition and our previous statements. For example, consider the statement A: All Russians are good soldiers. Statement A is false if only one Russian is not a good soldier. When we say that we believe the statement A to have a credibility level of 0.75, we normally mean to say that three out of four Russians are good soldiers. If we negate the statement, then we obtain $\bar{A}$: Some Russians are not good soldiers. If we wish to maintain the relation $b(A) + b(\bar{A}) = 1$, then we should set $b(A) = 0.25$ and interpret the statement A as

$\bar{A}$: One out of four Russians is not a good soldier.

Thus, the quantifier "some" means, in this example, one out of four.

If we start with the statement B: No Russian is a good soldier, then its negation is $\bar{B}$: Some Russians are good soldiers. The interpretation of "some" in this example is "one or more." If we assess the credibility of B as $b(B) = 0.75$, this is saying that approximately three out of four Russians are not good soldiers. Hence the "some" in $\bar{B}$ means "one out of four." The other quantifiers, each, any, one, etc., are handled similarly. ■■

Example 3.13 (Deduction and Beliefs). Recalling our previous discussions in Section 3.5 on the deductive form of reasoning, we now illustrate how deductive conclusions relate to conclusions drawn on beliefs. First let us consider the deductive form,

$$\begin{array}{l} A \rightarrow B \\ A \text{ is true} \\ \hline \therefore B \text{ is true} \end{array}$$

It was previously stated that our confidence in the conclusion B is estimated from our belief in the implication $A \rightarrow B$ and in the statement A. If either the argument $A \rightarrow B$ is valid or the premise A is true, but not both, the belief one places in the conclusion B approximates the lesser of the degrees of belief placed in the statement or argument–i.e., if the argument is valid but the premise is uncertain, we have,

$$\textbf{Pattern 1} \quad \begin{cases} b(A \rightarrow B) = 1 \\ \quad b(A) = \text{uncertain} \\ \quad b(B) \geqq b(A) = \text{uncertain} \end{cases}$$

In this case the belief in the consequence exceeds the belief in the premise.

If the argument is uncertain but the premise is true, we have

$$\textbf{Pattern 2} \quad \begin{cases} \quad b(A) = 1 \\ b(A \rightarrow B) = \text{uncertain} \\ \quad b(B) = b(A \rightarrow B) = \text{uncertain} \end{cases}$$

In this case the belief in the consequence approximates the belief in the validity of the argument. An example of pattern 1 follows:

A:
1. All G class submarines carry ballistic missiles.
2. Submarine sighted within 200 mi of Oahu, Hawaii is a G class submarine.

B: Ballistic missiles are within 200 mi of Oahu, Hawaii

From the statements A and B above, it is obvious that,

$$b(A \rightarrow B) = 1$$

If there is any doubt in the premise A, e.g., if the word "sighted" was changed to "probably sighted," then the statement B should be changed to read "ballistic missiles are probably...," and the value assigned to $b(B)$ modified accordingly to conform with $b(B) \geqq b(A)$. Thus the belief that ballistic missiles are within 200 mi of Oahu is at least as great as the belief that a submarine of the G class was sighted within the range specified. That the belief in B may be greater than the belief in A is explained through noting that submarine classes other than the G class may also carry ballistic missiles and may operate within 200 mi of Oahu.

3.18 PLAUSIBILITY SCALES

Pattern 1 can be diagrammed on paired plausibility scales, Figure 3.12. Since $b(A \rightarrow B) = 1$, it is placed on the extreme right of the A scale. Since $b(A)$ is uncertain, it is placed on the A scale at some position between T and F or 0 and 1. The position conforms with the value assigned to our belief in A. The pattern states that the belief $b(B)$ should be placed somewhere to the right of (higher value than) $b(A)$. The mark x for the value of $b(B)$ on the B scale indicates the relation $b(B) \geqq b(A)$ is satisfied. The difference $b(B) - b(A)$ is given by $b(\bar{A})\ b(A \rightarrow B)$.

Example 3.14 (Belief Patterns). There are two aspects of the second pattern that should be noted. The argument $A \rightarrow B$ may be uncertain for two different reasons. First, A may often (but not always) imply B. In this sense, there exists some probability for the occurrence of B given the occurrence of A. Second, it may be hypothesized that A implies B in the sense that *all* observations to date have shown that the presence of B may be inferred from the presence of A. However, the validity of this hypothesis may require further testing before it can be asserted with certainty.

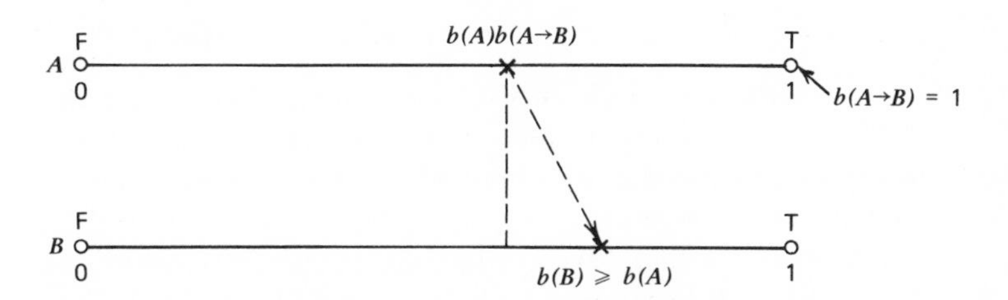

Figure 3.12 Plausibility scales.

In this example, statements A and B of Example 3.11 are modified to illustrate the principles involved in pattern 2.

A:
1. Approximately 30% of Zulu (Z) class submarines are fitted with new weapons.
2. Submarine sighted within 200 mi of Oahu, Hawaii, is a class Z submarine.

B: New weapon systems are within 200 mi of Oahu, Hawaii.

From the statements A and B above, it is noted that, $b(A) = 1$, $b(A \to B) =$ uncertain. There is doubt that the submarine sighted, as stated in A, is a Z class submarine fitted with new weapons. This uncertainty is transferred to the conclusion, so that $b(B) =$ uncertain. In the above example we would evaluate $b(A \to B) = 0.30$ so that $b(B) \cong 0.30$.

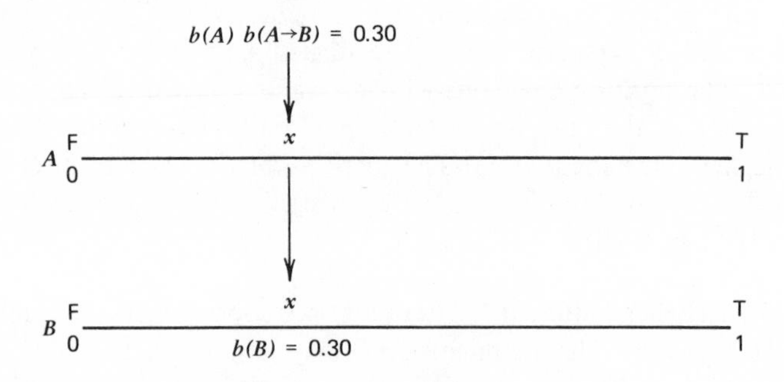

Figure 3.13 Plausibility scales.

Pattern 2 can also be diagrammed on paired plausibility scales as indicated in Figure 3.13. Since $b(A \to B) = 0.30$, this value can be fixed on the A scale as can $b(A) = 1$. The placement of $b(B)$ on the B scale is established as in pattern 1; $b(B)$ should be placed at the value $b(B) = 0.30$, since $b(\bar{A}) = 0$ and the term $b(\bar{A})\ b(\bar{A} \to B)$ does not make a contribution to the value of $b(B)$ as it did in pattern 1. ■■

In the above discussion, the events or statements A and $\bar{A}$ were viewed as two mutually exclusive events. In most cases we have more than two events, which can occur in conjunction with the event B. In such cases the requirement remains that the set of events $A_1, A_2, \ldots, A_n$ be mutually exclusive. Further, only one event A_i of the set of events can occur in conjunction with the event B at any time. For this occurrence both $b(A_1)$ and $b(A_i \to B)$ will be known. When these conditions are satisfied, our degree of belief in B can be expressed as

$$b(B) = b(A_1) \cdot b(A_1 \to B) + b(A_2) \cdot b(A_2 \to B) + \cdots + b(A_n) \cdot b(A_n \to B)$$

or

$$b(B) = \sum_{i=1}^{n} b(A_i) \cdot b(A_i \to B)$$

To illustrate the application of the above relation to a practical situation consider the following argument:

A_1: The submarine sighted is one of ten operational H class submarines, 5 of which are armed with ballistic missiles.

A_2: The submarine sighted is one of 50 operational J class submarines, 20 of which are armed with ballistic missiles.

A_3: The submarine sighted is one of 40 operational N class submarines, no one of which is armed with ballistic missiles.

A_4: No other class of submarine is operating in the area under surveillance.

B: Sighted submarine (class not identified) is armed with ballistic missiles.

The belief that the sighted submarine is armed with ballistic missiles can be expressed as

$$b(B) = b(A_1)\ b(A_1 \to B) + b(A_2)\ b(A_2 \to B) + b(A_3)\ b(A_3 \to B) + b(A_4)\ b(A_4 \to B)$$

where A_4 asserts that any sighted submarine will be of the H, J, or N class. We assume that the information provided in A_1, A_2, A_3, and A_4 is accurate– i.e., $b(A_1) = \frac{10}{100}$, $b(A_2) = \frac{50}{100}$, $b(A_3) = \frac{40}{100}$, wherein $b(A_1) + b(A_2) + b(A_3) = 1$. If the sighted submarine is of the H class, then $b(A_1 \to B) = \frac{5}{10} = \frac{1}{2}$, since five of ten carry BMs. For the J class, $b(A_2 \to B) = \frac{20}{50} = \frac{2}{5}$ since 20 out of 50 carry BMs. For the N class, $b(A_3 \to B) = 0$, since N class submarines do not carry missiles. These values can now be substituted in the expression for $b(B)$

$$\begin{aligned} b(B) &= b(A_1)\ b(A_1 \to B) + b(A_2)\ b(A_2 \to B) + b(A_3)\ b(A_3 \to B) \\ &= \frac{1}{10} \cdot \frac{1}{2} + \frac{1}{2} \cdot \frac{2}{5} + \frac{2}{5} \cdot 0 \\ &= \frac{1}{20} + \frac{4}{20} \\ &= \frac{25}{100} \end{aligned}$$

This last value agrees with our intuition since only 25 out of 100 possible submarines are armed with ballistic missiles. To obtain this value for $b(B)$

necessitated quantifying the degree of belief in A_1 and A_2. This was accomplished by directly estimating values for these factors. They may be obtained by weighing other data pertinent to the problem. Previous operational characteristics of the classes plus a knowledge of the locations of some of the units, could, for example, influence the analyst's evaluation of $b(A_1)$ and $b(A_2)$.

3.19 BELIEF EVALUATION

The objective of this section is to

1. Introduce new measurement intervals for source reliability, statement credibility, and argument plausibility.
2. Enlarge on the relationship between source reliability and statement credibility.
3. Compute a table of values for $b(B)$ whose entries are dependent on values for $b(A)$, $b(A \rightarrow B)$, $b(\bar{A})$, and $b(\bar{A} \rightarrow B)$.

Measurement Interval

Previously, in Section 3.4, we used the averages of five unequal length intervals when assigning values to the credibility of statements, the reliability of sources, the plausibility of arguments, and the probability of events. These intervals were illustrated in Figures 3.3 and 3.4. In this section we will use a 0-1 scale subdivided into ten equal intervals. The change to ten equal intervals is justified for two reasons. First, the objective techniques of evaluation discussed in Section 3.12, when used properly, should result in higher accuracies than the subjective techniques that were used in the five unequal interval scale. Measurements made on the five unequal interval scale are easily converted to the most representative interval of the ten interval scale. Second, the ten interval scale simplifies the calculations in the development of the belief table and in applications which make use of the table.

When evaluating the credibility of statements or the plausibility of arguments, we can either assign values at interval end points, 0, 0.1, 0.2, ..., 1.0, or select any decimal number between 0 and 1. When we do the latter, it will be necessary to round the number to the nearest interval endpoint. Thus the value 0.83 is rounded to 0.80. The value 0.69 is rounded to 0.70. The value at the center of an interval is rounded to the lower endpoint of the next higher interval. The values 0.25 and 0.65 are rounded to 0.3 and 0.7, respectively. The 0-1 scale and its interval are illustrated in Figure 3.14.

Many substantive analysts do not like to work purely from numbers. To satisfy the desire of these analysts nominal values are assigned to the upper five intervals of the 0-1 scale. Only the upper five are named because this is where the techniques of belief evaluation are most frequently applicable. The names of the five upper intervals are: low (L), medium-low (M-L), medium (M), medium-high (M-H), and high (H). More will be said on the applicability of interval values to belief calculation in the sections which follow.

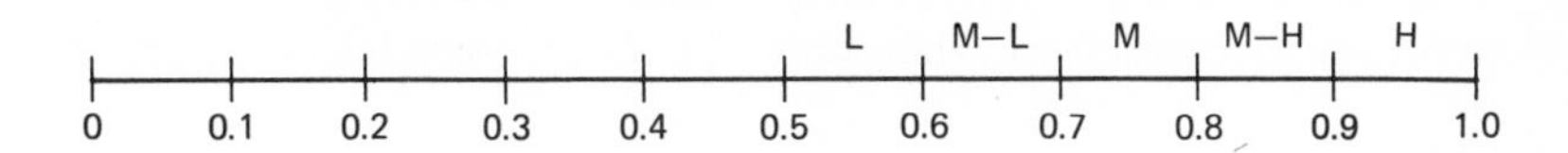

Figure 3.14 The 0-1 scale.

Source Reliability and Statement Credibility

In evaluating the credibility of statements, there are two applicable criteria which can be used. The first is the reliability of the source of the statement. The second is the opinions, judgements, or calculations employed by the analyst to evaluate the substantive content of the statement. The two criteria can be used in conjunction with each other if the correct and sufficient information is available.

First, we will consider the reliability of the source as a criterium for establishing the credibility of statements received from the source. To do so, we must enlarge on what we mean by the "reliability" of a source. And we must show how the reliability of a source and the credibility of its statements are related. Reliability is normally used in a combined sense of *repeatability, availability at full capacity,* and *accuracy.* Thus for a source to be considered reliable, it must be available and capable to report on each event accurately and should be able to repeat its performance over considerable periods of time. An electronic sensor such as radar can be available yet, due to slight but undetected decrease in output power, introduce errors in range measurements on targets at long range. A newspaper, witness testimony, trade journal, or literary magazine may be available, and highly repeatable, but each may repeat with views that are consistently biased to the right, left, or middle of the road. A human observer may not be 100% available (preferring a round of golf over attendance at the current air show) yet, when he is available, he is capable of repeatedly submitting reports of high accuracy.

From the examples it is clear that a source may be unreliable for any one, or a combination of, three reasons. First, the source may not be fully available to observe and hence to submit a complete report–i.e., the source may not have observed all there was to observe. Second, the source may be available but not fully capable in observing accurately that which it did observe. Third, the source may not be consistent in its ability to observe, e.g., its reports on

Monday (after a hard weekend) may not be as good as those submitted on Wednesday or Thursday. Whatever the reason for the reduced reliability, the submitted report contains statements which are not fully credible. The relationship between report or statement credibility and the reliability of the source will now be developed.

First, we assume that the three components of reliability can all be measured relative to a 0-1 scale. We abbreviate the three components of reliability as follows:

$$\begin{aligned} \text{availability} &= A \\ \text{capability} &= C \\ \text{variability} &= V \end{aligned}$$

Thus we have that,

$$\begin{aligned} 0 \leqslant A \leqslant 1 \\ 0 \leqslant C \leqslant 1 \\ 0 \leqslant V \leqslant 1 \end{aligned}$$

where $A = 0$ means that the source's reports or statements are highly incomplete, and $A = 1$ means essentially perfect completeness in the reports; $A = 1/2$ means that the source may have observed only 1/2 of that available for observation. When $C = 0$, then the source is not capable of making an accurate report. When $C = 1$, the source is completely capable of accurately reporting all there is to observe. When $C = 1/2$, the source will accurately report only 50% of those events it observes. When $V = 1$, the source has had a long record of repeatedly submitting reports of essentially the same performance (high or low). When $V \neq 1$, V measures the percent of the time the source has repeated its reporting performance.

Thus V can be used to measure the amount of bias and purposeful corruption the source and administrative system will introduce into the reporting of its observations. When $V = 0$, one admits that the variability of the source does not permit one to place any credence in the source's reports. The source is one which would submit contradictory descriptions of essentially the same event or occurrence to suit its own purpose even if these were separated by only a short period of time.

Using the three components of reliability as previously described, we define the reliability of a source R as the product of these three components,

$$\text{reliability} = A \cdot V \cdot C = R$$

Further, we state that the *credibility of statements* made by a source of reliability R will have approximately the same value as the reliability of the source. Hence, as previously indicated, the credibility $b(P)$ of a statement P from a source with reliability R will satisfy the following relation:

$$b(P) \cong R \ (P)$$

However, there are instances where the reporting source, called the second source, may have received the information reported from another source, called the first source. In its report the second source may have submitted an evaluation of the first source in accordance with the previously described credibility code (see Figs. 3.1 and 3.4). The analyst receiving the report from the second source must not only evaluate the reliability of the second source but also the ability of the second source to evaluate its source's credibility fairly and accurately. Thus, when a hierarchy of sources is involved, the problem of evaluating the credibility of reports and statements from the original source becomes more difficult since each source in turn can introduce inaccuracies, generalities, and subjective interpretations reflecting their own views.

When only a single source is involved and the source is reporting events, or observations on a firsthand basis, then the technique for evaluating the sources reliability, and hence the credibility of its statements, is a reasonably accurate and effective one. Its use is recommended (over a purely subjective one) in situations involving a single source or where each source in turn, in a hierarchy, has been evaluated by the receiving source, using the three components of reliability.

Example 3.15 (Credibility Evaluation). A report from source S contains the following statement P: Country X is shipping 30 of its newest fighter aircraft to country Y. Source S's previous reports are known to have an average accuracy of 0.8, i.e., it occasionally sends a report or parts of a report (about 1 in 5) which is off-the-mark. And source S is known to have missed events or important observations in several previous instances (about 1 in 10), i.e., source S was not present to observe the events. Source S is also known to vary only slightly in its ability to observe during a reporting period. What credibility value should be attached to statement P? In this example we have

$$\begin{aligned}
\text{availability} &= 0.8 = A \\
\text{capability} &= \frac{4}{5} = 0.8 = C \\
\text{variability} &= \frac{9}{10} = V
\end{aligned}$$

$$\text{reliability} = A \cdot C \cdot V = 0.8 \cdot 0.8 \cdot 0.9$$
$$R = 0.576 \approx 0.6$$

Thus we would evaluate

$$b(P) = R = 0.6$$

■■

Substantive Content and Statement Credibility

When the credibilities of statements are evaluated solely from the substantive content of the statement, i.e., the analyst is his own source of information, the analyst may use any of the previously described techniques of analysis such as analogy, cause-effect, historical, economic, social, and political trends, relative frequency. In so doing the analyst makes subjective or objective evaluations on the truth of the substantive content of the statement. For example, if a statement P forecasts a simple event, then the analyst attempts to evaluate the probability that the event will occur as stated. For example, if statement P: Country X will launch an offensive against country Y, then the analyst makes a subjective or objective evaluation of the probability that P will occur.

If the statement is a complex statement which involves events, dates, levels of activities, personalities, etc., then the evaluation of the credibility of the statement becomes a little more difficult. However, the evaluation can be simplified if the following steps are taken: First, the compound statement should be decomposed into simple statements combined by the logical connectives and, or, not, and exclusive or. Second, the credibilities of each simple statement can be evaluated independently of the other simple statements. Third, the independence, mutual exclusiveness, and exhaustiveness of the events contained in the simple statements are determined. And finally the credibilities of the simple statements are combined in accordance with the rules of combination for beliefs discussed in Sections 3.9-3.13. For example, the statement P: Country X will launch a spring offensive using 50,000 troops, preceded by a massive infiltration effort, against country Y, consists of the following four simple statements:

P_1 : X will launch offensive against Y.

P_2 : X will launch offensive in spring.

P_3 : X will use 50,000 troops.

P_4 : X will precede offensive by a massive infiltration effort.

The statement P is the conjunction of the four simple statements P_1, P_2, P_3, P_4, or

$$P = P_1 \wedge P_2 \wedge P_3 \wedge P_4$$

Thus if the four simple statements are independent, we would have,

$$b(P) = b(P_1) \cdot b(P_2) \cdot b(P_3) \cdot b(P_4)$$

However, statements P_1 and P_4 may not be independent. In that case we use the conditional form for the belief (or credibility) calculation

$$b(P) = b(P_1)\ b(P_3)\ b(P_4/P_2)\, b(P_2)$$

Other examples can be decomposed and evaluated accordingly.

Calculation of Credibilities

We are now in a position to use the 0-1 scale in evaluating the credibility of a statement Q from the credibility of a statement P and its substantive connection to Q through the argument $P \to Q$. The following three quantities will be evaluated relative to the 0-1 scale:

1. Plausibility of argument, $b(P \to Q)$.
2. Credibility of statements $b(P)$ and $b(Q)$.
3. Reliability of source R_S.

Table 3.1 below gives the numerical and nominal values to the five upper most intervals of the 0-1 scale. Abbreviations (or mnemonics) are included

Table 3.1. Values for the plausibility of argument $b(P \to Q)$, the credibility of statements $b(P)$ and $b(Q)$, and the reliability of source R_S.

Numerical values	Nominal values	Abbreviation
0.9-1.0	High	H
0.8-0.9	Medium-high	M-H
0.7-0.8	Medium	M
0.6-0.7	Medium-low	M-L
0.5-0.6	Low	L

in the last column of the table. Descriptions of the five intervals of argument plausibility follow:

Plausibility of high range (H): The argument is solid and there is little doubt that statement Q follows from statement P. The plausibility range is 0.9 to 1.0.

Plausibility of medium-high range (M-H): The argument is slightly lacking in its finality. There are isolated instances where statement Q does not follow from statement *P*. The plausibility range is 0.8-0.9.

Plausibility medium range (M): The argument is good but the substantive ties between premise and conclusion are lacking in about 25% of the cases that are considered. The plausibility range is 0.7-0.8.

Plausibility medium-low range (M-L): The argument is just above average. The substantive ties between premise and conclusion require more cohesion or reasonable statistical justification. The plausibility range is 0.6-0.7.

Plausibility low range (L): The argument is average where one finds only slightly above half the instances supporting the substantive tie between premise and conclusion. The plausibility range is 0.5-0.6.

In Table 3.1 one essential difference in interpretation between low-range plausibility and low-range credibility must be noted. With low-range credibility (0.5-0.6), we assert that we have insufficient information or no substantiation of information to decide on the truth or falsity of the statement. Hence we remain uncommitted and the statement remains unevaluated; the statement is neither true nor false. We may also receive information on the credibility of the statement and find that approximately half of it supports calling the information true, the remaining information supports calling the statement false. In this case the statement's credibility is evaluated 0.5 in the sense of a relative frequency of truthness.

The plausibility of an argument is determined either by its logical structure or by the substantive connections between premise and conclusion. The logical structure determines whether the argument is structurally valid (tautology) or just plausible. If the argument is not structurally valid, then the plausibility of the argument is determined by the technique of relative frequency, cause-effect, analogy, etc. Again, having no decisive information dictates that the plausibility of the argument be evaluated at the 0.5 level which is equivalent to saying that it is not known whether information favors or disfavors the argument. More information is required to evaluate the argument's plausibility toward 1 or 0.

Table of Credibilities

We indicated above we will not use the entire scale from 0 to 1.0. Certain parts of the scale are more significant than others in assessing credibilities and plausibilities. We assume that if $b(A \to B)$ or $b(A)$ is low, i.e., if we feel that our information about A is not very credible and that the substantive connection between statements A and B is not strong, we will not attempt to evaluate the credibility of B. Hence, we will only consider situations where the belief in the plausibility of the argument $b(A \to B)$ is at the 0.5 or above level. We will restrict our consideration to statements whose credibility level is below or above

0.5. If the credibility of A is below 0.5, then we will consider the statement $\bar{A}$ which will have a credibility level above 0.5 and consider the argument $\bar{A} \to B$ in place of $A \to B$.

We now show how to estimate the credibility of B from estimates of $b(A)$, $b(A \to B)$, and $b(\bar{A} \to B)$. All assumptions are listed in the following procedure:

1. Use the formula
$b(B) = b(A)\ b(A \to B) + b(\bar{A})\ b(\bar{A} \to B)$

Table 3.2

Range	Credibility of A, $b(A)$	Plausibility of $A \to B$, $b(A \to B)$	$b(B) = b(A)\ b(A \to B) + b(\bar{A})\ b(\bar{A} \to B)$ with $b(\bar{A} \to B) = a$, and				
			$a = 0$	$a = 0.1$	$a = 0.2$	$a = 0.3$	$a = 0.4$
	1.0	1.0 H	1.0	1.0	1.0	1.0	1.0
	1.0	.9 M-H	.9	.9	.9	.9	.9
91-100 = H	1.0	.8 M	.8	.8	.8	.8	.8
	1.0	.7 M-L	.7	.7	.7	.7	.7
	1.0	.6 L	.6	.6	.6	.6	.6
	.9	1.0 H	.90	.91	.92	.93	.94
	.9	.9 M-H	.81	.82	.83	.84	.85
81-90 = M-H	.9	.8 M	.72	.73	.74	.75	.76
	.9	.7 M-L	.63	.64	.65	.66	.67
	.9	.6 L	.54	.55	.56	.57	.58
	.8	1.0 H	.80	.82	.84	.86	.88
	.8	.9 M-H	.72	.74	.76	.78	.80
71-80 = M	.8	.8 M	.64	.66	.68	.70	.72
	.8	.7 M-L	.56	.58	.60	.62	.64
	.8	.6 L	.48	.50	.52	.54	.56
	.7	1.0 H	.70	.73	.76	.79	.82
	.7	.9 M-H	.63	.66	.69	.72	.75
61-70 = L-M	.7	.8 M	.56	.59	.62	.65	.68
	.7	.7 M-L	.49	.52	.55	.58	.61
	.7	.6 L	.42	.45	.48	.51	.54
	.6	1.0 H	.60	.64	.68	.72	.76
	.6	.9 M-H	.54	.58	.62	.66	.70
51-60 = L	.6	.8 M	.48	.52	.56	.60	.64
	.6	.7 M-L	.42	.46	.50	.54	.58
	.6	.6 L	.36	.40	.44	.48	.52

2. Use the relation
   $b(A) + b(\bar{A}) = 1$
3. Use values at 0.1 intervals from 0.5 to 1.0 for $b(A)$ and $b(A \rightarrow B)$. These values are selected as the upper limits of the five intervals. Thus 0.6 is the upper limit for the low range 0.5-0.6.
4. Ranges (or intervals) of 0.1 were selected as being matched to the accuracy of the judgmental evaluations and measurements used in determining values for $b(A)$ and $b(A \rightarrow B)$.
5. Use values at 0.1 intervals from 0 to 0.4 for $b(\bar{A} \rightarrow B)$. If values above 0.4 are required, one should consider using $\bar{A}$ as the premise in place of A and use the table accordingly.
6. The range or interval of values are listed in column 1 of Table 3.2.
7. Columns 2 and 3, of Table 3.2 contain the values for the credibility of A and the plausibility of the argument $A \rightarrow B$, respectively. There are 25 possible pairing of values at 0.1 intervals entered in columns 2 and 3.
8. Columns 4, 5, 6, 7, and 8 contain the computed values for $b(B)$ when $b(\bar{A} \rightarrow B) = 0, 0.1, 0.2, 0.3$, and 0.4, respectively.

We now show how to use the table to evaluate the credibility of the statement B as a function of the credibility of the statement A and the plausibilities of the arguments $b(A \rightarrow B)$ and $b(\bar{A} \rightarrow B)$. First, we use rounded values for $b(A)$, $b(A \rightarrow B)$, and $b(\bar{A} \rightarrow B)$ which agree with the entries in the table. Then we will show how to use the table when unrounded values are used and interpolate in the table as required.

Example 3.16 (Credibility Tables).

1. Evaluate $b(B)$ for $b(A) = 0.8$, $b(A \rightarrow B) = 0.9$, $b(\bar{A} \rightarrow B) = 0.2$.

We enter the table on rows 11-15 where $b(A) = 0.8$ in the second column. The value $b(A \rightarrow B) = 0.9$ appears in rows 2, 7, 12, 17, and 22. To match with $b(A) = 0.8$, we must use the entry on row 12; thus our answer is in row 12. The value $b(\bar{A} \rightarrow B) = 0.2$ is at the head of, and hence locates, the sixth column in the table. Moving down the sixth column to the twelfth row we read the value 0.76; hence 0.76 is the value of $b(B)$. The table contains the results of the following calculations:

$$\begin{aligned} b(B) &= b(A) \cdot b(A \rightarrow B) + b(A) \cdot b(\bar{A} \rightarrow B) \\ &= 0.8 \cdot 0.9 + 0.2 \cdot 0.2 \\ &= 0.76 \end{aligned}$$

2. Evaluate $b(B)$ for $b(A) = .75$, $b(A \rightarrow B) = 0.9$, $b(\bar{A} \rightarrow B) = 0.3$.

For $b(A) = .75$, we consider all possible values for $b(A) = 0.7$ and $b(A) = 0.8$. This restricts us to rows 11-20. For $b(A \to B) = 0.9$, we are further confined to two rows in the set of rows from 11-20. These are rows 12 and 17. In row 12, the value $b(\bar{A} \to B) = 0.3$ yields a value $b(B) = 0.78$ at column 7. In row 17 the value $b(\bar{A} \to B)$ yields a value $b(B) = 0.72$ at column 7. Thus when $b(A \to B) = 0.9$, we have the following:

for $b(A) = 0.8$	$b(B) = 0.78$
for $b(A) = 0.7$	$b(B) = 0.72$
for $b(A) = 0.75$	$b(B) = ?$

Since 0.75 is midway between 0.7 and 0.8, $b(B)$ for $b(A) = 0.75$ should be midway between 0.72 and 0.78. Hence $b(B) = 0.75$. The above process is called linear interpolation and is widely used with tables which give values at discrete intervals for continuous variables and functions.

3. Evaluate $b(B)$ for $b(A) = 0.9$, $b(A \to B) = 0.75$, and $b(\bar{A} \to B) = 0.2$.

For $b(A) = 0.9$, we restrict our attention to rows 6-10. For $b(A \to B) = 0.75$, we further restrict our attention to row 8, where $b(A \to B) = 0.8$, and row 9, where $b(A \to B) = 0.7$. At these two values we use the value $b(\bar{A} \to B) = 0.2$ to enter the sixth column and read the two values $b(B) = 0.74$ and $b(B) = 0.65$. Thus when $b(A) = 0.9$, we have the following:

for $b(A \to B) = 0.8$	$b(B) = 0.74$
for $b(A \to B) = 0.7$	$b(B) = 0.65$
for $b(A \to B) = 0.75$	$b(B) = ?$

Using interpolation, we find that

$$b(B) = 0.695$$ ▪▪

Up to this point we used a general formula to evaluate the credibility of the conclusion using the credibility of the premise and the plausibility of the argument. Now we use a simpler formula, based on the assumption that the argument $P \to Q$ is valid, to evaluate the credibility of the conclusion using the credibility of the premise and the plausibility of the argument. The formula,

$$b(Q) = b(P) \cdot b(P \to Q) + b(\bar{P}) \cdot b(\bar{P} \to Q)$$

used previously to evaluate $b(Q)$, can be simplified using the assumption that the argument $P \to Q$ is valid,

$$b(P \to Q) = 1$$

This allows us to write,

$$b(Q) = b(P) + b(\bar{P}) \cdot b(\bar{P} \to Q)$$

where

$$b(P) + b(\bar{P}) = 1$$

The formula states that one's belief in the credibility of the consequence $b(Q)$, in a valid argument, is as least as large as one's belief in the credibility of the premise $b(P)$. The increase in belief in the credibility of the consequence is determined by the product of one's disbelief in the credibility of the premise, $b(\bar{P})$ multiplied by (or weighted by) one's belief in the plausibility that the negation of the premise implies the consequence $b(\bar{P} \to Q)$.

The accompanying table gives values for $b(Q)$ in terms of sets of typical values for $b(P)$ and $b(\bar{P} \to Q)$:

$b(P)$	$b(\bar{P} \to Q)$	$b(Q)$
.9	.5	.95
.9	.4	.94
.9	.3	.93
.9	.2	.92
.9	.1	.91
.7	.5	.85
.7	.4	.82
.7	.3	.79
.7	.2	.76
.7	.1	.73
.5	.5	.75
.5	.4	.70
.5	.3	.65
.5	.2	.60
.5	.1	.55

The table indicates that when the credibility of the premise $b(P)$ is high (0.90), the contribution to the credibility of the conclusion $b(Q)$ from $b(\bar{P})$ and $b(\bar{P} \to Q)$ is not high. This is due to the fact that $b(\bar{P})$ is small (0.1). The first five entries in the table shows this relationship. When the credibility of the premise $b(P)$ is reduced from 0.9 to 0.7, then the contribution from the quantity $b(\bar{P}) \cdot b(\bar{P} \to Q)$ becomes more significant for larger values of $b(\bar{P} \to Q)$. The second five entries in the table illustrates this relationship. As the credibility of the premise is still further reduced from 0.7 to 0.5, then the term $b(\bar{P}) \cdot b(\bar{P} \to Q)$ can become a larger contributor to the credibility of the consequence.

The last five entries of the table shows a difference of 0.25 when $b(\bar{P} \to Q) = 0.5$ and $b(P) = 0.5$. We can rewrite the last formula as,

$$b(Q) - b(P) = b(\bar{P}) \cdot b(\bar{P} \to Q)$$

Written in this form we can see that the difference which exists between the credibilities of conclusion and premise is accounted for in the product of the two beliefs $b(\bar{P})$ and $b(\bar{P} \to Q)$.

Example 3.17 (Statement Credibility). Consider the following premise P and conclusion Q. The premise consists of two statements

$$P: \begin{cases} P_1: \text{Big guns have a range of 20,000 yd.} \\ P_2: \text{Big guns sighted approximately 10,000 yd of site } X. \end{cases}$$

Q: Big guns are within range of site X.

The condition $b(P \to Q) = 1$ is satisfied in this argument. We want to check the credibility of Q. The reports indicate that statement P_1 has credibility 1, but P_2 has credibility $b(P_2) = 0.75$. Now premise P is the conjunction of P_1 and P_2. For this example we accept the credibility of P to be the product of the credibilities of P_1 and P_2, which is 0.75. The formula

$$b(Q) = b(P) + b(\bar{P}) \cdot b(\bar{P} \to Q)$$

indicates that we need find $b(\bar{P})$ and $b(\bar{P} \to Q)$. The first quantity is simply,

$$b(\bar{P}) = 1 - b(P) = 1 - 0.75 = 0.25$$

The second quantity must be analyzed through the decomposition of $\bar{P}$ into the union of three events which are

1. $\bar{P}_1 \wedge \bar{P}_2$
2. $P_1 \wedge \bar{P}_2$
3. $\bar{P}_1 \wedge P_2$

For the three events, our beliefs are

1. $b[(\bar{P}_1 \wedge \bar{P}_2) \to Q] = 0$
2. $b[(P_1 \wedge \bar{P}_2) \to Q] = \text{maybe}$
3. $b[(\bar{P}_1 \wedge P_2) \to Q] = \text{maybe}$

Events 2 and 3 require further discussion. Event 2 denies P_2, i.e., big guns were not sighted approximately 10,000 yd of site X. This still leaves open the possibility that some guns are located outside of 10,000 yd and within 20,000 yd, the range of the big guns. Event 3 denies that big guns have a range of 20,000 yd. If the range is less than 20,000 yd but still above 10,000, the big guns will be within their range to site X. We estimate

2. $b[(P_1 \wedge P_2) \rightarrow Q] = 0.10$
3. $b[(P_1 \wedge P_2) \rightarrow Q] = 0.20$

$$b(\bar{P} \rightarrow Q) = b\left\{[(\bar{P}_1 \wedge P_2) + (P_1 \wedge \bar{P}_2) + (\bar{P}_1 \wedge \bar{P}_2)] \rightarrow Q\right\}$$

Since the events in the square brackets are mutually exclusive, we can write

$$\begin{aligned} b(\bar{P} \rightarrow Q) &= b[(\bar{P}_1 \wedge P_2) \rightarrow Q] + b[(P_1 \wedge \bar{P}_2) \rightarrow Q] + b[(\bar{P}_1 \wedge \bar{P}_2) \rightarrow Q] \\ &= b[(\bar{P}_1 \wedge P_2) \rightarrow Q] + b[(P_1 \wedge \bar{P}_2) \rightarrow Q] \\ &= 0.20 + 0.10 = 0.30 \end{aligned}$$

Using these estimates, we have

$$\begin{aligned} b(Q) &= b(P) + b(\bar{P}) \cdot b(\bar{P} \rightarrow Q) \\ &= 0.75 + 0.25 \cdot 0.3 \\ &= 0.825 \end{aligned}$$

Note that we have lessened our emphasis on the differences between premise and conclusion. When dealing simultaneously with two statements and the two relations $A \rightarrow B$, $B \rightarrow A$, we are most interested in the strength of the substantive tie between A and B. When we have information on the credibility of A and not on B, we seek to use the information on A and the substantive ties between A and B to gain some knowledge on the credibility of the statement B. If we have information on B and not on A, we interchange the roles of A and B (change notation) and proceed as we did in the previous case. If it is determined that $b(A \rightarrow B)$ is H and $b(B \rightarrow A)$ is L, then we prefer to label A the premise and B the conclusion. However, there is no convention which forces us to do this. If the credibilities of A and B are essentially equal, then the premise-conclusion relationship can go either way. If the credibility of A is exceeded by the credibility of B, then we may fall back on the statement that in a *valid* argument the credibility of the conclusion exceeds the credibility of the premise. This guides us in calling A the premise and B the conclusion. If the argument $A \rightarrow B$ is only plausible, $b(A \rightarrow B) \neq 1$, the latter guide is not applicable.

The key objective here is the evaluation of information based on its substantive ties to previously evaluated information. Name-calling is part of the game and while it does not contribute substantively to the objective, it does help to label and identify statements and arguments as part of the overall objectives.

Exercise 3.1. Determine the credibility of B for the combination of values given for $b(A)$, $b(A \rightarrow B)$, and $b(\bar{A} \rightarrow B)$.

(a) $b(A) = 0.7$, $b(A \rightarrow B) = 0.8$, $b(\bar{A} \rightarrow B) = 0.3$
(b) $b(A) = 0.85$, $b(A \rightarrow B) = 0.8$, $b(\bar{A} \rightarrow B) = 0.2$
(c) $b(A) = 0.9$, $b(A \rightarrow B) = 0.75$, $b(\bar{A} \rightarrow B) = 0.1$
(d) $b(A) = 0.9$, $b(A \rightarrow B) = 0.7$, $b(\bar{A} \rightarrow B) = 0.25$ ■■

Exercise 3.2. If $b(A) = 0.7$, and $b(A \rightarrow B) = 0.9$, at what values of $b(\bar{A} \rightarrow B)$ does $b(A) = b(B)$? ■

Exercise 3.3. What significance or confidence, if any, can be placed in the values for $b(B)$ when $b(A) = 0.6$ and $b(A \rightarrow B) \leqq 0.8$? ■

At this point an example should help clarify the steps to be taken in evaluating the credibility of a statement as a consequence based on the credibility of other statements taken as premises and the plausibility of arguments formed from the premise and consequence statements.

Example 3.18 (Premise Credibility). Let us assume that we have been provided with a hypothesis P and three units of information (evidence) which allegedly support the hypothesis. We list the hypothesis and three units of evidence as simple statements:

P: Airfield under construction in Siberia.

Q: Airfield construction experts on site.

R: Site meets geophysical and environmental airfield specifications.

S: Heavy construction equipment and personnel shelters on site.

The objective now is to evaluate the credibility of hypothesis P which we assume to be unknown. We want to use our knowledge of the credibilities of statements Q, R, and S and the plausibilities of the six possible arguments: $Q \rightarrow P$, $\bar{Q} \rightarrow P$, $R \rightarrow P$, $\bar{R} \rightarrow P$, $S \rightarrow P$, and $\bar{S} \rightarrow P$, to estimate the credibility of statement P, and we devise the following procedure:

Step 1. Consider P and one other statement, say Q. Form the arguments $\bar{Q} \rightarrow P$ and $Q \rightarrow P$. Using the methods of cause-effect analogy, relative frequencies, etc., estimate values for $b(Q \rightarrow P)$, $b(\bar{Q} \rightarrow P)$, and $b(Q)$. For this example we will use the estimates

$$b(Q \rightarrow P) = 0.9$$
$$b(\bar{Q} \rightarrow P) = 0.1$$
$$b(Q) = 0.8$$

■■

Step 2. Use the three values obtained in Step 1 to enter Table 3.2. In this example $Q = A$ and $P = B$ to conform with the table headings. The values

$b(Q \to P) = 0.9$ and $b(Q) = 0.8$ direct us to enter Table 3.2 on row 12. The value $b(\bar{Q} \to P) = 0.1$ tells us to read the value for $b(P)$ in row 12 and column 4 headed by $a = 0.1$. At that point we read $b(B) = 0.74$.■■

Step 3. Repeat Steps 1 and 2 for the two remaining statements R and S. For the first statement R, we estimate,

$$b(R \to P) = 0.7$$
$$b(\bar{R} \to P) = 0$$
$$b(R) = 1$$

For these values Table 3.2 yields $b(P) = 0.7$. For the second statement S, we estimate

$$b(S \to P) = 0.6$$
$$b(\bar{S} \to P) = 0.1$$
$$b(S) = 0.9$$

and Table 3.2 yields $b(P) = 0.58$. ■■

Repeating the three steps to estimate credibilities of P as a function of Q, R, and S, we obtain three values. We shall call these interim credibilities and denote them as

$$b_Q(P) = 0.74$$
$$b_R(P) = 0.70$$
$$b_S(P) = 0.58$$

The problem now reduces to the following: We have a table and a procedure to estimate interim credibilities for P when credibilities for Q, R, and S are known and the argument plausibilities $b(Q \to P)$, $b(R \to P)$, and $b(S \to P)$ can be estimated. We now need a procedure, model, or table which will utilize the independently arrived at interim credibilities $b_Q(P)$, $b_R(P)$, and $b_S(P)$ to estimate a joint credibility for P which reflects the contributions from all the items of information available.

The computation for the joint credibility of P is based on Bayes' rule. Bayes' rule will be covered in Chapter 6, hence it will not be discussed here. For the present purposes we will use the recursive form of Bayes' rule without explaining the rationale behind its use and the assumptions that must be satisfied for its proper use. Before we use Bayes' rule, let us repeat some notation used previously and introduce some new notation. The interim credibilities for P were estimated as

$$b_Q(P) = 0.74 \qquad b_R(P) = 0.70 \qquad b_S(P) = 0.58$$

We can interpret the interim credibility $b_Q(P)$ as the probability of P being true given the credibility of statement Q and the plausibilities of the arguments $Q \to P$ and $\bar{Q} \to P$. With this interpretation we can write

$$\overline{b_Q(P)} = 1 - b_Q(P) = 1 - 0.74 = 0.26$$
$$\overline{b_R(P)} = 1 - b_R(P) = 1 - 0.70 = 0.30$$
$$\overline{b_S(P)} = 1 - b_S(P) = 1 - 0.58 = 0.42$$

Now let us denote the odds that P is true rather than $\bar{P}$ by the symbol θ (theta). If we use a subscript Q, then θ_Q is the odds that P rather than $\bar{P}$ is true, restricted to the information provided by the statement Q. Similar statements can be made for θ_R and θ_S. The three odds are given by

$$\theta_Q = \frac{b_Q(P)}{\overline{b_Q(P)}}$$
$$\theta_R = \frac{b_R(P)}{\overline{b_R(P)}}$$
$$\theta_S = \frac{b_S(P)}{\overline{b_S(P)}}$$

Bayes' rule (recursive form) states that the joint odds is given by

$$\theta = \theta_Q \cdot \theta_R \cdot \theta_S$$

For the information P, Q, R, and S, we can now compute the interim and joint odds as follows:

$$\theta_Q = \frac{0.74}{0.26} = 2.84$$
$$\theta_R = \frac{0.70}{0.30} = 2.33$$
$$\theta_S = \frac{0.58}{0.42} = 1.37$$

The joint or final odds are

$$\theta = \theta_Q \cdot \theta_R \cdot \theta_S = 2.84 \cdot 2.33 \cdot 1.37 = 9.05$$

Thus the odds that P is true over $\bar{P}$ is approximately 9 to 1. To convert odds to probabilities, we use the formula

$$\text{Probability} = \frac{\text{odds}}{\text{odds} + 1}$$
$$= \frac{9}{9 + 1} = 0.9$$

Hence the credibility (probability) that P is true is 0.9. Since we are assuming that probabilities and credibilities can be used interchangeably, we can state that the final estimate for the credibility of P based on all information provided is 0.9. ■■

Exercise 3.4. Follow the methods used in Example 3.18. Estimate the credibility of P based on the statements Q, R, and S where,

$$b(Q) = 0.80$$
$$b(R) = 0.90$$
$$b(S) = 0.75$$

and

P: Runways are being extended at several military airfields.

Q: Many landing accidents due to short runways have occurred.

R: Higher capability fighter or bombers requiring longer runways are being produced and will soon be operational.

S: Commercial cargo aircraft requiring longer runways are being produced and will soon be operational. ■■

Simplicity of Examples

Before proceeding, a comment may be in order about the simplicity of our examples. Simplicity is of great value in explaining the basic principles of evaluation, but in practice much greater complexity is, of course, the rule. We have been considering simple implications of the type $P \rightarrow Q$. In practice analysts work with far more complicated implications. It is worth noting that for statements P, Q, and R, the plausibility of $P \rightarrow R$ and that of $Q \rightarrow R$ may both be very low, while the plausibility of $(P \wedge Q) \rightarrow R$ may be extremely high. The example with oxygen and kindling temperatures as necessary conditions is an obvious one. Other situations are not as obvious. For example, if P, Q, and R are as follows:

P: The aircraft has U.S. markings.

Q: The aircraft is a Harrier.

R: The aircraft belongs to the Marine Corps.

then neither $P \rightarrow R$ nor $Q \rightarrow R$ has a plausibility as good as even, yet $(P \wedge Q) \rightarrow R$ has a very high plausibility. It could happen that the credibility of R obtained through $(P \wedge Q) \rightarrow R$ is greater than the credibility of R obtained through $P \rightarrow R$ and $Q \rightarrow R$ considered independently and not conjoined. Hence, analysts should not only check cause-effect relationships for simple statements but also compound statements conjoined by and or or. It is through conjoining of seemingly unrelated statements and deriving highly credible consequences that analysts can gain intuitive insights on new relations and intelligence of eventual significance.

REFERENCES

Copi, Irving M. *Symbolic Logic.* Macmillan, New York: 1965.

Copi, Irving M. *Introduction to Logic.* Macmillan, New York: 1968.

Dulles, Allen. *The Craft of Intelligence.* Harper & Row, New York: 1963.

Kent, Sherman. *Strategic Intelligence.* Archon Books, Hamden, Conn.: 1965.

Polya, G. *How to Solve It, A New Aspect of Mathematical Method.* Doubleday Anchor Books, Garden City: 1945.

Polya, G. *Patterns of Plausible Inference.* Princeton Univ. Press, Princeton, N.J.: 1954.

Quine, W. V. O. *Elementary Logic.* Ginn, Boston: 1941.

Chapter 4 / Descriptive Statistics

4.1 INTRODUCTION

Prior to describing the basic concepts and applications of statistics, some of the essential differences between probability and statistics will be discussed. Probability theory deals with situations where the composition of the universe or sample population is known and the objective is to deduce the *probable* composition of a sample taken from that universe. The simplest example of this is the game of craps. Two die are used. Each die has six faces numbered from 1 to 6. The universe consists of the numbers from 2 to 12 which are the sums of the numbers which appear on the top faces of the two die after the die are tossed and come to rest on a flat surface. Probability considers or states the chances that a specific number or some one of a set of numbers might result after a single toss of two die.

Statistics, on the other hand, has the same objective, i.e., to specify the chances of obtaining a specific result, but approaches the problem from the opposite direction. Statistics deals with situations where a sample of the universe is known and the objective is to deduce the makeup of the universe or to state the chances that the sample came from any of a set of universes. For example, in the game of craps, the sample would assume the form of a set of numbers resulting from a series of throws of the dice. From the makeup of the sample, statistics considers and states the chances that each of a set of hypothesized universes (pairs of die with different properties such as loaded, true, or wrongly marked) was the universe employed in obtaining the sample.

Statistics is also concerned with the design of experiments for the collection of information, the organization of information, its display through charting or graphing, its summarization, and its interpretation. In this chapter we will consider the descriptive aspects of statistics and a few of its uses in drawing inferences (generalizations) from specific examples.

The findings of an analyst have their justification or verification in statistical (historical) fact. Thus, if the analyst is to convince others of the quality of his findings, it is important that he summarize, describe, and interpret his facts in an effective manner. Where applicable, descriptive statistics provides the analyst with one means of achieving these aims.

The descriptive aspects of statistics seek to describe three important properties of collections of data. The first concept or aspect of descriptive statistics is that of "central tendency." Here the commonly used terms of average (mean), median, and mode most frequently apply. We speak of the average fire power of a combat unit, the median cost of foreign aid, or the modal class of ethnic structures. The second aspect of descriptive statistics is that of "dispersion." Here the commonly used terms of variance and standard deviation most frequently apply. We speak of uniformity in the productive abilities of our allies, or the variability in support provided to us by our allies, or we speak of the deviation in the accuracy or range of our offensive or defensive weapon systems. The third concept of descriptive statistics is that of "associations." We say that the production of heavy weapons is associated (correlated) with the production of iron. Or we may say that the defense budget of Russia is highly correlated with the defense budget of the United States (or vice versa). The association might be causal or independent.

Briefly then, descriptive statistics is used to assess and describe sets of observations and to judge their significance to specific analysis goals. Two typical problems which can be analyzed and answered by statistical description and inferential procedures are stated here:

1. We know that the Russians are deploying a new weapon to its 100 combat divisions. We have information which indicates that 7 out of 20 divisions have the new weapon deployed. What can we say about the number of divisions in the 100 having the new weapon deployed?
2. From a sampling of 1000 Cubans in urban areas it was determined that 350 favored Castro, 500 distrusted him, and 150 were noncommittal. A sampling of 600 Cubans in rural areas indicated a split of 325 favoring, 175 distrusting, and 100 noncommittal. The population figure of Cuba is 3,250,000, divided by 57% urban and 43% rural. What confidence can we have that Castro is favored in rural Cuba, in urban Cuba, and in all of Cuba?

The Analyst's Universe and Sampling Procedures

Social, diplomatic, political, military, demographic, and legal substantive analysts study international situations, nations, ethnic groups, individuals, economies, geography, military forces, political philosophies, etc. Most of the information employed in their studies is historical in nature. The objectives of the studies normally tend to stress assessments of current strengths, weaknesses, philosophies, alignments, accessibilities, attitudes and to predict trends or changes which might occur in the future. The information for these studies come from a spectrum of sources. Some are observations of other analysts, some from the open literature, some from closed literature (where accessible), some from informers, witnesses, defectors, and agents. The key point being made here is that the substantive analyst does not have complete control over the selection of his information sources or their reliability and credibility. The universe or population from which he draws his sample is not a well-defined one. His sampling procedures are of necessity opportunistic, yet he still is required to do any of the following three things with his information:

1. Give a current description or assessment of the universe (functional area) under his cognizance based on the sample at hand.
2. Give a prediction or prescription of the appearance of the universe under his cognizance at some future date.
3. Predict the response of his universe to actions (stimuli) anticipated, i.e., answer questions of the type: If we do this, what will they do in return?

Thus the substantive analyst who desires to use statistics in his analysis faces a different and more difficult situation than does an experimenter who has control of his universe, his sampling procedures, his measuring devices, and his analytic procedures. Yet with all of these differences, the substantive analyst will still find that descriptive statistics allow him to gain full utilization of information, its characterization, its evaluation, and its selection. For example, in information collection, correct sampling procedures can reduce costs through the removal of redundant, nonrelevant, and inaccurate information at source. Proper selection of information at the source helps reduce storing, transmitting, processing, analyzing, and destroying this unneeded information.

A *universe* or population is a *set* of things, events, people, objects, etc., having a common measurable or observable characteristic. A *sample* of a universe is any *subset* of members of the universe. Samples can be drawn randomly from the universe, i.e., without regard to any specific property possessed by members of the universe. For example, one could draw names from a list of Russian diplomats with each name having an equal chance of being drawn. Draws, or samples, are assumed to be independent of each other; the results of one draw does not influence the results of any other draws. Sampling can be done with respect to a well-defined subset of the universe. For example, the list of names of Russian diplomats can be restricted to those above a stated diplomatic rank, or those assigned to specified geographical areas. In selecting samples from a universe, two procedures are usually employed. The first process is selection with replacement; the name drawn is recorded and returned to the list of names. It has the same chance as any other name being selected again. The second process is selection without replacement, i.e., the name drawn is recorded and removed from the list of names. It has a zero chance of being selected again.

Sampling in substantive analysis is performed when an analyst selects information sources in his research. Two diplomats being queried may be reporting or interpreting the same or different information or opinions. The interrogation or sampling of individuals may result in their reporting the sighting of different ships (sampling without replacement) or the same ship (sampling with replacement). During the Cuban missile crisis, the interrogation of Cuban refugees resulted in an unmanageable amount of redundancy and duplication. It It was extremely difficult to determine the existence, number, or size of missile emplacements in Cuba due to the uncertainty in the interrogees' objectivity and the lack of independence and exclusiveness of their reports.

When an analyst checks information sources, he decides whether the sources are reporting or concluding on the basis of the same or different information

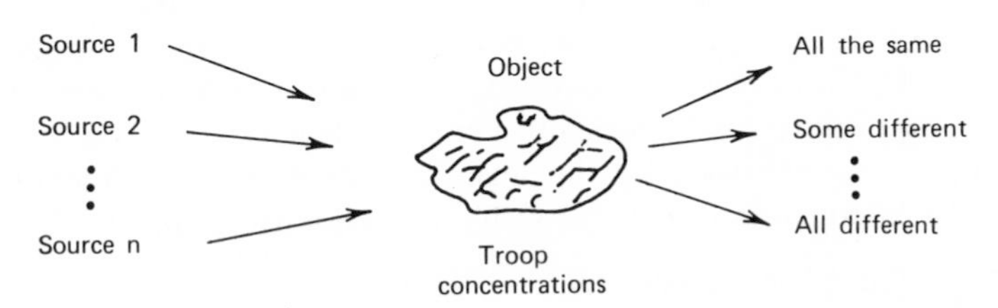

(samples). For example, several sources may be reporting on an object, say troop concentrations in a particular area. Their reports are not explicit and hence it is difficult to determine whether the reports all cover the same, some different, or all different concentrations.

Substantive analysts should be aware of the fact that they work with three types of universes. The first type of universe is inaccessible to the analyst and can only be observed through filtering media such as reports, agents, on covert information. The second type of universe is directly observable and can be studied and measured under control of the analyst. The third type of universe is only partly accessible and directly observable. In this situation the analyst must be extremely careful and ascertain whether that part of the universe accessible to him is representative of the entire universe.

The substantive analyst seldom if ever has the opportunity to work directly with the universe he is attempting to analyze. Rather, the analyst must utilize and evaluate measurements and observations obtained by others. Thus he has only partial information on the universe (limited sample). Further, the sampling techniques used by others to acquire more information for the analyst on the universe are usually suspect, as is the credibility of the information or measurements gathered in through the sampling process. Thus, when an analyst states that he has done his research, i.e., sampled his information sources, he usually means that he has read a list of reports, interpreted some photographs, used historical sources, questioned other experts, and sampled the political or social theme for the day. From the information extracted from these samples he will provide any of the following:

1. Description of the universe (verbal with some graphics).
2. Prediction of the future "shape" of the universe.
3. If asked, give some indication of how the universe would respond to some supplied stimulus.

It should be noted that the "sampling" process alluded to above and utilized by most substantive analysts today has evolved over years of slowly changing practices. Little, if any, formal evaluation of the information sources and their collection methods has been attempted in the past. Hence most analysts, with little control over these processes, are limited in their analysis by the quality of information this sampling process provides. A good substantive analyst must know how to identify good information sources. He must know how to control their output and be aware of the many statistical processes and methods available to interpret, present, and verify the information provided.

Example 4.1 (Sampling Considerations). An order of battle (OB) analyst is asked to estimate the size and strength of a troop concentration in a specific area. The analyst and his information collection sources do not have direct access to the area under consideration. The analyst, in order to study the area and the troop concentrations, decides that he must obtain the following sample to gain the information he requires:

Aerial photographs of area.

Past reports of troops in area.

Identification of any units found to have men or equipment in the area.

Copies of signals and communications into and out of the area.

Interrogation of friendly citizens who have recently been in the area.

Reports from covert sources operating in the area.

Reports in the open press on units which might be in the area.

Whereabouts of key military, political, or diplomatic personnel who might be associated with the activities planned in the area.

Political or military objectives for the area and sectors bordering the area.

The above constitute a good portion but not all of the resources available to the analyst. From all of these sources he arrives at a profile of the number of personnel present, the units in the area, their strength in fire power, logistics, fuel, and any other statistic required. Note that the analyst's sampling techniques may count the same personnel, equipment, or support activity many times. Hence the sampling is done with replacement. It is up to the analyst to determine where the reports are duplicative and hence unnecessary. Also note that to obtain a good statistical coverage of the area (if not total) the analyst must design his information requests (i.e., instruct his collection sources) so that all characteristics requiring analysis will obtain sufficient coverage. For example, the location and size of fuel and ammunition depots are important. They are usually camouflaged or well-concealed. The analyst must devise sampling techniques (coverage) which will assure obtaining information on the size and location of these resources. Covert operations, infiltration, prisoners, or defectors may be the only source for this type of information.■■

Example 4.2 (Sampling Considerations). It is desired to know the production rate of a particular model of missile currently being produced in several factories in country X. The location of all factories is not known. One would like to know the average length of time for the missiles to be produced at each factory, delivered to an operational site, and made operational at the site. The location of all operational sites is not known.

To answer these questions requires a tremendous effort at information collection sampling. The statistical analyst would call the plan of information collection his design of experiment. The information collection design would seek to uncover information using good statistical practices which would guarantee sufficient coverage with a level of redundancy restricted to that required to remove errors and deception in reporting the information. Specifically, many different production units would be appropriately sampled. The delivery time to the different operational sites and the time required to make missiles operational at each site would also be randomly (if possible) and independently sampled. The objective should be to obtain a good estimate of the times involved while expending a minimal collection (sampling) effort.■■

Example 4.3 (Sampling Considerations). An analyst wishes to determine whether defoliation was effective in certain operations in a combat area. The analyst interviews a cross-section of combat troops who have taken part in operations where defoliation was employed and where it was not. The analyst's universe consists of all combat personnel who have taken part in both types of operations. There are too many to be interviewed. Hence he selects a sample of these troops consisting of infantrymen, artillery, helicopter pilots, bomber pilots, etc. The number he selects from each category of combat troops for questioning very much determines the efficacy and accuracy of the statistics and opinions obtained from the interviews.■■

Briefly then, analysts work mainly with universes that are finite. Only a few of their queries deal with infinite universes like the *time* required to perform certain operations. Finite universes will be large enough in some instances to be considered infinite (voters in a large state, rifles in the U.S. Army). Most universes will be inaccessible in part or as a whole. For example, in attempting to determine which factory produced operational missiles, it is not possible to gain access to all missile sites and inspect each and every missile for factory markings. Yet it is reasonable and possible to gain access to a sample of missiles and make the necessary inspections. Statistics will help answer many questions on sampling procedures and size and provide descriptions which are more objective than impressionistic. Specifically, statistics will identify the relationships between universe size, sample size, and applicability of estimating procedures. Not all of the observations and measurements information analysts seek to compile and study admit to statistical description or analysis. But many do. Much work needs to be done in determining those parameters of substantive information which can be expressed statistically. Too little effort has been expended by information analysts to gain experience in this field. Before any advances in applications can be made in this area, from the point of view of the information analyst, the analyst has to learn as much as he can about statistical description and analysis. With this knowledge and skill he can seek the help of statistical experts and guide their efforts to solve problems of interest to the substantive analyst. This chapter on statistical description and analysis is designed to provide the substantive analyst with the skills in statistical description necessary for him to accomplish his collection and analysis of information more efficiently and effectively.

4.2 USE OF SETS, LOGIC, AND PLAUSIBLE REASONING IN STATISTICAL ANALYSIS

Sets deal with the categorization of elements by specific properties. A set is defined by listing all its elements with known properties (or more precisely, listing all elements known to the lister), or by specifying properties which determine whether an element belongs to the set. The latter form of set definition is

$$X = \{x \mid x \text{ has property } P\}$$

When one ascertains with certainty whether an element under observation has property P, then sets are well-defined. However, in the fields of intelligence, international politics, foreign affairs, social science, and most pseudoscientific endeavors, one seldom possesses sufficient information to determine with certainty whether the element has property P. Good analysts will have some information on the properties possessed by the elements and will usually venture a guess whether x has property P. For example, let the set X be all those countries that will have a GNP of 100 billion by 1980.

$$X = \{x | x \text{ is a country having a GNP of \$100 billion by 1980}\}$$

This is not a well-defined set. Some might call it a fuzzy set. Now let us see where uncertainties might enter in defining this fuzzy set. One can readily ascertain those countries with GNPs which currently greatly exceed $100 billion. Unless something like a major natural disaster or war afflicts these countries, most analysts will assign these countries as members of the set X. Now what about those countries which currently have GNPs close to $100 billion? In the years until 1980, will their development and growth take them or keep them over the 100 billion mark? This is not an easy question to answer, even for the best international economic expert. Yet in some sense it must be answered. Now how will this question be answered by various types of analysts who may have to provide this answer? First, let us take the cocktail circuit analyst. Here the answer provided by the cocktail analyst will be quite loose and stated as a belief such as "If country x doesn't get caught in a sizable conflict (hic) with its neighbors y, then it is possible (hic) that the GNP of country x will be at least $100 billion by 1980." An international politics expert, in answering the question, views the GNP of country x as a major contributor to its role in the power struggle among nations. As such he spends a considerable amount of time studying the economic growth of country x. His answer to the question will be highly qualified with economic and political conditions which must be met if country x is to achieve a $100 billion GNP. He will make an assessment that runs something like this: "Country x will achieve a $100 billion GNP by 1980 only if economic, social, and political developments A, B, C, ... are achieved in the intervening period. If all are achieved, then there is a *70-90% chance of success.*"

A professional economist who is expert on the economy of country x and dedicates virtually all his professional career on keeping abreast of economic and related social and political developments in country x would answer the question after referring to or recalling such analysis as business cycles, trend analysis, correlation and regression, production predictions, inflation, etc. Yet he would still state that country x will achieve a $100 billion GNP by 1980 with a 85-90% chance of success only if certain conditions and trends continue to exist in the intervening period.

The three answers have only one thing in common. Each expresses an opinion on whether country x will achieve a $100 billion GNP by 1980. The cocktail circuit estimate is totally useless. It simply states what we already know about everything, i.e., the event is not impossible. Yet the statement "it is possible that..." has appeared in estimates prepared by some of the best experts in their professional field. Such usage by a professional analyst is inexcusable, except in polite conversation.

The international politics expert stated that the event was possible and he utilized his information to restrict the realm of possibility to a range of values. The amount of information he uses and the reliance he places in his information is reflected in the range of values he assigns to the possibility that country x will reach a \$100 billion GNP by 1980.

The economic expert supposedly utilizes all information available and his prediction should be the most accurate. Further, the size of the range of values he places around his estimate (confidence limits) will normally be smaller and reflects the utilization of the most information with a high degree of reliance.

Briefly then, any analyst can answer, "it is possible that," to virtually every question asked of him and not be in error. However, this is not what analysts are paid for. Rather every analyst should be able to state a level of possibility for any event he is assessing and he should be able to give a range over which his assessment should be correct at specified confidence levels. For example, he should be able to state that the chance that country x will achieve a \$100 billion GNP by 1980 ranges from 75 to 90% and he can state this with a 95% chance of being correct. (The numbers in this statement are fictitious, but the form of the statement is real.) He will not always be able to make such assessments quantified to the level indicated. However, when he is able to, he should be employing the above form of the statement using the elements of descriptive statistics, inferential statistics, and probability to form his best estimates.

Clearly then, formally defined sets are used in situations wherein elements belonging to the set can be clearly identified. Elements of the set can be objects, truth sets of statements, or events. When it is only probable that some elements belong to the set, the set becomes a fuzzy set. With fuzzy sets the question reduces to: what is the chance (probability) that the element or the event belongs to the set? When we are dealing with objects, we are questioning whether the object has the properties required for membership. When we are dealing with statements (algebraic, logical) we are asking whether the statement is true (or if the statement achieves some level of credibility). When we are dealing with events, we are asking whether the event will or will not occur, subject to specific conditions. Chapter 3 indicated how an analyst can assign ranges of values to his observations using subjective, conventional, or objective measures. This chapter will indicate to the analyst how he can use statistics to describe and assess individual sets of events, statements, or objects.

4.3 FREQUENCY DISTRIBUTIONS

In this section frequency distributions will be described, their applications illustrated, and their construction from basic data sets demonstrated. Frequency distributions utilize the breakdown of data sets or information into units such as classes, scores, or intervals. Each class, score, or interval is to be considered as a set. For example, an interval defines the set of all measurements on elements of a prespecified universe which lie between two values (the end points of the interval). Another example is the class (set) of all American males who are high school graduates. Note, in order for an element to be a member of the interval or class, the element must pass a test or measurement, e.g., lie between two values (a, b), or graduate from high school. As indicated in the previous section, the classes, scores, or intervals must be clearly defined if fuzziness in determining membership is to be avoided. Many of the shortcomings of statistical description can be traced back to fuzzy definitions of classes, scores, and intervals and hence to disagreements as to who or what belongs to each. For example, in political polls, the important classes are party affiliation, support of a spectrum of policy options, etc. Depending on who is conducting the poll and how the poll taker defines the various parties or policy options and selects his sample, a large variance in results is possible. If the issues are not clearly defined, then opinions on those issues will vary in accordance with the different interpretations of those issues. Hence when dealing with statistics it is important that extreme care be exercised in defining the units in an unambiguous manner. Use of the principles and concepts of sets applied to the definition of statistical classes or categories will reduce the possibility of working with fuzzily defined classes.

Information from a collection of measurements or observations increases in value to potential users when analyzed, interpreted, and presented in an understandable (to the user) manner. When the collection of measurements or observations is small, the information content can be readily summarized and communicated. When the collection is large, some expedient is required to consolidately summarize, and present the information content. One expedient used in the analysis and interpretation of data is called a frequency distribution. A frequency distribution is a means of codifying, ordering, and displaying information obtained from sets of observations, measurements, opinions, and judgments in a graphical form in order to facilitate communicating important characteristics of the information and to allow for ease in further processing and analysis.

To construct a frequency distribution, one must identify the concept, person, or thing to be studied and then collect, observe, or measure certain properties or characteristics of variables associated with the concept, person, or thing. For example, if the objective is to measure air-traffic density over an airport, then a variable to be observed and measured is the number of takeoffs and landings by hour, minute, or day. If one wanted to study some features of a national or ethnic group of people, one would observe or measure characteristics such as height, weight, eye and hair color, sex, age, etc. Note that in the former example, the variable assumes numerical values, the number of takeoffs and landings. In the latter example, some variables assume numerical values (weight, height, age) and others assume name values (hair and eye color, sex). All variables to be discussed in this section will assume either name or numerical values.

When a variable or characteristic assumes values which are designated by names, these variable values are called *NOMINAL values.* Examples of variables which assume nominal values are:

Military grades–recruit to general.

Color–blue, yellow, red, and all combinations.

Tank size–light, medium, heavy.

Aircraft–private, commercial, military

Budget items–city, state, federal.

Budget items–health, agriculture, commerce.

When a variable or characteristic assumes values which are designated by numerics, these variable values are called *NUMERICAL values.* Examples of variables which assume numerical values are:

Height, age, firepower, budget costs, rank in size, or scores, tank size (by tonnage), aircraft (by range, capacity, load), and any variable which can be measured in some physical unit of measurement.

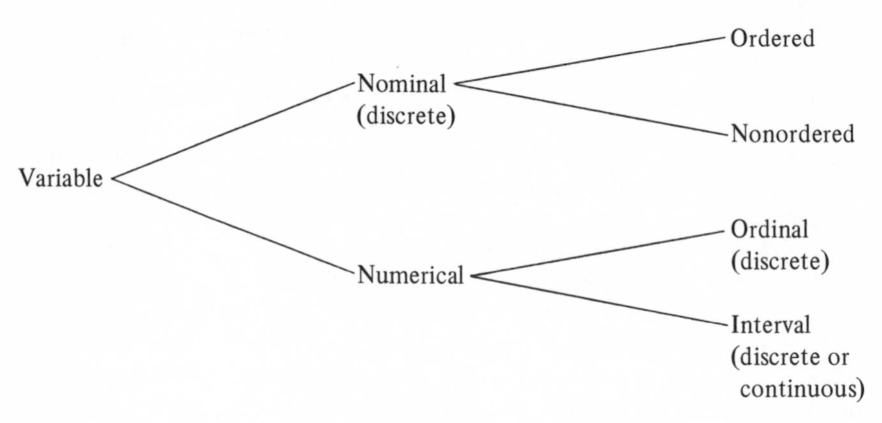

Figure 4.1

Figure 4.1 illustrates that variables take on either nominal or numerical values. Variables which take on nominal values like color and sex do not have a numerical order associated with them, i.e., if we were to list all the colors, there would be no standard or convention which would guide us to list the colors in any preferred order, notwithstanding the fact that the colors are related on the artist's color wheel. Other variables which take on nominal values, like military grades, have a natural or accepted order to them, i.e., from the lowest to the highest grade, or vice versa. Note here that although enlisted and military grades have nominal values, they can also be ordered or ranked through numerical values such as E-1 to E-9 and 0-1 to 0-10. In most instances variables which assume nominal values will be considered as being nonordered. In other instances interest will lie in the order or rank of the variable values. In this case the variable values will be considered as ordered or ranked values.

Variables which assume numerical values do so either through direct measurement or by ranking in accordance with a set of relative measurements. For example, a country's rank in the international power structure is obtained not solely by its economic, political, and military power (the measured quantities), but also by the relative power level or rank achieved by other nations considered in the international structure. Power differences between two successive nations in the order of rankings will not be uniform from the head to the bottom of the order, as illustrated in Table 4.1. The power level difference (two) between ranks 1 and 2 differs from that between ranks 3 and 4 (six) although the difference in ranks is the same. Variables whose values are assigned or measured in accordance with a rank or sequenced order are called *ORDINAL values*.

Table 4.1

Rank	Power level	Difference
1	98	
2	96	2
3	93	3
4	87	6
5	86	1
6	82	4
7	82	0
8	77	5
9	43	34
10	37	6

Some variables assume values derived from measurements employing standard units of measurement (time, length, weight, volume, quantity). In most instances the unit of measurement can be made as fine as one considers necessary. For example, the weight of a ship is measured in tons while the weight of gold is measured in thousandths of an ounce. With these measurements it is possible to assign a scale on which the units of measurements on the scale are equal (equal interval) and the fineness of measurement is limited only by the accuracy of the measuring device and the observer. Variables whose values can be measured on an equal interval scale are called *INTERVAL values.*

Variables like the number of rivers or rank on a test can assume only discrete (clearly separated) values. There may be 2, 5, or 7 rivers, but never 3½ or 6.79 of them. Ranks are first, second, third but never 1.5 or 3.14. Variables like height and weight can be measured into the nearest tenth, hundredths, or thousandths of a unit. These variables are called continuous variables. As indicated in Figure 4.1, nominal and ordinal variables are always discrete variables. Interval variables can be either discrete or continuous.

The reasons for discussing the various types of variables will now become apparent. First, there are differences in the construction of the frequency distributions associated with each type of variable. Second, the measures of central tendency and dispersion differ among the variable types. Third, not all types of variables are amenable to further statistical processing and analysis.

Prior to describing the construction of a frequency distribution, it is useful to become familiar with parts of the frequency distribution and the terminology which is employed in describing them. Frequency distributions are used to summarize graphically sets of observations or measurements taken of a set of variables. For our initial purposes we shall restrict ourselves to one variable. Recall that variables can assume nominal or numerical values. With nominal variables, our observations or measurements will be associated with a set of names. With interval variables, our observations or measurements will be numerical values associated with some measurable property or characteristic. If our variable is the *nationality* of representatives in attendance at a U.N. session, then our observations will specify the nationality of all representatives in attendance. If our variable is the height of attendees at a U.N. session, then our measurement will be the height in feet and inches of the attendees.

In constructing a frequency distribution, the x axis of a standard x-y coordinate system is used to record the names or scale the numerical values assumed by the variable under observation. For the first example discussed above, the names of all countries with representatives at the U.N. will be labeled and spaced on the x axis. The distance between names and the length of x axis will be determined by the number of names to be recorded and the space required to make legible the names recorded on the x axis. Also, the order of the names on the x axis outside of protocol, will be unimportant in constructing the frequency distribution. In the second example, the possible range of heights of all attendees will be labeled and spaced on some length of the x axis. The length of x axis and the distance between units of heights will be determined by the expected range of heights and the smallest unit of measurement determined by the accuracy of the measuring instrument (inch, quarter inch, etc.) and the interval length selected for the smallest unit of measurement.

To continue our examples, consider the nationalities and height of 30 U.N. representatives selected at random from five nations: United States, Russia, France, England, and China. The countries and heights of the selected representatives are as follows: France, 5 ft 8 in.; England, 5 ft 9 in.; U.S., 5 ft 10 in.; Russia, 5 ft 7 in.; Russia, 5 ft 6 in.; England, 5 ft 10 in.; France, 6 ft 1 in.; France, 6 ft 0 in.; China, 5 ft 4 in.; U.S., 6 ft 3 in.; China, 5 ft 7 in.; England, 5 ft 10 in.; Russia, 5 ft 11 in.; U.S., 5 ft 8 in.; China, 5 ft 7 in.; France, 5 ft 9 in.; U.S., 6 ft 1 in.; Russia, 5 ft 11 in.; England, 5 ft 9 in.; China, 5 ft 5 in.; U.S., 5 ft 10 in.; France, 5 ft 10 in.; U.S., 6 ft 1 in.; France, 5 ft 9 in.; Russia, 5 ft 7 in.; U.S., 5 ft 10 in.; France, 5 ft 10 in.; U.S., 6 ft 0 in.; Russia, 5 ft 9 in.; U.S., 5 ft 7 in.

Before we consider constructing the frequency distribution for these two variables, let us learn how to prepare a frequency table. A frequency table lists all possible values a variable can assume and the number of times (the frequency of occurrence) this value is assumed by the variable. The frequency tables for the two variables considered in this example are those of Tables 4.2 and 4.3. Table 4.2 tells us that nine representatives were from the United States, four from China, etc. Table 4.3 tells us that seven representatives were 5 ft 10 in. and only one was 5 ft 4 in. Naturally, one could have culled the same information from the original listing, but only after a scan and counting of that list. The frequency table is a simple way of keeping track of all counts or measurements taken on a variable of interest. Once completed, other users of the information can readily observe the same information without having to repeat the scan and counting procedure.

Table 4.2

Nationality	Frequency
United States	9
Russia	6
France	7
China	4
England	4

Table 4.3

Height	Frequency	Height	Frequency
5' 4"	1	6' 0"	2
5"	1	1"	3
6"	1	2"	0
7"	5	3"	1
8"	2	4"	0
9"	5	5"	0
10"	7		
11"	2		

In the two examples cited above, we have two different scales as follows:

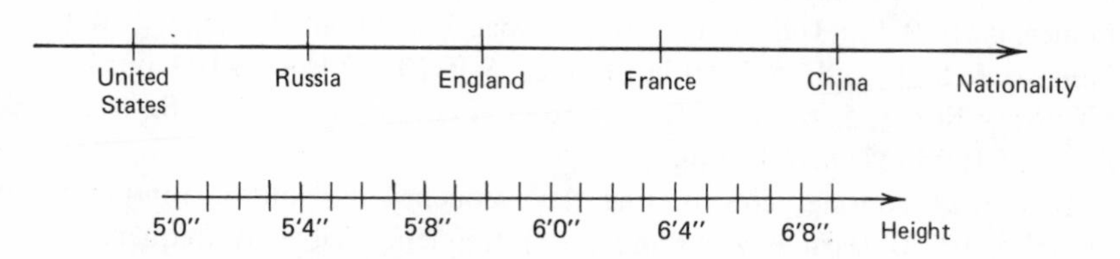

On the first scale above, the names of five nationalities are equally spaced and labeled along the x axis. On the second or height scale, the smallest unit of height measurement is recorded to be 1 in. On the height scale, 1/8 in. of the x axis was selected to represent 1 in. of height measurement.

Once one has a frequency table it is a simple thing to construct a frequency distribution. For the five nationalities, one labels the names of the nationalities

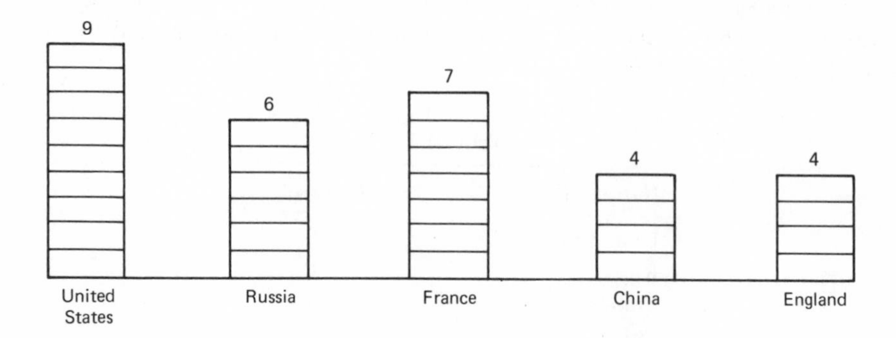

Figure 4.2 Nationalities of United Nations representatives.

separated by equal distances along the x axis of the x-y plane. In the positive direction of the y axis, one counts a number of spaces above each nationality, equal to the frequency count recorded in the frequency table for that nationality. The scale selected for the count depends on the highest frequency in the frequency table and the space (or length) available above the x axis. The resulting frequency distribution is illustrated in Figure 4.2. Note again, the order of the nationalities is of no significance.

The construction of a frequency distribution for the heights of U.N. representatives presents some additional problems in scaling and in positioning. In the frequency distribution for a variable assuming nominal values, the selection of scales, spacing, and labeling is governed by graphic and esthetic considerations. When the variable assumes numerical values, the construction of the frequency distribution depends on the difference between the greatest and least value assumed by the variable (called the *range* of the variable), the smallest unit of measurement considered significant, the number of intervals required, and the length of the x axis available to represent the range of the variable. For interval data, a guide to the number of intervals to use is given by $N = 1 + 3.3 \log n$, where N = number of intervals, n = number of data points. For $n = 10$, $N = 4$ or 5, for $n = 100$, $N = 7$ or 8. For the example above, the heights varied from a least height of 5 ft. 4 in. to a greatest height of 6 ft. 3 in. The difference is 11 in. The least significant unit of measurement in height is 1 in. Hence, if only 6 in. of the x axis were available to represent the height difference of 11 in. then the scaling of 1/4 in. of the x axis to represent 1 in. in actual height measurement would be satisfactory and would result in the scale illustrated in Figure 4.3. Note that the heights included on the scale are those recorded in the frequency table. No attempt is made to refine the scale to include accuracy in measurements which were not present in the original measurements. We could have recorded heights to 1/4 in. by adding three equally spaced marks between successive inch marks. This would have resulted in additional work which yielded no additional information for our effort.

To construct the frequency distribution for heights, we proceed as we did in constructing the frequency distribution for nationalities. First, we construct and scale a segment of the x axis, as in Figure 4.3. However, this time it is necessary to leave sufficient space on the paper over the x axis to construct the frequency

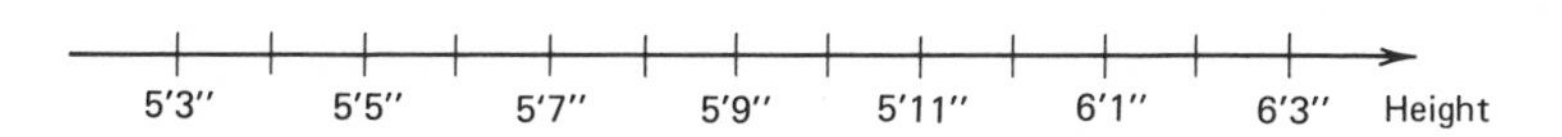

Figure 4.3

bars. The frequency table provides the range of heights and the number of individuals who were measured at each height. Before the frequency bars can be constructed in an acceptable manner, two measurements must be observed or made. The first is the largest frequency to be included in the frequency distribution. The second is the length of paper available to represent the largest frequency over the x axis. In the example, the largest frequency is 7 and it occurs at the height 5 ft. 10 in. If 2 1/2 in. are available above the x axis, and if 1/4 in. is selected to represent each frequency count, then 1 3/4 in. are needed to represent the highest frequency of occurrence (7) as illustrated in Figure 4.4. Placing dots over each other separated by 1/4 in. for each measurement at the heights appearing on the frequency table, the dot frequency diagram or distribution results.

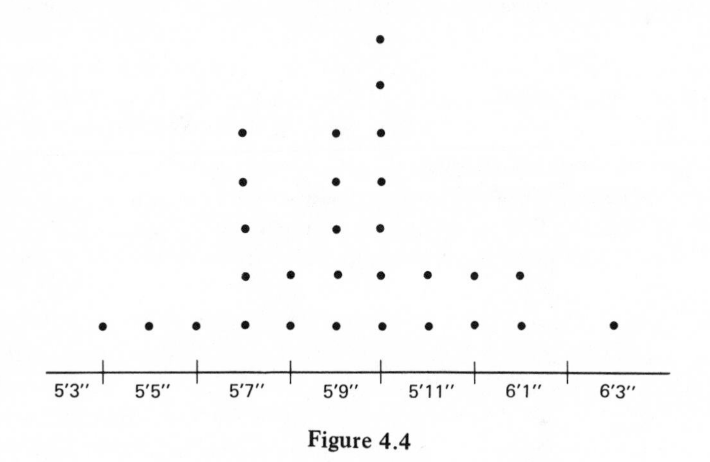

Figure 4.4

The construction of a frequency distribution, dot or bar, becomes a simple task if the following steps are practiced and used as a guide:

Step 1. Identify variable. Determine whether the variable assumes nominal or numerical values. If nominal, proceed to Step 2. If numerical, proceed to Step 8.■■

Step 2 (Nominal Variable). Identify or establish all the names the variables can assume. For example, if the variable is religion, how many religions are to be considered? If the variable is tank size, are the names (categories) light, medium, heavy sufficient?■■

Step 3. Construct a frequency table containing all names assumed by the variable and record the number of occurrences of each name in the observations.■■

Step 4. Observe the highest frequency in the frequency table and use this to estimate the bar length or dot spacing needed in the frequency distribution (bar graph).■■

Step 5. Draw a horizontal line and mark equally spaced points along the line, one mark for each name assumed by the variable, as noted in the frequency table. Label names at marks.■■

Step 6. Determine a bar width (not required if a dot frequency distribution is constructed) and center a bar over each name marking on the horizontal line. Normally, bar widths are taken as half the distance between marks on the horizontal line.■■

Step 7. Using the frequency table (Step 3), draw vertical bars (or dots) over each name in agreement with the scale determined in Step 4 and the count contained in the frequency table for the variable name.■■

Step 8 (Numerical Variable). Decide whether the variable values are ordinal or interval. If ordinal, proceed to Step 9. If interval, proceed to Step 14.■■

Step 9 (Ordinal Variable). Establish the order of the values and the range covered by the variable.■■

Step 10. Construct a frequency table containing all values (scores) assumed by the variable and record the frequency of occurrence of each value (score).■■

Step 11. Observe the highest frequency in the frequency table and use this to estimate the bar length or dot spacing needed in the frequency distribution (bar graph).■■

Step 12. Draw a horizontal line and scale equally spaced points along the line, one mark for each value (score) assumed by the variable as noted in the frequency table. Label values (scores) of variable at marks.■■

Step 13. Using the frequency table (Step 10) draw vertical bars (or dots) centered over each mark in the count contained in the frequency table for the variable values. Frequency bars may be constructed contiguously or separated by spaces.■■

Step 14 (Interval Variable). Since the variable assumes interval values, establish the range of the variable values, the number of intervals and the size of the intervals. Use the approximation $N = 1 + 3.3 \log n$. ■■

Step 15. Construct a frequency table containing all intervals. Using the data or measurements record the frequency with which the variable values fall in each interval.■■

Step 16. Observe the highest frequency in the frequency table and use this to estimate the bar length needed in the frequency distribution. ▪▪

Step 17. Draw a horizontal line and scale a sufficient number of intervals along the line to cover the range of the variable as noted in the frequency table. Label each interval with the values it represents. ▪▪

Step 18. Using the frequency table (Step 15) draw vertical bars over each interval in agreement with the scale determined in Step 17 and the count contained in the frequency table for the interval values. Frequency bars will be drawn contiguously except where intervals do not possess any counts. Variable values falling on the ends of intervals, by convention, will be assigned to the higher interval. Midpoints of class intervals will be the representative value for the class interval. For example, for the interval 3.5-4.5, the end point 3.5 will belong to the interval, the endpoint 4.5 will belong to the next higher interval 4.5-5.5. The midpoint 4 will represent all values falling in the interval 3.5-4.5.▪▪

Example 4.4 (Nominal Values).

Step 1. Identify variable. In this example the variable will be the counter-actions that one nation can take against another. The values assumed by the the variable are nominal values listed in the order of intensity of action and will be as follows:

	Counteraction	Frequency
DP	Diplomatic protest	40
UNA	United Nations action	10
SDR	Sever diplomatic relations	23
BBS	Boycott, blockade, or seizure	17
TM	Troop movements	7
GW	Guerrilla warfare	10
LCW	Limited conventional war	6
LSW	Large-scale war	12
AOW	All-out war (nuclear)	0

Step 2. The values are nominal and we have identified all the names the variables can assume in the frequency table included in Step 1.▪▪

Step 3. For the sake of simplicity, it is assumed that international counter-actions over the previous 200 yrs. have been analyzed and totaled as given in the frequency table of Step 1.▪▪

Step 4. Observe that the highest frequency in the table is 40 occurring for the value diplomatic protest. If we provide but 5 in. to the bar length, then each inch of the bar will represent a frequency of 8.■■

Step 5. The horizontal line will have nine equally spaced marks, one for each name assumed by the variable, as indicated. Note the abbreviations for the names of the values assumed by the variables and contained in the frequency table.■■

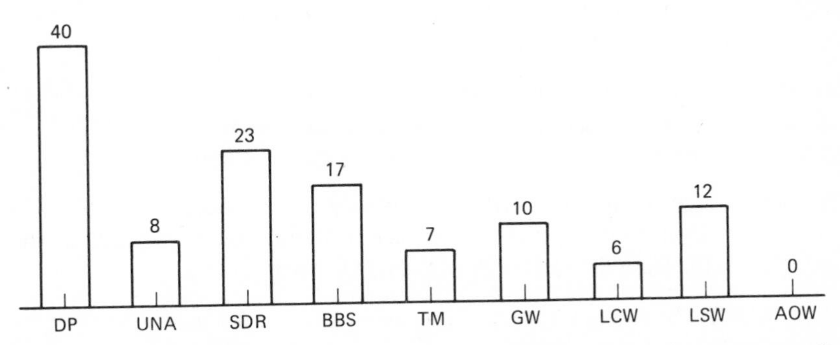

Step 6. The bar width is selected as being one-half the distance between marks, as illustrated in Step 5.■■

Step 7. Using the frequency table, the frequency bars are drawn centered on the marks and of heights scaled to the counts given in the frequency table for each variable value. The bars are illustrated in Step 5.■■

Example 4.5 (Ordinal Values)

Step 1. Identify variable. In this example the variable will be the ranking of nations according to their ability to wage war as scaled and ordered in the following frequency table:

Level or rank	Action	Number of nations
1	Limited guerrilla	48
2	Protracted guerrilla	35
3	Limited conventional	40
4	Full-scale conventional	25
5	Limited nuclear	5
6	Full-scale nuclear	2

Nations are scored at the highest level of ability. The values assumed by the variables are the ranks. Associated with each rank is a name descriptive of what the rank implies. For example, Russia and United States are ranked at level 6, and France, Britain, China, Canada, and Israel are ranked at level 5 because this is the highest level achievable by these nations. Note that the values are numerical. ■■

Step 8. The values assumed are ordinal (although they can also be considered as nominal values).■■

Step 9. The frequency table of Step 1 identifies the range and order of the values assumed by the variable.■■

Step 10. The frequency table is contained in Step 1. It was assumed that 155 nations of the world had the capacity to wage war. Each was evaluated in accordance with its potential to wage war at the six levels indicated.■■

Step 11. Observe that the highest frequency in the frequency table is 48 occurring at level 1. If we provide but 6 in. to the maximum frequency, then each inch of the bar will represent a frequency of 8.■■

Step 12. The horizontal line will have six equally spaced marks, one for each level or rank, as indicated.■■

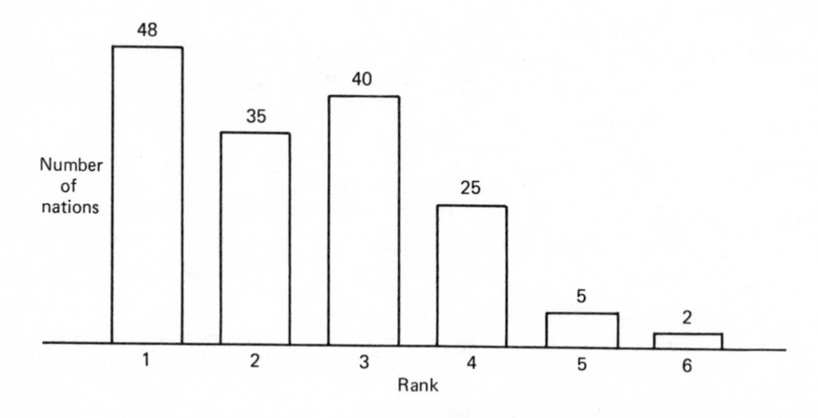

Step 13. The bar width was selected as being one-half the distance between adjacent marks, as illustrated in Step 12. The frequency table of Step 1 is used to construct bars of correct length over the marks representing the levels or ranks achieved by the nations. In this frequency distribution, the frequency bars are spaced (not contiguous) to indicate that the differences between successive ranks are not uniform on the scale shown. For example, the difference in war

potential between levels 1 and 2 is considerably less than the difference between levels 4 and 5.■■

Example 4.6 (Interval Values).

Step 1. Identify Variable. In this example the variable will be the error in accuracy of a rifle fired at 100 yds. The measurement of error is in units of 1/8 in. from center of target. The measurement is on 52 shots aimed from a fixed firing device. The variable assumes numerical values. Hence we proceed to Step 8.■■

Step 8. The variable values are interval values. Hence we proceed to Step 14. The following measurements are distance in inches from the center of target: 1/8, 3/8, 3/4, 1/4, 1/8, 1/8, 1/4, 3/8, 1/8, 1/8, 1/4, 3/8, 1/2, 3/8, 1/8, 5/8, 1/8, 1/4, 1/8, 1/2, 1/8, 3/8, 1/4, 1/4, 7/8, 5/8, 1/8, 1/8, 3/8, 1/4, 1/4, 1/8, 1/4, 1/8, 1/4, 1/8, 1/8, 3/8, 1/2, 3/8, 1/8, 1/4, 1/4, 1/8, 3/8, 1/4, 1/8, 1/2, 3/8, 1/4, 1/8, 1/4.■■

Step 14. The range of the variable is 0 to 3/4 in. The width of the intervals should be selected as 1/8 in. wide or larger. Why? We shall select 1/8 in. The center of the intervals shall be at the 1/8 in. points and the ends of the intervals at the 1/16 in. points.■■

Step 15. To construct the frequency table we list the intervals as identified above in Step 14.

Interval	Frequency
1/16 - 3/16	19
3/16 - 5/16	15
5/16 - 7/16	10
7/16 - 9/16	4
9/16 - 11/16	2
11/16 - 13/16	1
13/16 - 15/16	1

Step 16. The highest frequency in the table is 19. Thus if we select a scale of 1/4 in. for each frequency count, we will need at least 4-3/4 in. for the frequency bar height.■■

Step 17. The horizontal line will have seven intervals representing the misses from 0 to 7/8 in. Thus, if we select a scale of 1/2 in. for each interval, we will need a length of 4 1/2 in. to represent the seven intervals and one additional

1/2 in. on each end of the scale. The line and its labels are given below. Note the centers of the intervals are at 1/8 in. intervals.■■

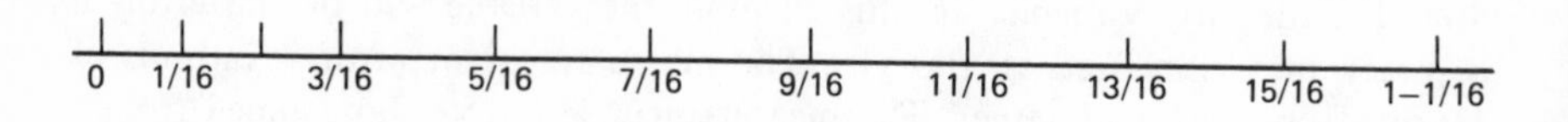

Step 18. The frequency distribution is illustrated in Figure 4.5.

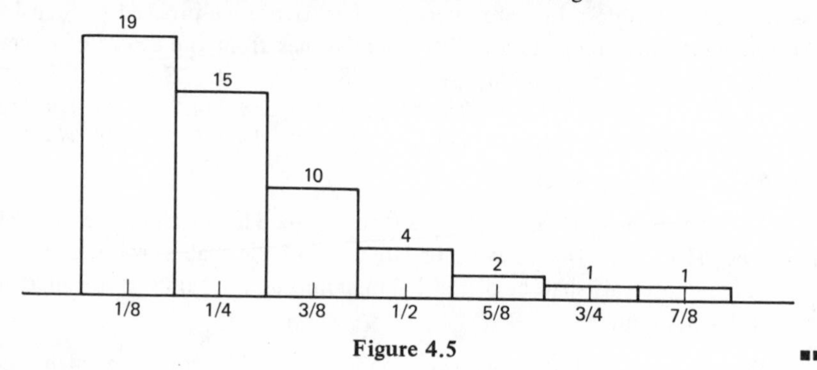

Figure 4.5 ■■

Exercise 4.1. Classify observations or measurements on the following variables as continuous or discrete, nominal, ordinal or interval:

(a) Number of ethnic groups by nations of the world.
(b) Size of ethnic groups within a specified nation.
(c) Amount in dollars of principal imports and exports for a specified nation.
(d) Involvement in major conflicts by nations of world.
(e) Population by major nations of world.
(f) Iron production by major producers of the world.
(g) Ages of population within a nation.
(h) Heights of population within a nation.
(i) Average daily temperature by month in a specified city.
(j) Crime rates by types in a specified country.
(k) Voting records of general population by parties in a specified country.
(l) Voting records of specified religious group by parties in a specified country.■■

Exercise 4.2. A sampling of heights to the nearest inch of the adult population of country X is given by the following frequency table. Construct a frequency distribution from the frequency table. Use eight intervals.

Height, in.	62	63	64	65	66	67	68	69	70	71	72	73	74	75	76	77
Frequency	1	4	7	10	14	20	27	32	20	13	7	2	0	2	1	1

(a) What interval has the highest frequency?

(b) What is the range of the variable? In other words, what is the difference in height between the shortest and tallest male?

Frequency and Percentage Polygons

There are other graphic forms used to display the information contained in a frequency table. The midpoints of the tops of the frequency bars can be joined to form a *frequency polygon.* Figure 4.6 illustrates the relationship between the bar graph and the frequency polygon. Note that one interval of frequency zero is added to each end of the frequency distribution in order to completely close the frequency polygon on the x axis.

The frequency polygon has two properties of interest. First, it is an approximation to a smooth curve which follows the shape of the frequency distribution. As the number of intervals increase and their widths decrease, the frequency polygon will appear more like a smooth curve. With continuous variables, polygonal or smooth curve approximations to the frequency distribution are commonly used. Second, the area under a frequency polygon is the same as the area under the bar graph. This is an important property because the area under the bar graph is proportional to the number of samples

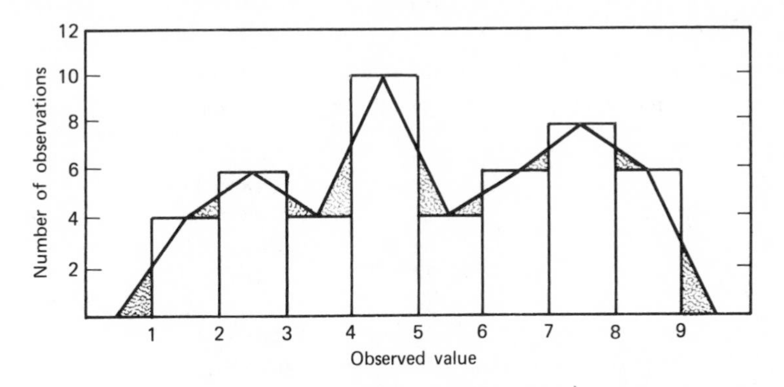

Figure 4.6 Bar graph and frequency polygon.

used to construct the bar graph. The area under any bar is given by the product of the width of the bar (length of interval) and the frequency count for that interval. In Figure 4.6 the area under the bar for the interval between 7 and 8 is:

$$\begin{aligned} \text{(frequency count) times (interval length)} &= \text{area} \\ 8 \times 1 &= 8 \end{aligned}$$

In the above example, the area under the frequency bar is equal to the frequency since the length of its interval base is unity. From Figure 4.6 it is observed that every line joining two midpoints of the tops of adjacent bars identifies two triangular areas. The triangles are equal in area. One area is added to the bar graph by the line. The area added is shaded in Figure 4.6. The other area is cut off from the bar graph. The area removed is left unshaded.

The frequency polygon can be constructed directly from the frequency table. The polygon is constructed by joining midpoints of adjacent intervals at heights equal to the frequencies of the intervals.

Relative and Cumulative Frequency Distributions

The information contained in the frequency distribution illustrated in Figure 4.6 can be presented graphically in other meaningful ways. The resulting graphics are called the relative and the cumulative frequency distributions. To describe the construction of the relative and the cumulative frequency distribution, let us reconstruct the frequency table for the frequency distribution in Figure 4.6. This is given in Table 4.4. Additional columns (c), (d) and (e) are included in the frequency table.

Table 4.4

(a) Interval	(b) Frequency	(c) Cumulative Frequency	(d) Relative Frequency	(e) Cumulative Relative Frequency
1 - 2	4	4	4/48 = .083	.083
2 - 3	6	10	6/48 = .125	.208
3 - 4	4	14	4/48 = .083	.291
4 - 5	10	24	10/48 = .208	.499
5 - 6	4	28	4/48 = .083	.582
6 - 7	6	34	6/48 = .125	.707
7 - 8	8	42	8/48 = .166	.873
8 - 9	6	48	6/48 = .125	.998

Columns (a) and (b) constitute the frequency table for the frequency distribution in Figure 4.6. Column (c) is the sum of the interval frequencies of column (b) up to and including the last interval considered. The cumulative frequency up to the first interval is the frequency for the first interval (4). The cumulative frequency up to the fourth interval (4-5) is the sum of the frequencies for the first four intervals, 4 + 6 + 4 + 10 = 24. The cumulative frequency for the last interval is 48 which is the total number of observations or measurements included in the frequency table.

Column (d) entries are obtained from column (b). The entries of column (b) are expressed as decimal fractions of the total frequency (48) and entered in column (d). For example, interval 7-8 has a frequency of 8 and a relative frequency of 8/48 = 1/6 = 0.166. Column (e) is the cumulative sum of the relative frequencies of column (d). Note that the last entry in column (e) should be the value 1. However, due to round-off in computing the decimal fractions, the sum may not be exactly 1.

Relative frequencies can be graphed along with, or in place of, absolute frequencies. When comparing two frequency distributions wherein the number of measurements (total frequencies) do not agree, the relative frequency method of presenting the information allows the two distributions to be compared directly without accounting for the differences in the number of observations or measurements considered. Figure 4.7 is a graphic presentation of columns (a), (b), and (d) of Table 4.4.

From Figure 4.7 it is possible to read the percentage of observed values that fall in any interval. For example, the interval 6-7 contains 12.5% of the values

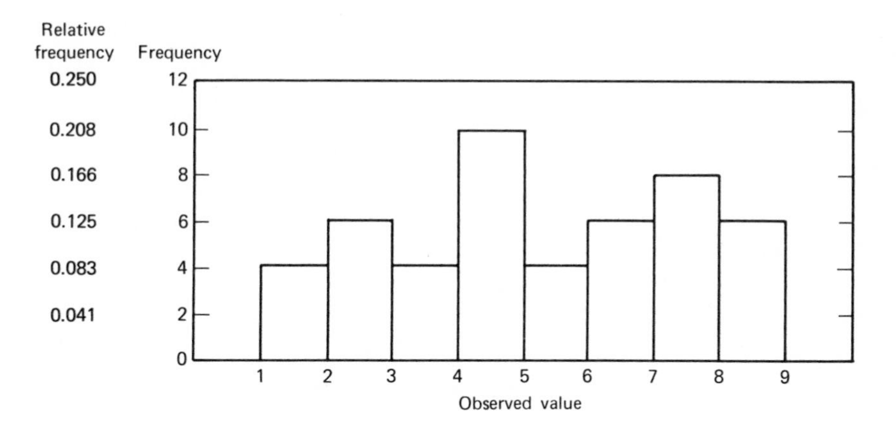

Figure 4.7

observed. Interval 4-5 has 20.8% of the values observed. In many applications it is useful to know the number or percentage of observations having values below a given value, between two given values, or above a given value. If the observed values of Figure 4.7 are interpreted as performance scores for an athletic event, then one could be interested in knowing how many or what percentage of athletes had scores below 6, between 4 and 8, or above 6. These figures can be obtained from Figure 4.7 with a little work. However, questions of this type are asked so frequently in statistics that some additional effort is warranted in the construction of a graph from which questions of the type can be answered directly. To do this we utilize Table 4.4 and graph columns (c) and (e).

In Figure 4.8 we graph the cumulative frequencies of column (c) using the upper end point of the intervals as the abscissas (x axis) and the cumulative frequency for all preceeding intervals as the ordinates (y axis). Successive points from left-to-right are connected by line segments. One interval to the left of the first interval is included in the graph. This interval contains no measurements and is the interval 0-1, i.e., the frequency count of interval 0-1 is zero. For the upper end point of each interval we can obtain the number or percentage of measurements which do not exceed the value given by the end point. For

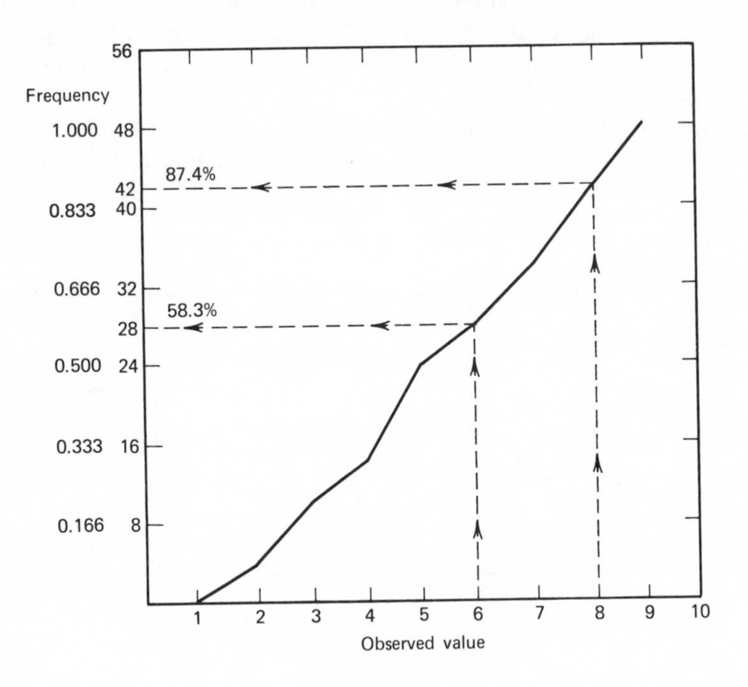

Figure 4.8

example, for the value 6, the cumulative frequency is 28 and the percentage frequency is 58.3%; 28 measurements or 58.3% of the measurements have values *less* than 6. These values were obtained by starting at the observed value and proceeding vertically on the graph until the polygonal line is reached. From the point of intersection one proceeds horizontally to the ordinate axis and reads the values of frequency and relative frequency at that point. Since there are 48 observations, 20 or 41.7% of the measurements have values *greater* than or equal to 6. If we are interested in the number or percentage of observations with values between 6 and 8, we simply read the differences in total count and percentages to obtain our answers. For example, 42 and 28 observations have values below 8 and 6, respectively. Hence $42 - 28 = 14$ have values between 8 and 6. The percentage calculation follows the same format, the percentages being 87.4 and 58.3 to yield a difference of $87.4 - 58.3 = 29.1$. These values are recorded on Figure 4.8.

When one is describing a set of observations or measurement, one is sometimes interested in knowing how many or what percentage of observations fall above or below a specified value. This information is important when the measurements are scores or ranks of a set of individuals, teams, or nations. For example, we rank nations by size of army, navy, imports, exports, production of raw materials, production of finished goods, etc. We are interested in the number of nations whose armies do not exceed a certain head count or firepower. The question of ranking in the production of fissionable materials is an oftentimes used criteria. College aptitude tests rank applicants by percentiles. The most frequently used percentiles are quarters, fifths, and tenths. If you scored in the ninetieth percentile, your test score exceeded the test scores of 90% of individuals taking that test. If you scored in the upper quintile, your score was better than the scores of at least 80% of those taking the same tests.

Exercise 4.3. The frequency table for the ages of 243 employees of company X in age intervals of 5 yr follows. For this frequency table construct the following:

Frequency distribution.

Cumulative frequency distribution.

Percentage frequency distribution.

Cumulative percentage frequency.

(a) How many employees are of ages between, but not including, ages 32 and 48?

(b) What is the median age (50% younger, 50% older) for the 243 employees?

(c) What percentage of employees will retire in 10 yr? Retirement day is the employee's sixty-eighth birthday.

(d) How many employees are younger than 37 yr old?

(e) What percentage of employees have passed their forty-seventh birthday?

Age	f_i	Age	f_i
18-22	53	43-47	34
23-27	22	48-52	26
28-32	30	53-57	15
33-37	25	58-62	6
38-42	30	63-67	2

■■

Summary

The objectives of the discussions on descriptive statistics up to this section are two. The first is to increase the analyst's awareness of the benefits to be gained through the application of statistics to his analysis problems. The second is to guide him through the practice necessary to gain the requisite skill for successful application to his specific problems. The analyst should feel that he can perform all of the seven operations listed below with a minimum of assistance. If not, the analyst should review the text where required before proceeding to the next section.

1. Selecting and ordering variable values (names, properties, intervals) to be used in collecting the observations or measurements.
2. Identifying, abstracting, observing, measuring, and organizing information for statistical analysis in terms of nominal-ordinal- or interval-valued variables.
3. Converting a set of observations or measurements to a frequency table.
4. Constructing a frequency distribution (histogram) from the frequency table.
5. Computing cumulative frequencies and percentages from frequency tables and expanding frequency tables to include same.
6. Constructing cumulative polygons for frequency counts or percentages.
7. Answering questions on frequencies, cumulative frequencies, and percentiles.

4.4 STATISTICAL DESCRIPTORS

Frequency distributions serve a useful purpose in ordering, classifying, and presenting the information contained in a set of observations or measurements.

However, their construction usually involves a considerable amount of effort. Once the construction is completed it should not be necessary to reproduce the distribution in full to convey to others the information it contains or to utilize it for further analysis. Hence a need exists for identifying some properties of frequency distributions which characterize, or provide a good summary of, the information they contain. Our descriptions up to this point have used three such properties: the *range* (greatest minus least value of variable, number of values for nominal variable), the *median* (value of variable at which 50% of observations fall below and 50% above), and the *percentiles.* For example, the median gives us some idea where the "middle" of the distribution is located, and the range gives us some idea how far from the "middle" the distribution extends. The difference between the first and last deciles, quintiles, or quartiles could be used to give some idea of the variability or scatter of the distribution.

The *median, range,* and *differences of percentiles* are useful but not exacting descriptors for some types of distributions. They are not applicable to the description of frequency distributions for nominal-valued variables. For nominal variables, a new descriptor of typical, modal, or central value is needed. This descriptor is called the *mode* or *modal* value and is discussed later in this chapter. For continuous-valued variables, other more precise descriptors amenable to additional computation and analysis are needed. For these reasons another measure or descriptor of the "center," "middle," or "cluster point" of a distribution is introduced. It is called the *mean* or *arithmetic mean* of the distribution. Also, the variability is described or measured by the *standard deviation* of the distribution. We first present the methods one employs to find the arithmetic mean. Then through examples we illustrate its utility in describing sets of observations or measurements.

Arithmetic Mean (Ungrouped Data)

As previously stated, the arithmetic mean is one measure or descriptor of central tendency or the middle of a set of observations or measurements. The arithmetic mean of a set of measurements $x_1, x_2, \ldots, x_n$ is the sum of the individual measurements x_i divided by the number of measurements n. The formula is

$$\bar{x} = \frac{x_1 + x_2 + \cdots + x_n}{n}$$

If we denote the index set by I, then

$$I = \{1,2,\ldots,n\}$$

and we can abbreviate the formula to

$$\bar{x} = \frac{1}{n} \sum_{i \in I} x_i$$

When it is understood that the summation is to be carried out for the complete set of observations, i.e., over the index set, the identification of the index set is dropped. The formula for the mean becomes

$$\bar{x} = \frac{1}{n} \sum x_i$$

Example 4.7 (Mean Value).

1. The mean of four measurements $x_1 = 2$, $x_2 = 5$, $x_3 = 2$, $x_4 = 3$, is computed by adding their values and dividing by 4.

$$\bar{x} = \frac{1}{n} \sum x_i = \frac{x_1 + x_2 + x_3 + x_4}{n} = \frac{2 + 5 + 2 + 3}{4} = 3$$

2. The annual salaries of five employees in one department are \$3600, \$3800, \$4500, \$5100, and \$50,000. What is the mean salary (arithmetic mean), and what is the median salary?

   The mean or average salary is,

$$\bar{x} = \frac{3600 + 3800 + 4500 + 5100 + 50{,}000}{5} = \frac{68{,}000}{5} = 13{,}600$$

   The median salary is \$4500 (middle value in order of amount).

3. If the department was seeking to hire another employee, what salary figure, the mean or the median, would best represent the department's typical salary?

   Since one salary (\$50,000) is considerably above any of the others, the arithmetic mean, which considers all observations equally, is pulled out of the typical salary range for the department. The median salary (\$4500) is more typical in this example, because it reflects the salary which equals or exceeds half the salaries paid by the department and is equal to or less than the remaining half of the salaries. In this sense it is the "middle" salary for the department.

4. If everyone in the department received a pay raise of \$1000 a year, what would be the department's average salary?

$$\bar{x} = \frac{4600 + 4800 + 5500 + 6100 + 51{,}000}{5} = \frac{73{,}000}{5} = 14{,}600$$

   Note that the average salary also increased by \$1000, the amount each salary increased.

5. Compute the arithmetic mean for each of the following sets of observations:

$$x_1 = 2, \quad x_2 = 5, \quad x_3 = 6, \quad x_4 = 3$$

$$y_1 = 8, \quad y_2 = 6, \quad y_3 = 1, \quad y_4 = 5$$

$$\bar{x} = \frac{2 + 5 + 6 + 3}{4} = 4$$

$$\bar{y} = \frac{8 + 6 + 1 + 5}{4} = 5$$

6. Compute the arithmetic mean for the combined set of observations of item 5.

$$z_1 = x_1 + y_1, \quad z_2 = x_2 + y_2, \quad z_3 = x_3 + y_3, \quad z_4 = x_4 + y_4$$

Compare this with $\bar{x} + \bar{y}$.

$$z_1 = 10, \quad z_2 = 11, \quad z_3 = 7, \quad z_4 = 8$$

$$\bar{z} = \frac{10 + 11 + 7 + 8}{4} = 9$$

$$\bar{x} + \bar{y} = 4 + 5 = \bar{z} = 9$$

■■

Items 4, 5, and 6 in Example 4.7 illustrate two useful relationships to remember when dealing with arithmetic means of sets of observations. The first relationship states that if a constant value is added or subtracted from every measurement, then the average measurement is increased or reduced by the same constant value, respectively. The second relationship states that if corresponding members of two sets of measurements are added to form a new set of measurements, the mean of the new set is equal to the sum of the means of the original two sets.

Rule 1. If a measurement set, $x_1, x_2, \ldots, x_n$ has mean $\bar{x}$, then the measurement set obtained by adding a constant k to each measurement, i.e., $x_1 + k$, $x_2 + k$, ..., $x_n + k$, has mean $\bar{x} + k$.

Rule 2. If $x_1, x_2, \ldots, x_n$ has mean $\bar{x}$, and $y_1, y_2, \ldots, y_n$ has mean $\bar{y}$, then $x_1 + y_1, x_2 + y_2, \ldots, x_n + y_n$ has mean $\bar{x} + \bar{y}$.

Rule 1 is very useful in that it shows us how to "translate" any set of measurements to another set of measurements which has an arithmetic mean of 0. In Example 4.7 (5) the x set has a mean of 4. If we add -4 to each measurement, the resulting measurement set -2, 1, 2, -1 has a mean of 0. In the sequel we will see that computational procedures for computing the standard deviation of a measurement set are simplified when the measurement set is translated to have an arithmetic mean equal to 0.

Rule 2 is useful when one is interested in parts as well as the whole of a set of observations. A typical example would be the observation of a number of flights at an airport for every day of a month. One could be interested in airport usage during the hours 6 AM to 2 PM and could label the number of flights $x_1, x_2, ..., x_{30}$ for the 30 days of the month. For the hours 2 PM to 10 PM the number of flights could be labeled $y, y_2, ..., y_{30}$. Thus, the corresponding sums $x_1 + y_1, x_2 + y_2, ..., x_{30} + y_{30}$ would be the number of flights for the hours 6 AM to 10 PM for the 30 days observed. If the x_i had the mean $\bar{x}$, and the y_i had the mean $\bar{y}$, then the day totals $x_i + y_i$ would have the mean $\bar{x} + \bar{y}$.

It should be noted that if the two sets of 30 observations were combined into one set of 60 observations, the mean of the resulting set of observations would be,

$$\bar{z} = \frac{\bar{x} + \bar{y}}{2}$$

The arithmetic mean has some other useful properties. For example, suppose that 30 manufacturing plants produce an average of 60 units per day. What is the daily total output of the 30 plants? In this example $n = 30$ and $x = 60$. We know that

$$\bar{x} = \frac{1}{n} \Sigma x_i$$

$$n\bar{x} = \Sigma x_i$$

where x_i is the production of ith plant, and Σx_i is the total daily production. Hence $n\bar{x} = 30 \times 60 = 1800$ units/day. Simply then, the total production can be obtained by multiplying the average production (mean) by the number of producers (measurements).

Example 4.8. Suppose that the production record of four plants are as follows:

Plant	Average output/day, units	Length of time, days
1	60	20
2	110	30
3	92	25
4	60	25

What is the average daily production for all the plants?

Solution: To find the average daily production of all plants, we must find the total production of all plants and divide this by the total number of days required to produce this amount. From the table we compute the total production from each plant by the formula $n\bar{x}$. We construct another table to record the computations.

Plant	Average output/ day, units	Length of time, days	Plant production $= n\bar{x}$
1	60	20	1,200
2	110	30	3,300
3	92	25	2,300
4	60	25	1,500
Totals		100	8,300
Average $= \frac{8,300}{100} = 83$			

The plant productions are $n_1\bar{x}_1$, $n_2\bar{x}_2$, $n_3\bar{x}_3$, and $n_4\bar{x}_4$, where the subscripts designate the four different plants and n_i is the number of days plant i produced the average $\bar{x}_i$ number of units. The total production of all plants is given by

$$\Sigma\, n_i x_i = n_1\bar{x}_1 + n_2\bar{x}_2 + n_3\bar{x}_3 + n_4\bar{x}_4 = 8300$$

The total number of days the plants produce their product is given by

$$n = n_1 + n_2 + n_3 + n_4 = 100$$

Hence the average for all plants is $\bar{x} = \frac{1}{n}\, \Sigma\, n_i x_i = 83$.

Arithmetic Mean (Grouped Data)

In the previous section each measurement was considered as a distinct element of a set of measurements. The arithmetic mean was computed by summing the measurement values and dividing by the number of measurements. The formula used to compute the mean is repeated here

$$\bar{x} = \frac{1}{n} \sum_{i=1}^{n} x_i$$

It is possible to simplify the calculation of the arithmetic mean of a set of measurements by grouping the measurements into subsets, classes, or intervals.

To do this, the range of the measurements is subdivided into intervals or classes as suggested for the construction of frequency tables or distributions. We will have two types of frequency tables or distributions; one for measurements which result in repeating discrete values and one for measurements which result in interval values. An example of a frequency table for each type is shown in Figure 4.9 (*a*) and (*b*).

Value, x	Frequency, f
4	2
5	4
6	7
7	8
8	6
9	5
10	1

(*a*) discrete value

Interval, x	Frequency, f
1- 3	1
4- 6	4
7- 9	6
10-12	9
13-15	5
16-18	3
19-21	2

(*b*) interval value

Figure 4.9

In Figure 4.9(*a*), the variable value $x = 6$ occurs with a frequency of 7. In Figure 4.9(*b*), the variable assumes values between 13 and 15 with a frequency of 5.

To compute the arithmetic mean of the measurements assuming *discrete values,* we can proceed as before, i.e., add the numbers 4, 4; 5, 5, 5, 5; 6, 6, 6, 6, 6, 6, 6; etc.; to obtain $\Sigma\, x_i = 229$ and then divide by the number of measurements, which in this case is the sum of the entries in the frequency column,

$$n = 2 + 4 + 7 + 8 + 5 + 5 + 1 = 33$$

The mean of these values is given by

$$\bar{x} = \frac{1}{n} \Sigma\, x_i = \frac{1}{33} \cdot 229 = 6.94$$

An alternate way of computing the mean is to multiply the variable value by the frequency with which it occurs. This replaces summing the variable value with itself in accordance with the number of times it occurs. The following table indicates the operations required to compute the mean for repeated discrete values:

Variable value, x_i	Frequency, f_i	Sum of values, Σx_i	Product, $x_i \cdot f_i$
4	2	4 + 4	$4 \cdot 2 = 8$
5	4	5 + 5 + 5 + 5	$5 \cdot 4 = 20$
6	7	6 + 6 + ··· + 6 + 6	$6 \cdot 7 = 42$
7	8	7 + 7 + ··· + 7 + 7	$7 \cdot 8 = 56$
8	6	8 + 8 + ··· + 8 + 8	$8 \cdot 6 = 48$
9	5	9 + 9 + 9 + 9 + 9	$9 \cdot 5 = 45$
10	1	10	$10 \cdot 1 = 10$
Totals	$n = 33$	229	$\Sigma x_i \cdot f_i = 229$

Arithmetic mean $= \dfrac{229}{33} = 6.94$

The special formula used in computing the mean $\bar{x}$ for grouped discrete measurements is

$$\bar{x} = \frac{x_1 f_1 + x_2 f_2 + \cdots + x_k f_k}{f_1 + f_2 + \cdots + f_k} = \frac{1}{n} \Sigma x_i f_i$$

where x_i is the ith measurement value and f_i is the frequency of occurrence of the ith measurement value x_i.

With *interval values,* the calculation of the arithmetic mean is identical to that for discrete values. The only difference is that we use the interval midpoint as the value in the interval which represents all measurement values which fall in the interval. This of course, is an approximation. All internal values are *rounded off* to the value given by the midpoint of the interval. For example, in Figure 4.9(*b*), the midpoints of the intervals are 2, 5, 8, 11, 14, 17, and 20. The measurement value 5.5, is rounded off to the value 5. The measurement value 13 is rounded off to 14. Rounding off introduces some error in the calculation for the mean. But there is little reason to be concerned with this error, for in most cases it will be small when compared with the errors which may be introduced in accomplishing the original measurements. If one requires very high accuracy in the calculation of the arithmetic mean, then one can readily use the original measurements without grouping and compute the mean as with discrete and distinct measurement values. To compute the mean for the measurements given in Figure 4.9(*b*), one constructs the following computation table and completes the indicated operations:

Interval, i	Midpoint, x_i	Frequency, f_i	Product, $x_i \cdot f_i$
1	2	1	2
2	5	4	20
3	8	6	48
4	11	9	99
5	14	5	70
6	17	3	51
7	20	2	40
Totals		30	330

$$\bar{x} = \frac{x_1 f_1 + x_2 f_2 + \cdots + x_k f_k}{f_1 + f_2 + \cdots + f_k} = \frac{330}{30} = 11$$

$$\bar{x} = \frac{\text{sum of product column}}{\text{sum of frequency column}}$$

Example 4.9 (Mean for Grouped Data). Find the average age of death for the following sample of 8000 deaths occurring in 1964:

Age group, i	Midpoint, x_i	Number of deaths (frequency), f_i	Product, $x_i \cdot f_i$
55-59	57.5	62	3,565
60-64	62.5	190	11,875
65-69	67.5	530	35,775
70-74	72.5	1,090	79,025
75-79	77.5	1,800	139,500
80-84	82.5	1,970	162,525
85-89	87.5	1,570	137,375
90-94	92.5	600	46,500
95-99	97.5	188	18,330
Totals		8,000	634,470

Solution: A death in age group 70-74 is interpreted to mean that the age of the person who died could have been exactly 70 (died on his or her seventieth birthday) or one day before his or her seventy-fifth birthday. Hence the midpoint of the interval is 72-1/2 and the interval represents a 5 yr period.

The products of each interval midpoint and frequency for the nine intervals are given in the last column. The sample size (number of observations or measurements) is 8000. Hence the mean is

$$\bar{x} = \frac{\text{sum of product column}}{\text{sum of frequency column}}$$

$$= \frac{634470}{8000}$$

$$= 79.3$$

■■

Example 4.10 (Data Interpretation). A sampling of annual percentage rates of profit after taxes on stockholders' equity across 34 major U.S. industries in 1970 yielded the following set of observations: 6.3, 6.1, 6.8, 9.9, 9.1, 8.3, 8.6, 7.0, *4.3,* 10.7, 5.9, 7.9, 6.9, 14.2, 10.0, 10.3, 11.5, 8.5, *17.6* (drugs), 11.0, 11.0, 7.1, 10.8, 7.1, 10.8, 10.2, 8.8, 10.5, 15.7 (tobacco), 5.1, 9.3, 7.0, 11.2, and 9.4. Construct the frequency table, frequency distribution, frequency polygon, and accumulative frequency polygon. Compute or determine the mode, median, and mean for the measurement set. How many U.S. industries return 10% or better on stockholder's equity?

Solution: The first thing to do is to classify the variable by type. It is clear that the variable values are continuous and of interval type. The next thing to determine is the range of the measurements. The least value is 4.3% and the greatest value is 17.6%. Hence the range is 17.6 – 4.3 = 13.3. To cover both ends of the measurements, we select the range starting from 4% and ending at 18%. There are 34 data elements or measurements. The use of the formula 1 + 3.3 log n for the number of intervals indicates that 6 intervals will suffice. However, for ease of calculation and recording we shall pick seven intervals of length 2. With these decisions made we can now construct the frequency table as indicated below and tally the 34 measurements given previously:

Interval	Midpoint	Tally	Frequency
4- 5.99	5	111	3
6- 7.99	7	~~1111~~ 1111	9
8- 9.99	9	~~1111~~ 111	8
10-11.99	11	~~1111~~ ~~1111~~ 1	11
12-13.99	13		0
14-15.99	15	11	2
16-17.99	17	1	1

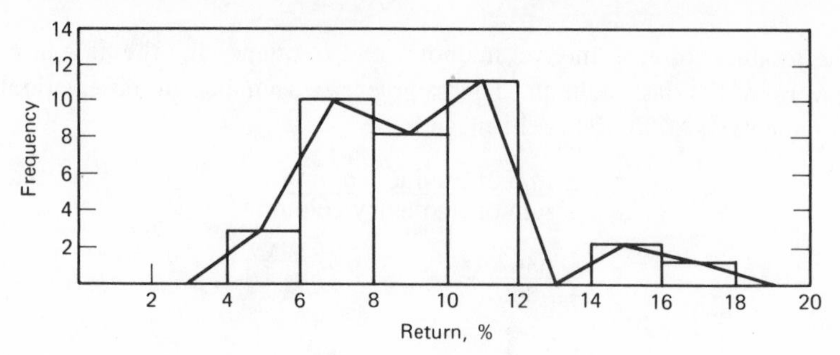

Figure 4.10 Frequency polygon.

The frequency distribution is constructed from the frequency table and is illustrated in Figure 4.10.

The frequency polygon is constructed by connecting the midpoints of the tops of adjacent frequency bars starting with an interval of zero frequency inserted at the start of the frequency distribution and ending with a zero frequency interval inserted at the end of the frequency distribution. This is illustrated in Figure 4.10.

To obtain the cumulative frequency polygon we expand the frequency table to include the cumulative frequency and then plot these values at the end points of intervals associated with the successive accumulations. The assumption here is that the measurements are distributed uniformly in the intervals.

Interval	Midpoint	Frequency	Cumulative frequency
4- 5.99	5	3	3
6- 7.99	7	9	12
8- 9.99	9	8	20
10-11.99	11	11	31
12-13.99	13	0	31
14-15.99	15	2	33
16-17.99	17	1	34

To determine the number of industries which return 10% or better on stockholders' equity, we use the cumulative frequency polygon, Figure 4.11. Entering the frequency polygon at the value $x = 10\%$, we proceed vertically to the point of intersection on the graph. From that point we proceed horizontally to the y axis where we read the value 20. This value is interpreted to mean that 20 industries had a return of less than 10% (remembering that the upper value of an interval belongs to the next higher interval). Since there are

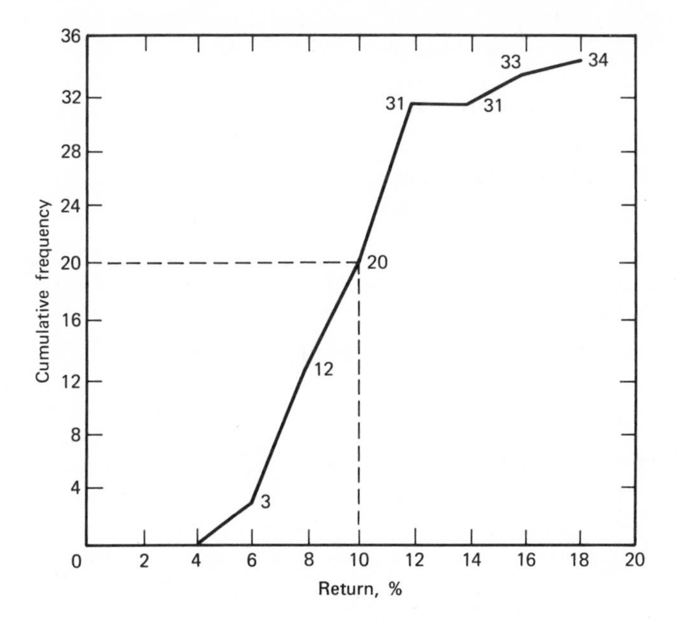

Figure 4.11 Cumulative frequency distribution.

34 industries in the sample, the remaining 14 industries had a return of 10% or better on stockholders' equity.

The calculation of the mean is relatively straightforward when we complete the following computation table:

Interval, i	Midpoint, x_i	Frequency, f_i	Product, $x_i \cdot f_i$
4.0- 5.99	5	3	15
6.0- 7.99	7	9	63
8.0- 9.99	9	8	72
10.0-11.99	11	11	121
12.0-13.99	13	0	0
14.0-15.99	15	2	30
16.0-17.99	17	1	17
Totals		34	318

$$\text{Mean} = \frac{318}{34} = 9.35$$

Exercise 4.4 Presidents of the United States are considered to be healthy and long-lived. The age at death of past presidents are as follows: 67, 90, 83, 85, 73, 80, 78, 79, 68, 71, 53, 65, 74, 64, 77, 56, 66, 63, 70, 49, 56, 71, 67, 58, 60, 72, 67, 57, 60, 90, 63, 78, 46. What is the average age at death of former U.S. presidents? What is the median age at death for former U.S. presidents? Construct the frequency distribution, frequency polygon, and percentage frequency polygon.■■

Exercise 4.5 The 1970 profits after taxes as percent of sales of manufacturing corporations by 23 industry groups are listed by the FCC and SEC as follows: 4.1, 4.7, 3.9, 5.4, 3.8, 4.4, 6.6, 4.7, 3.5, 4.8, 7.8, 3.8, 2.6, 5.2, 2.9, 2.3, 4.8, 4.7, 6.5, 10.0, 10.1, 3.8, and 2.6. Construct the frequency distribution and frequency polygon. What was the average corporation profit after taxes as percent of sales?■■

Exercise 4.6 The following figures are the estimated expenditures per pupil in average daily attendance in public elementary and secondary day schools, listed alphabetically by state (and D.C.) for the year 1969-1970.

503	688	1,137	1,108	645
1,416	964	874	835	775
915	706	1,019	1,420	636
632	959	1,105	675	709
1,067	847	534	764	716
798	1,037	842	804	1,034
966	920	982	617	822
1,106	693	649	1,022	880
(D.C.) 1,372	746	877	1,056	706
923	816	856	1,010	988
				931

(a) What is the range of the average state expenditure?

(b) What is the median of the average state expenditure?

(c) What is the average state expenditure per student?

(d) If there were 46 million students in the United States, in 1969-1970, can you determine from the information provided the yearly cost for elementary and secondary education in the United States for 1969-1970?

Mode for Nominal Values

The interval, ordinal, or nominal class whose frequency is not exceeded by any other interval, ordinal, or nominal class, respectively, is called a *modal* class.

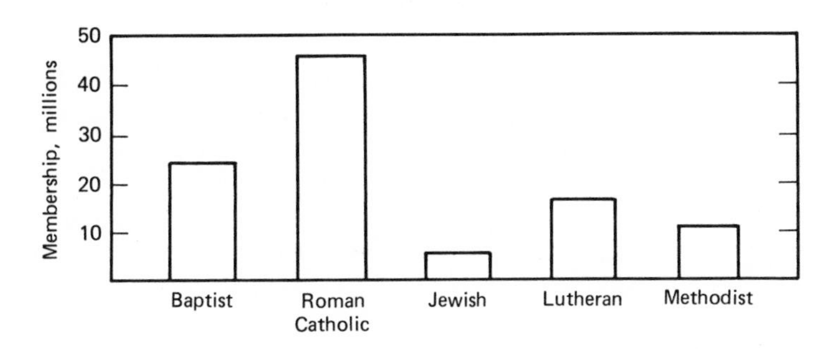

Figure 4.12 Religious membership in the United States, 1971.

The modal class indicates where the distribution tends to cluster. In a frequency distribution for a nominal variable, the nominal classes can be arranged in any order; they do not have a natural order of arrangement from lowest to highest, smallest to largest, coolest to hottest, darkest to brightest, etc. For example, the frequency distribution for the largest religious bodies in the United States in 1971 would appear as in Figure 4.12. There is no "center" to this distribution. However, one class is exceeded by no other class. It indicates a tendency to cluster at that nominal value.

Some distributions will have two or more classes with approximately equal frequencies. Such distributions will be called *multimodal.* When a distribution

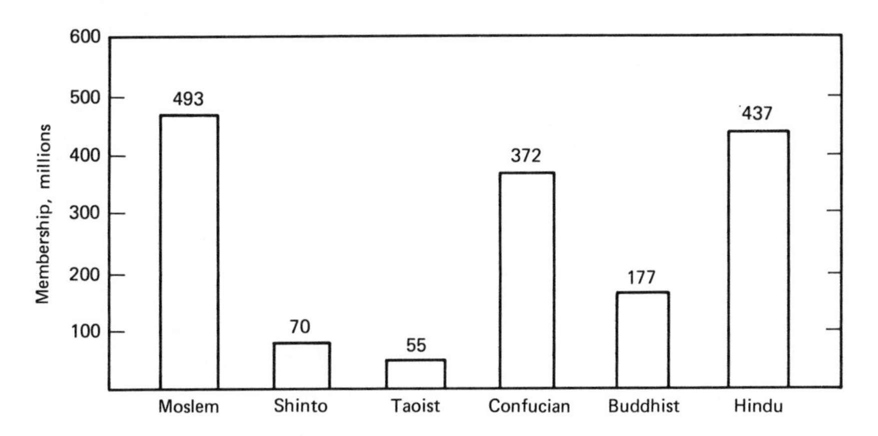

Figure 4.13 Non-Judeo-Christian religions–membership in the world, 1968.

has two classes with approximately equal frequencies, the distribution will be called *bimodal.* A distribution with only one modal class will be called *unimodal.*

An example of a multimodal distribution is one which depicts the non-Judeo-Christian religious population of the world, as of 1968, Figure 4.13.

From the frequency distribution it is clear that the Moslem religion has the largest membership of the non-Judeo-Christian denominations. Hence it is definitely a modal class of the distribution. Some judgement is required to decide whether the Hindu or Confucian classes should also be called modal classes. Most statisticians would call the Hindu class a modal class. Not as many would call the Confucian class a modal class. Since these three classes greatly exceed the frequency counts of the other classes, one would not be in error in calling all three modal classes.

The utility associated with identifying modal classes decreases as the number of modal classes increases. For the identification of one dominant class is normally more useful in analysis than the identification of two or more classes of equal dominance. With one modal class the separation of classes is clear. With two or more modal classes, the selection of one modal class over another must be based on some characteristic, property, or requirement other than that utilized in the distribution.

Mode for Ordinal Values

The modal class can be employed to identify clustering in distributions for ordinal or interval variables. For example, consider the distribution of military enlisted personnel in the nine enlisted grades E1-E9. The grades have a rank

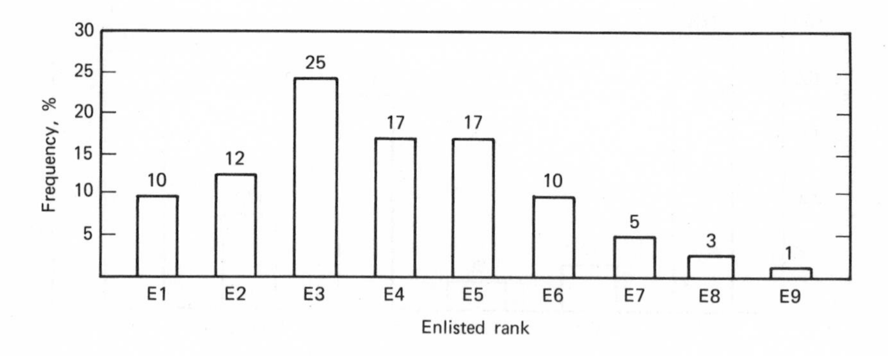

Figure 4.14 Percent of frequency distribution

ordering and are considered as values for an ordinal variable. However, military grades cannot be considered an interval-valued variable. Since the total number of enlisted personnel may vary considerably with time, the more static percent frequency distribution across the nine grades will be employed. The distribution, based on 1970 data, is illustrated in Figure 4.14.

The frequency distribution clearly is unimodal. The modal class is E3–the rank of private first class. There are more enlisted men of rank private first class than any other rank. Note that with ordinal values such as rank, the modal class will assume a position relative to the other classes. This is not true for nominal-valued variables where the classes, modal or otherwise, can be placed in any position in the frequency distribution.

Some distributions for ordinal-valued variables are markedly bimodal. For example, the sensitivity of patients to drug dosage in some instances, are bimodal. The typical distribution follows and indicates a clustering around both a high and low dosage level:

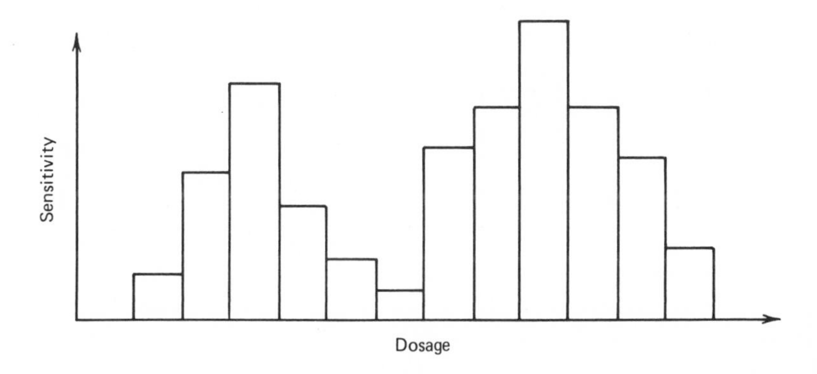

The modal class on the left of the distribution indicates that some people's reaction threshold is low (due to allergy or other reasons) while the larger percentage require a much larger dose to obtain a reaction. In the distribution, the high-frequency class on the left is called a modal class because its frequency exceeds by some margin the other classes in its immediate neighborhood. Multimodal distributions indicate that the sample is not homogeneous with respect to the property being measured. In the above example, there appears to exist two different groups, one with a generally low sensitivity, the other with a high sensitivity to the drug being administered. It indicates that pretesting for sensitivity is necessary prior to determining whether to administer the drug and what dosage level to recommend.

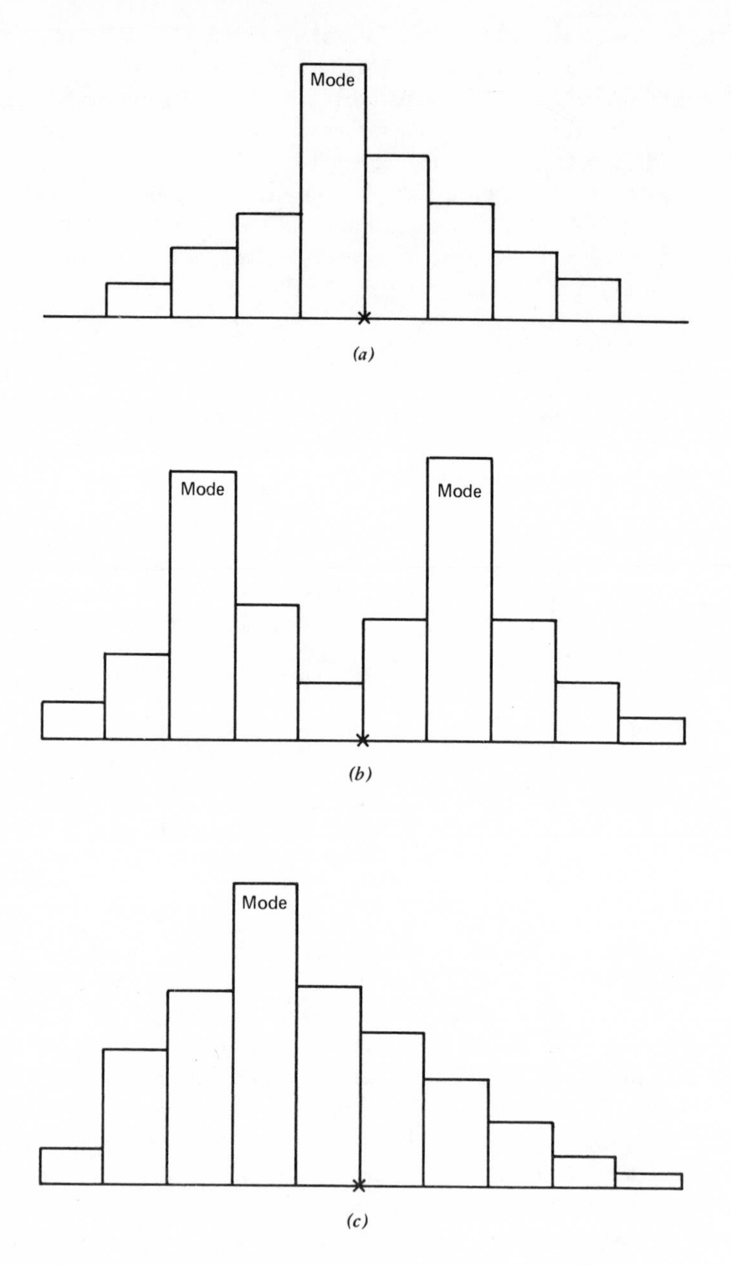

Figure 4.15. (*a*) Unimodal distribution, (*b*) bimodal distribution, and (*c*) skewed right distribution.

Mode for Interval Values

As with nominal and ordinal values, the modal class for *interval* values is that class whose frequency is exceeded by no other class in the frequency distribution. The *midpoint* of the modal class or interval is called the *mode* of the distributions.

Modes for interval-valued distributions are numerical values around which the distribution tend to cluster. If the distribution is unimodal, then the central tendency of the distribution is positioned at the mode. Unimodal distributions whose mode is not at the center of the range of the distributions are called *unsymmetrical* or *skewed.* If the mode is positioned to the left of center in the range of the distribution, the distribution is said to be skewed to the right. If the mode is positioned to the right of center, the distribution is said to be skewed to the left. Figure 4.15 illustrates distributions having the properties introduced above. Distributions for interval-valued variables are utilized.

It should be noted that the mode of a distribution is dependent upon the number and size of the intervals used in constructing the distribution. The two distributions illustrated in Figure 4.16 are from the same set of measurements. Intervals of different lengths are used. This results in moving the position of the modal class. The frequency table for the distributions gives the two sets of intervals employed and is shown in Table 4.5.

In distribution *A*, the three high-frequency intervals could be combined into one modal class with mode equal to 40.5. The two top frequency classes of distribution *B* could be combined into one modal class with mode equal to

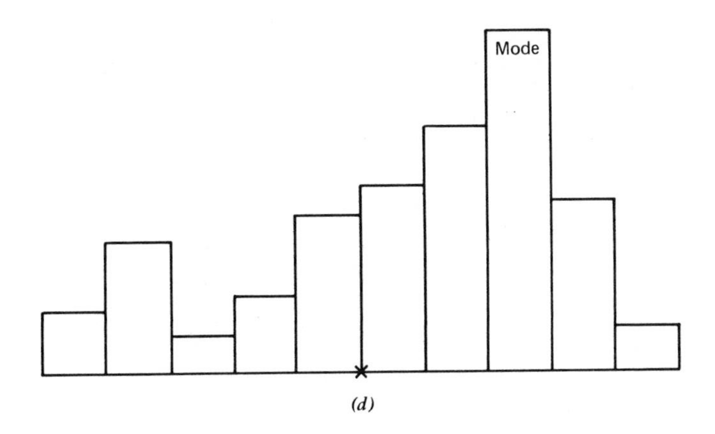

(d)

Figure 4.15 (continued). (*d*) Skewed left distribution. *Note:* x = range center.

38. In either case, the modes of the resulting distributions remain unequal. Hence the mode is not a property of the set of measurements alone. It is also dependent on the size of intervals used in constructing the frequency distribution. The mean and the median, when considered for interval (grouped) data, also vary with the selection of interval location and size. When calculated from ungrouped data, the mean and the median are unique.

When it is necessary to determine the mode of a distribution for grouped data more accurately than the midpoint of the modal class, the following formula is frequently used:

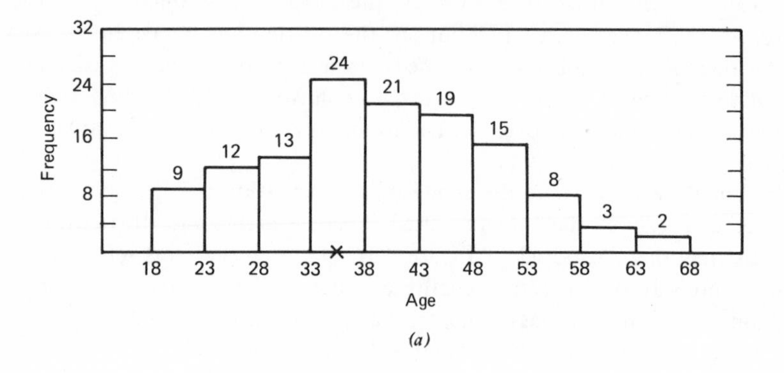

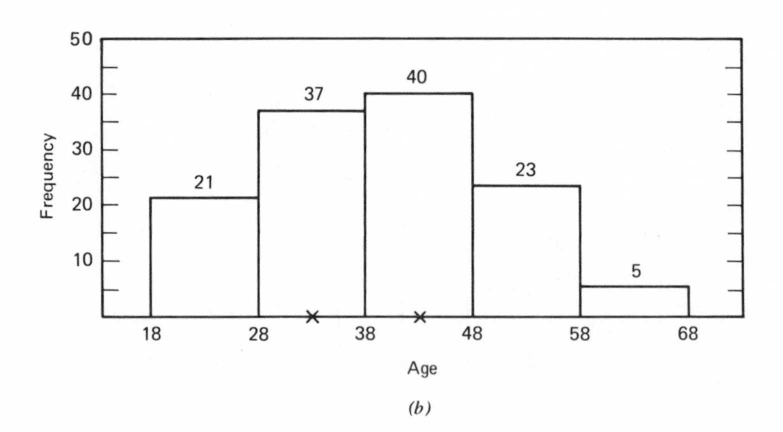

Figure 4.16 (*a*) Distribution *A* and (*b*) distribution *B*.

$$\text{mode} = M + \frac{1}{2}\left(\frac{f_U - f_L}{2f_M - f_U - f_L}\right) \cdot C$$

where

M = midpoint of modal class
C = length of class interval
f_L = frequency of class below modal class
f_M = frequency of modal class
f_U = frequency of class above modal class

We illustrate the use of this formula by computing the mode for distribution A as given in Figure 4.16. From the figure we read,

$M = 35.5$, $C = 5$, $f_L = 13$, $f_M = 24$, $f_U = 21$

Using the formula, we compute,

$$\text{mode} = 35.5 + \frac{1}{2}\left(\frac{21 - 13}{2 \cdot 24 - 21 - 13}\right) \cdot 5$$

$$= 35.5 + 1.42$$

$$= 36.92$$

A quick look at the distribution indicates that the correction is in the right direction.

Table 4.5. Frequency table for different intervals.

Distribution A		Distribution B	
Class (age)	f_i	Class (age)	f_i
18-23	9	18-28	21
23-28	12		
28-33	13	28-38	37
33-38	24		
38-43	21	38-48	40
43-48	19		
48-53	15	48-58	23
53-58	8		
58-63	3	58-68	5
63-68	2		

Exercise 4.7 Use the formula

$$\text{mode} = M + \frac{1}{2} \cdot \left(\frac{f_U - f_L}{2f_M - f_U - f_L}\right) \cdot C$$

to compute the mode for distribution B in Figure 4.16. ■■

Exercise 4.8 Determine the mode of distribution B without using the formula. Compare your result with that obtained in Exercise 4.8 and explain the difference–if any. ■■

Median for Ordinal Values

The use of modal classes to indicate clustering in a distribution of nominal values is a rational one. Since the value assumed by nominal variables do not normally have an order associated with them, the frequency of occurrence of any nominal value is the only indication of clustering not dependent on the arrangement of values. The modal class is also useful to indicate central tendency or clustering in a distribution of ordinal values. However a more natural and rational concept to use with ordinal values is that of the *median*. As defined previously, the *median* of a set of values which have been arranged in order of magnitude, from lowest to highest, is the value of the variable which occurs at the midpoint or midcount of the sequence. For example, the median of the ordered sequence of numbers 2, 5, 9, 10, 11, 13, 21, 31, is a number between 10 and 11. For there are eight numbers in the sequence and the median occurs between the fourth and fifth numbers. If the sequence is rewritten as 2, 5, 9, 10, 10, 13, 21, 31, then the number 10 is the median of the sequence. If the original sequence is enlarged to contain an odd number of elements in the sequence, say 2, 5, 9, 10, 11, 13, 21, 31, 35, then the median of the sequence is the number 11, which occurs at the fifth position in a sequence of nine. Note in the last sequence, four numbers lie below the median and four numbers lie above the median.

The median of a sequence of numbers arranged in order and having an even number of elements is obtained by averaging the two values occurring at the middle of the sequence. If the two values occurring at the middle are 21 and 27, then the median is

$$\frac{21 + 27}{2} = 24$$

The median of a sequence consisting of an odd number of elements is the value of the element occurring as the middle element of the sequence. Hence, the median is an indicator of the middle of the sequence or distribution in the sense that half of the measurements have values above it. The median *does not* indicate the center of the *range* of the sequence or distribution. For a symmetrical distribution, the median equals or approximates the center of the range of the distributions. For example, the sequence 2, 5, 9, 10, 11, 13, 21, 31, 35, has a range of 35 − 2 = 33. The numerical middle of the range is

$$\frac{35-2}{2} = \frac{33}{2} = 16.5$$

The median of the sequence is the *value* of the middle element of the sequence, which is 11.

Example 4.11 (Median Value). To illustrate how one determines the median of a distribution for ordinal values we utilize the frequency table for test scores received by 150 students on a 20 question test, each correct answer given a grade value of 5.

Grade	Number achieving grade	Cumulative frequency
100	2	2
95	8	10
90	12	22
85	18	40
80	20	60
75	32	92
70	20	112
65	10	128
60	10	138
55	8	146
50	4	150

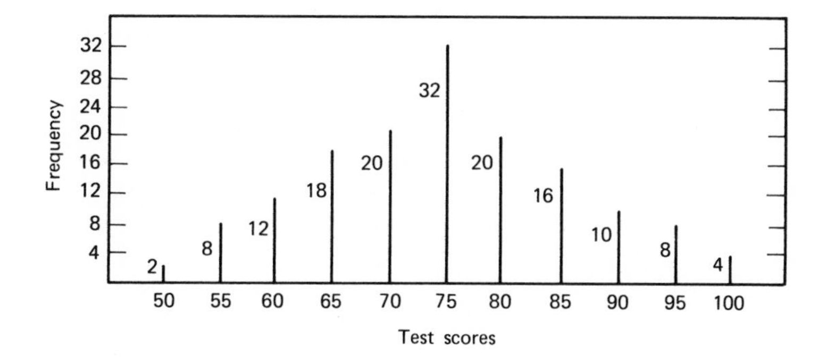

There are 150 test scores. This is an even number. The median score is the the score which was achieved at the frequency count between $n/2 = 75$ and $n/2 + 1 = 76$. The score of 80 or above was achieved through the sixtieth frequency count and the score of 75 was achieved from the sixty-first through the ninety-second frequency count. Hence the score of 75 was achieved at both the frequency count of 75 and 76. Note that 60 students had grades above 75 and 58 students had grades below 75. Hence the score 75 does not satisfy the definition of being the median score precisely—i.e., an equal number of scores in the distribution lie below it and above it. The next section will show how to handle this slight discrepancy by considering the scores as interval values. However, since the distribution is essentially symmetrical, the score 75 is a close enough approximation for most practical purposes.■■

Median for Interval Values

The median, which marks the fiftieth percentile, is a natural indicator of central tendency for ordinal-valued variables. The median can also be used for interval-valued variables. When used with interval-valued variables, the median is the sample value of the variable positioned in the middle of all sample values when these sample values are arranged in order from the least to the greatest value.

When the interval valued measurements have been grouped into classes to form a frequency distribution, the median will fall within a class interval. The class in which the median is located is called the *median class*. Normally it is sufficiently accurate to say that the midpoint of the median class interval is the median. However, a more accurate procedure, which will now be illustrated by means of an example, is available and is generally used.

Example 4.12 (Locating the Median). Consider the frequency table, frequency distribution, and frequency polygon used in Example 4.9. These are repeated here for ready reference.

Interval	Midpoint	Frequency	Cumulative frequency
4- 5.99	5	3	3
6- 7.99	7	9	12
8- 9.99	9	8	20
10-11.99	11	11	31
12-13.99	13	0	31
14-15.99	15	2	33
16-17.99	17	1	34

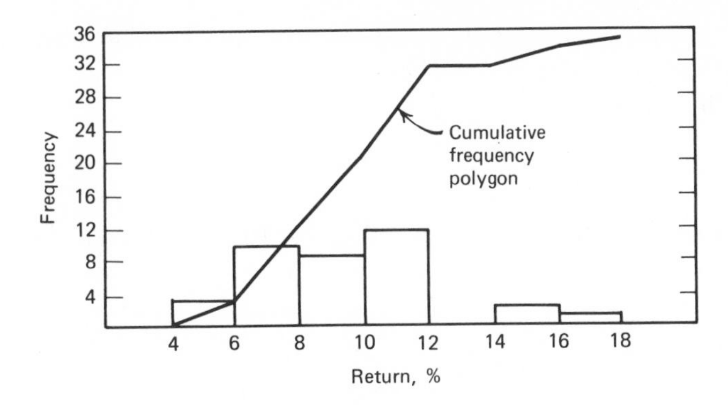

In this example we could start with the original 34 measurements and list them in order of magnitude as follows: 4.3, 5.1, 5.9, 6.1, 6.3, 6.8, 6.9, 7.0, 7.0, 7.1, 7.1, 7.9, 8.3, 8.5, 8.6, 8.8, *9.1, 9.3,* 9.4, 9.9, 10.0, 10.2, 10.3, 10.5, 10.7, 10.8, 10.8, 11.0, 11.0, 11.2, 11.5, 14.2, 15.7, 17.6. Since the sequence has an even number of elements, there are two middle values which occur at $\frac{34}{2}$ = sample 17 and $\frac{34}{2} + 1$ = the sample 18. The average of these two values is, by definition, the median of the distribution, i.e.,

$$\frac{9.1 + 9.3}{2} = 9.2$$

Thus, by ordering and counting, we obtain the median value as 9.2.

The median value can be obtained directly from the frequency distribution or cumulative frequency polygon. Using the frequency distribution to obtain the median is more difficult and will be described first. The first determination is the class interval containing the "middle" value. In this example, the middle values are the seventeenth and eighteenth measurements when the measurements are listed in the order of increasing values. Counting the frequencies from the first interval on the left, we accumulate frequency counts as follows:

interval 4-6 has 3

intervals 4-8 have 3 + 9 = 12

intervals 4-10 have 3 + 9 + 8 = 20

The seventeenth and eighteenth samples thus are in the third interval designated by the end points 8 and 10. The length of the interval is two percentage points. The frequency counts at the end points of the third interval are 12 and 20. Thus the measurement values 8 and 10 are associated with the frequency counts 12 and 20, respectively. Figure 4.17 illustrates this relationship. We seek the measurement value that is associated with the pseudofrequency count 17.5. To find this value, we use the technique known as interpolation. The value of the pseudofrequency count 17.5 is positioned

$$\frac{17.5 - 12}{20 - 12} \times 100 = \frac{5.5}{8} \times 100\%$$

of the distance into the interval from 12 to 20. The median value is positioned the same percentage distance into its interval from 8 to 10, i.e., the median is located at

$$\text{median} = 8 + \frac{5.5}{8} \times 2$$

$$= 8 + 1.37 = 9.37$$

Thus 9.37 is the median percentage return on investment for the 34 industries sampled. The median for ungrouped data was previously computed as 9.2.

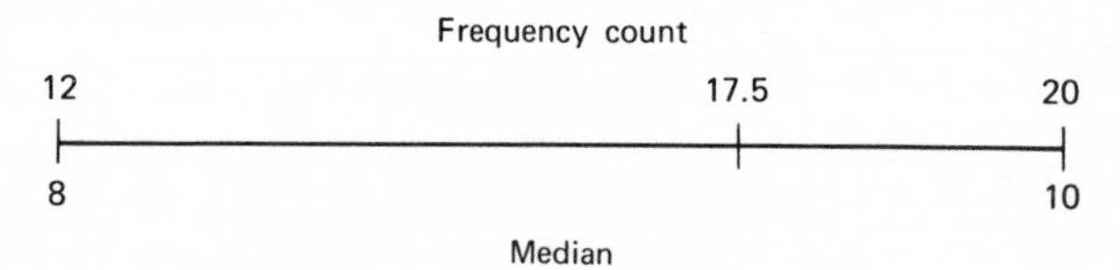

Figure 4.17 Measurement value.

In the interpolation above, it is assumed that the eight measurements in the interval were distributed uniformly across the interval. This assumption allowed us to use a linear form of interpolation in the interval.

Locating the median through the use of the cumulative frequency polygon is much easier than above. In this example we enter the frequency polygon graph at the pseudofrequency count 17.5. This point is on the y axis. We proceed horizontally from this point to the point where we intersect the graph of the polygon. From the point of intersection we proceed vertically downward to a point where we intersect the x axis, which is 9.37. This is the median value for the rate of return on stockholder's equity.

There is a formula for computing the median for interval-valued variables. To use it one must have values for the following:

N = number of samples = total frequency count

a = value of variable at lower boundary of median interval

b = value of variable at upper boundary of median interval

f_a = cumulative frequency count at lower boundary of median interval

f_b = cumulative frequency count at upper boundary of median interval

f_M = median frequency $= \dfrac{N+1}{2}$

M = median

The formula is

$$M = a + \frac{\dfrac{N+1}{2} - f_a}{f_b - f_a}(b - a)$$

■■

Example 4.13 (Median Formula). Using the median interval of the last example we have

$f_a = 12$ $\quad$ $\dfrac{N+1}{2} = 17.5$ $\quad$ $f_b = 20$

$a = 8$ $\quad$ M $\quad$ $b = 10$

Substituting directly into the formula,

$$M = a + \frac{\dfrac{N+1}{2} - f_a}{f_b - f_a}(b - a)$$

$$= 8 + \frac{17.5 - 12}{20 - 12}(10 - 8)$$

$$= 8 + \frac{5.5}{8}(2)$$

$$= 9.37$$

■■

Utility of Mean, Median, and Mode

The following chart indicates where and when each of the three measures of central tendency can be used:

Type of variable	Measure of central tendency
Nominal	Mode
Ordinal	Mode and median
Interval	All three

With *nominal variables* only the *mode* is available as an indicator or descriptor of clustering, typical value, or central tendency. With *ordinal variables* both the *mode* and *median* are available.

The mode is useful to indicate the score or value achieved most frequently. It indicates clustering or popularity but it is not a reliable measure of central tendency in ordinal-valued distributions. The median value splits the frequency distribution into two equal areas. In this sense it is a measure of the center of the distribution with respect to number of samples or measurements contained in the distribution. It is not a reliable indicator of the center of the range of an ordinal-valued distribution, except in the case where the distribution is approximately symmetrical. With *interval-valued* distributions, the mean is used almost exclusively except in the case where the distribution contains some extreme values or is strongly skewed. In the latter case the mode and preferably the median is employed to indicate the "center" of the distribution.

An interesting relationship exists between the relative positions of the median, mode, and mean. The relationship is illustrated by means of the two distributions given in Figure 4.18 (*a*) and (*b*). When a distribution is skewed to the right, as illustrated in Figure 4.18 (*a*), the mode is located to the left of the median which in turn is located to the left of the mean. When the distribution is skewed to the left as in Figure 4.18 (*b*), the positions of the mode, median, and mean are reversed. Numerically, the median is located approximately one third of the way from the mean to the mode, i.e.,

$$\text{mean} - \text{mode} = 3\,(\text{mean} - \text{median})$$

When the distribution is symmetrical, the mean, median, and mode coincide.

Example 4.14 (Mean, Mode, and Median Relationships). Figure 4.19 shows the distribution for the period in years of major comets (except Halley's). We wish to determine the mean, median, and mode period for the distribution and to check the accuracy of the estimate mean – mode = 3 (mean – median).

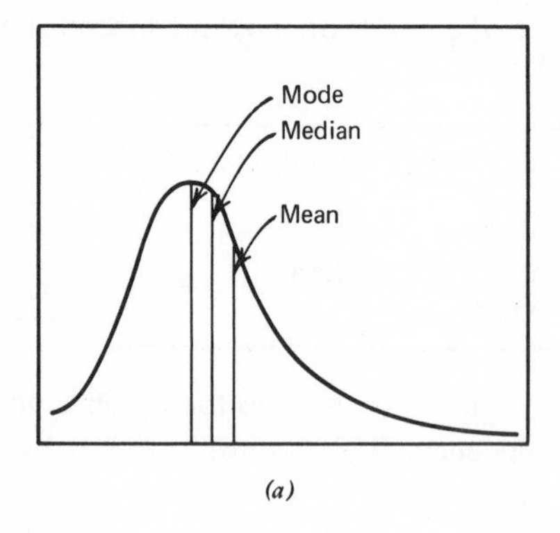

(*a*)

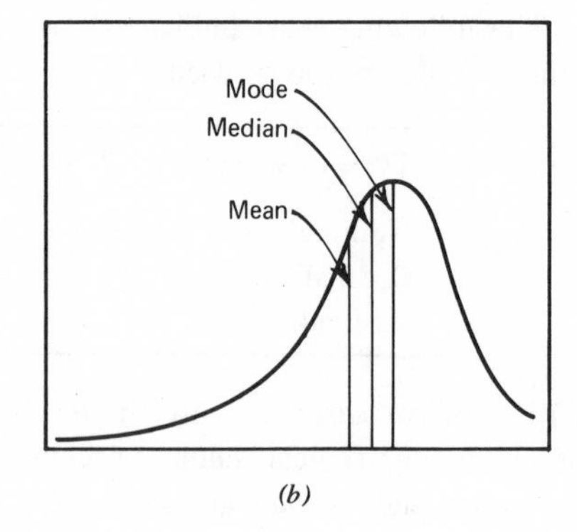

(*b*)

Figure 4.18

Solution: To compute the mean, median, and mode, we construct the computation table and perform the operations indicated by the column headings. Doing this we obtain the following table of values:

Interval midpoint	Frequency f_i	Cumulative frequency	Products of $x_i f_i$
3.5	2	2	7.0
4.5	1	3	4.5
5.5	6	9	33.0
6.5	12	21	78.0
7.5	7	28	52.5
8.5	3	31	25.5
9.5	1	32	9.5
10.5	1	33	10.5
11.5	1	34	11.5
12.5	0	34	0
13.5	1	35	13.5
Totals	$N = 35$		245.5

The mean is computed from the table by the formula

$$\bar{x} = \frac{\text{sum of product column}}{\text{sum of frequency column}} = \frac{245.5}{35} = 7.0$$

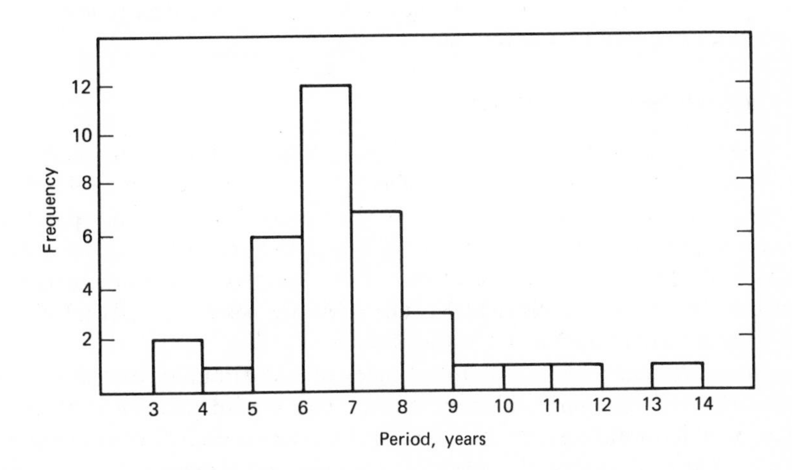

Figure 4.19 Distribution of comet periods.

Since the distribution has an odd number of observations ($N = 35$), the median occurs at the $\frac{(N + 1)}{2}$ = observation 18. Checking the cumulative frequency column we see that the eighteenth observation occurs in the fourth interval. The fourth interval with the frequency counts associated with its end poins $a = 6$ and $b = 7$ follows:

$f_a = 9$ $f_M = 18$ $f_b = 21$

$a = 6$ M $b = 7$

We use the formula and the indicated values to compute the median.

$$M = a + \frac{f_M - f_a}{f_b - f_a} (b - a) \qquad \text{where } f_M = \frac{N + 1}{2}$$

$$= 6 + \frac{9}{12} (1) = 6.75$$

Thus the median is 6.75. The mode is the midpoint of the modal class (fourth interval) which is 6.5. Thus the mode is situated to the left of median (which is 6.75) and the mean (which is 7).

The median is at the midpoint between the mode and median. For a normally skewed distribution we expect the relation

$$\text{mean} - \text{mode} = 3 (\text{mean} - \text{median})$$

to hold. Thus the distribution in the example is skewed slightly less than the amount anticipated by the formula.

Dispersion and its Measurement

When an analyst uses the word *dispersion* in context, it is usually with reference to the spread, or scatter of people, opinions, geographical areas, or things. Basic to the meaning of the word dispersion is a reference point from which something is being dispersed. In descriptive statistics, the dispersion is used to describe the scatter or spread of the samples or measurements which make up the frequency distribution. The reference point is usually taken as the arithmetic mean or median of the distribution.

There are many ways that the samples of a distribution can be dispersed and still have the same arithmetic mean. Thus a good measure of dispersion must be able to differentiate between, or characterize, each of the ways samples can be scattered around the mean of a distribution. Figure 4.20 shows some of the ways that the same number of samples of a distribution can be scattered and still retain the same arithmetic mean but with quite distinct ranges and symmetries.

Distributions (*a*)-(*f*) all have the same mean, $\bar{x} = 8$, and the same number of samples, $N = 20$. Distribution (*a*) has the same range (7) as (*b*) but the samples of (*b*) are more dispersed than those of (*a*). The frequency bars of (*e*) agree in magnitude with those of (*a*) but are dispersed over a range of 13 units (1.5-14.5). Distributions (*c*) and (*d*) have the same dispersion but are skewed in opposite directions. We will note that our measure of dispersion will not differentiate between the distributions of (*c*) and (*d*), but as previously stated, the location of the median and mode relative to the mean will identify the differences in skewness for (*c*) and (*d*).

All will agree that distribution (*a*) is less dispersed than (*b*) and (*e*) and that (*b*) is less dispersed than (*e*). Also (*c*) and (*d*) appear to be less dispersed than (*a*), (*b*), and (*e*).

From (*a*) and (*b*) one can decide that the range of a distribution is not a reliable measure of dispersion. For (*a*) and (*b*) have the same range but have different dispersions. From (*c*) and (*d*), the location of the range of values is

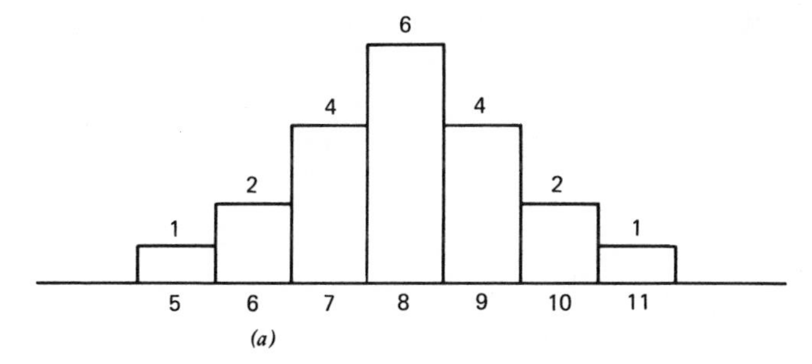

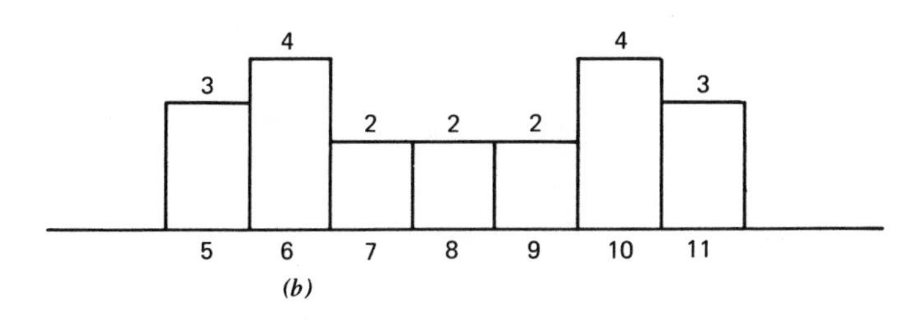

Figure 4.20 Dispersion.

also not a reliable measure of dispersion. For (*c*) and (*d*) have the same dispersion and range yet their ranges are located differently.

We could try to sum the deviations of the samples from the mean–i.e., compute $f_i\ (x_i - \bar{x})$ for each frequency class and sum for all the frequency classes. When we do this, we discover that the sum of deviations is 0,

$$\Sigma f_i\, (x_i - \bar{x}) = 0$$

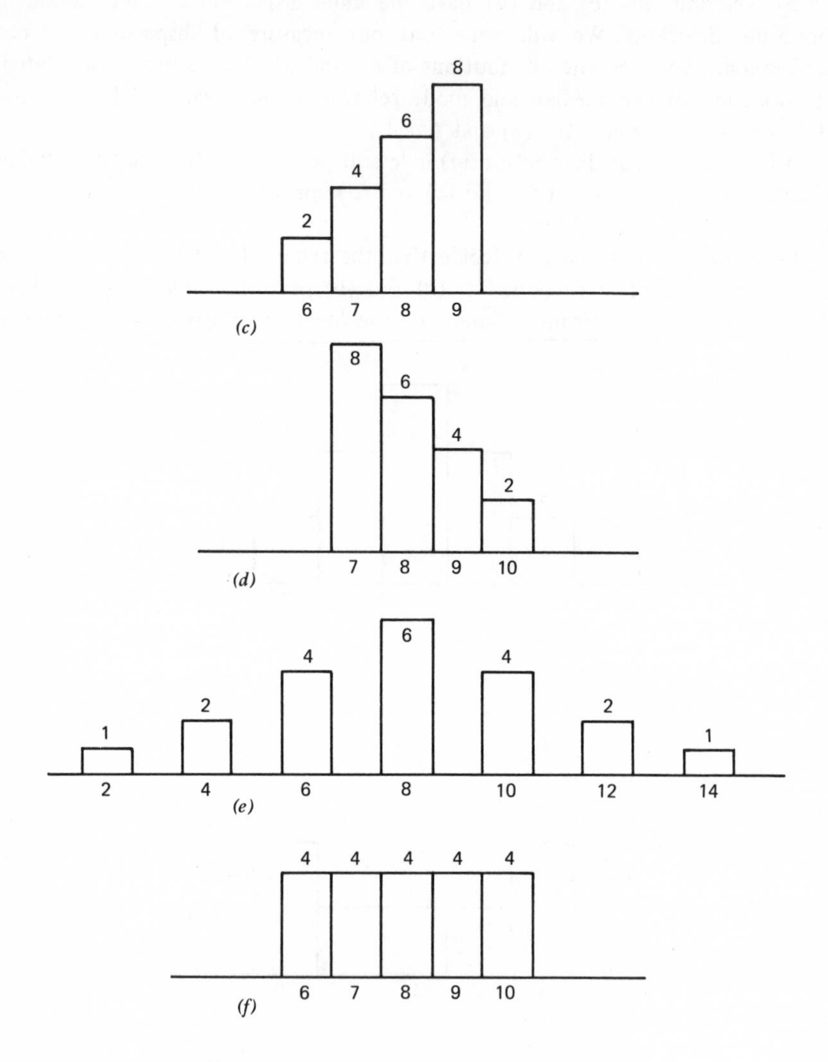

Figure 4.20 (continued) Dispersion.

We could try to sum the absolute values of the deviations, i.e., compute $f_i\,|x_i - \bar{x}|$ for each frequency class and sum over all the classes. The absolute value of $|x_i - \bar{x}|$ is taken as the magnitude of the difference without regard to sign—e.g., $|-3| = 3$ and $|3| = 3$. This measurement for dispersion holds up rather well for examples (*a*)-(*e*). However, when we consider distribution (*f*), we observe that although (*f*) is less dispersed than (*a*) or (*b*), its absolute deviation is slightly greater than that calculated for (*a*) but less than that calculated for (*b*). The values are

$$(a) = \Sigma\, f_i\, |x_i - \bar{x}| = 22$$
$$(b) = \Sigma\, f_i\, |x_i - \bar{x}| = 38$$
$$(f) = \Sigma\, f_i\, |x_i - \bar{x}| = 24$$

Thus even for example (*f*) we find that the absolute deviation is a reasonably good measure for dispersion. However, there is a better measure from the point of view of differentiating between distributions with different dispersions, increasing computational tractability, and providing additional significance in advanced analyses of distributions.

A measure of the dispersion, variation, or scatter of the distribution's samples from its arithmetic mean is the arithmetic mean of the squared deviations of the samples. This quantity is called the variance and is denoted by σ^2 (sigma squared). For ungrouped variable values $x_1, x_2, \ldots, x_N$, the variance is

$$\sigma^2 = \frac{1}{N} \sum_{i=1}^{N} (x_i - \bar{x})^2$$

where $\bar{x}$ is the arithmetic mean of the set of values $x_1, x_2, \ldots, x_N$. The arithmetic mean of the same set of values is

$$\bar{x} = \frac{1}{N} \sum_{i=1}^{N} x_i$$

The reader should note the similarities and differences between the two formulas. In computing the variance, it is necessary to compute the mean of the distribution and then average the squared deviations $(x_i - \bar{x})^2$ from the mean. A sample calculation of the variance of the samples 2, 3, 4, 5, 6, 7, 8, is given below; the mean is

$$
\begin{aligned}
\bar{x} &= \frac{1}{N} \Sigma x_i \\
&= \frac{2+3+4+5+6+7+8}{7} \\
&= \frac{35}{7} \\
&= 5
\end{aligned}
$$

The deviations of the samples from the mean, $x_i - \bar{x}$, are −3, −2, −1, 0, 1, 2, 3. Their squares are 9, 4, 1, 0, 1, 4, 9. The sum of the squares is 28. The mean of the squared deviations is

$$
\begin{aligned}
\sigma^2 &= \frac{1}{N} \Sigma (x_i - \bar{x})^2 \\
&= \frac{9+4+1+0+1+4+9}{7} \\
&= \frac{28}{7} \\
&= 4
\end{aligned}
$$

The sequence of samples 2, 3, 4, 5, 6, 7, 8 has a mean of 5 and a variance of 4. When computing the variance for ungrouped data, it is possible to simplify the computational procedure. This is done by using the following equivalent formula for the variance:

$$
\sigma^2 = \frac{1}{N} \Sigma x_i^2 - (\bar{x})^2
$$

This formula states that the variance is the mean of the squares of the samples minus the square of the mean of the samples. For the last example, we find that the mean of the squares is

$$
\begin{aligned}
\frac{1}{N} \Sigma x_i^2 &= \frac{1}{7} (2^2 + 3^2 + 4^2 + 5^2 + 6^2 + 7^2 + 8^2) \\
&= \frac{1}{7} (203) \\
&= 29
\end{aligned}
$$

and the square of the mean is

$$
\begin{aligned}
(\bar{x})^2 &= (5)^2 \\
&= 25
\end{aligned}
$$

Hence the variance is

$$\sigma^2 = \frac{1}{N}\Sigma x_i^2 - (\bar{x})^2$$

$$= 29 - 25$$

$$= 4$$

Exercise 4.9 Use both formulas to determine and compare the variance for the following sets of numbers.

(a) 2, 4, 6, 8, 10, 12, 14, 16

(b) 2, 2, 6, 6, 12, 12, 16, 16 ■■

Variance for Grouped Data

When the measurements or observations are ungrouped, each measurement contributes to the calculation for the mean and variance. When there are a large number of measurements, this individualized computational procedure becomes excessively time-consuming. Grouping the measurements into intervals reduces the amount of computation required to obtain the mean and variance. If the interval size is selected properly, the error introduced through grouping can be quite small and in most instances negligible. Grouping of measurements also provides a presentation which can be described and interpreted more readily. Grouping also simplifies the process of collecting and recording (tallying) the measurements.

When observations or measurements are grouped into intervals, the number of measurements tallied in the ith interval is designated by f_i. The center (or midpoint) of the interval is designated by x_i. The interpretation is that f_i measurements have value x_i. In the tallying it will be noted that some measurements in the interval have values below x_i and some have values above x_i. The assumption is that these errors, some positive and some negative, will balance each other in each interval. While this is a good assumption it is not entirely correct. For those requiring more accuracy in determining the mean and variance for grouped data, a correction factor may be introduced. This correction factor will be explained at the end of this section.

To calculate the variance for grouped observations or measurements, the data are arranged in a frequency table. The table will identify the class intervals, their midpoints x_i, the class frequencies f_i, and the mean $\bar{x}$ calculated from the x_i and f_i. The table is enlarged to include a column for the deviations $x_i - \bar{x}$ and their squares $(x_i - \bar{x})^2$. The last column contains the product

$f_i\,(x_i - \bar{x})^2$. The table now contains all the quantities necessary to compute the variance using the formula for grouped data

$$\sigma^2 = \frac{1}{N} \sum_{i=1}^{k} f_i\,(x_i - \bar{x})^2$$

Example 4.15 (Computing the Variance). To illustrate the computation of the variance from grouped data, we calculate the variance for the percent return on stockholders' equity measurements given in Example 4.9. Recall that the mean of the distribution is $\bar{x} = 9.35$. The frequency table containing all the required columns is completed as follows:

Interval midpoint, x_i	Frequency, f_i	Deviation, $x_i - \bar{x}$	Squared deviation, $(x_i - \bar{x})^2$	Product, $f_i\,(x_i - \bar{x})^2$
5	3	4.35	18.92	56.76
7	9	2.35	5.52	49.68
9	8	0.35	0.12	0.96
11	11	−1.65	2.72	29.92
13	0	−3.65	13.32	0.00
15	2	−5.65	31.92	63.84
17	1	−7.65	57.52	57.52

Totals $\quad N = \Sigma f_i = 34, \quad \bar{x} = 9.35, \quad \Sigma f_i\,(x_i - \bar{x})^2 = 258.68$

Computations: $\quad \sigma^2 = \frac{1}{N} \Sigma f_i\,(x_i - \bar{x})^2 = \frac{258.68}{34} = 7.60$

Note that the variance is obtained from the table by the following operation:

$$\text{variance} = \frac{\text{sum of product column}}{\text{sum of frequency column}}$$

The computation for grouped data can be simplified through the use of the following equivalent formula:

$$\sigma^2 = \frac{1}{N} \sum_{i=1}^{k} f_i x_i^2 - (\bar{x})^2$$

This formula states that the variance can be expressed as the difference between the mean of the squares and the square of the mean. The table to employ to calculate the variance by this formula is as follows:

Interval midpoint, x_i	Frequency, f_i	Product, $f_i x_i$	Second product, $f_i x_i^2$
5	3	15	75
7	9	63	441
9	8	72	648
11	11	121	1,331
13	0	0	0
15	2	30	450
17	1	17	289
Totals	34	318	3,234

Computations: $\bar{x} = \frac{318}{34} = 9.35$ $\qquad (\bar{x})^2 = 87.42$

$$\frac{1}{N} \Sigma f_i x_i^2 = \frac{3234}{34}$$

$$= 95.11$$

$$\sigma^2 = \frac{1}{N} \Sigma f_i x_i^2 - (\bar{x})^2$$

$$= 95.11 - 87.42$$

$$= 7.69$$

To complete the table the third column entries are the products of the first and second columns. The fourth column entries are the products of the first and third columns $x_i\,(x_i\,f_i) = x_i^2\,f_i$. The second, third, and fourth columns are totalled. The total of the third column $\Sigma\, f_i\, x_i$ is divided by the total of the second column $(\Sigma\, f_i)$ to obtain the mean $(\bar{x})$. The mean is squared to obtain $(\bar{x})^2$. The total of the fourth column $(\Sigma\, f_i\, x_i^2)$ is divided by the total of the second column $(\Sigma\, f_i = N)$ to obtain the mean of the squares $[(1/N)\, \Sigma\, f_i\, x_i^2]$. To complete the computation, the square of the mean $(\bar{x})^2$ is subtracted from the mean of the squares $[(1/N)\, \Sigma\, f_i x_i^2]$ to obtain the variance (σ^2).

Transformation of Variables

It is possible to transform the measurements of a distribution in such a way as to simplify the calculation of the mean and variance. The two transforms normally applied to sample values are the addition to, and multiplication by,

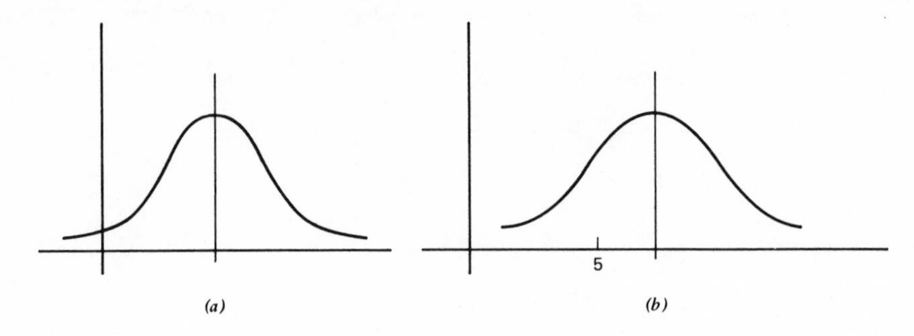

Figure 4.21 Translation of mean: (*a*) $\bar{x} = 5$ for samples $x_1, x_2, \ldots, x_n$ and (*b*) $\bar{x} + k = 7$ for samples $x_1 + k$, $x_2 + k$, ..., $x_n + k$, $k = 2$.

constants. These operations, illustrated in Figures 4.21, 4.22, and 4.23, affect the values of the arithmetic mean and variance as follows:

1. If a constant k is added to each member of the set of values $x_1, x_2, \ldots, x_n$ which has mean $\bar{x}$, the new set of values $x_1 + k, x_2 + k, \ldots, x_n + k$, will have mean $\bar{x} + k$–i.e., translating the samples translates the whole distribution and its mean. The variance of the distribution is not changed by this transformation.
2. If a constant m is used to multiply each member of the set of values $x_1, x_2, \ldots, x_n$ which has variance σ^2, the new set of values $mx_1, mx_2, \ldots, mx_n$ will have variance $m^2\sigma^2$–i.e., expanding or contracting the samples expands or contracts the whole distribution and its variance. The mean of the distribution is changed from $\bar{x}$ to $m\bar{x}$.

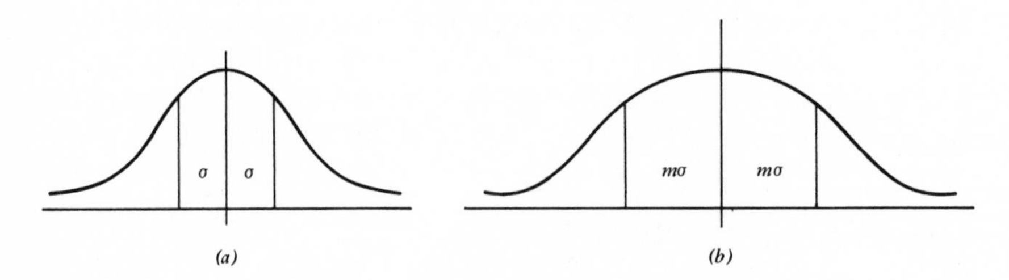

Figure 4.22 Expansion of variance: (*a*) $\bar{x} = 5$ for samples $x_1, x_2, \ldots, x_n, \sigma^2 = 4$, and (*b*) $m\bar{x} = 10$ for samples $2x_1, 2x_2, \ldots, 2x_n$, $m = 2$ and $m^2\sigma^2 = 16$.

3. If the distribution samples are modified by both a multiplier m and an additive constant k, the new sample set is $mx_1 + k, \ldots, mx_n + k$, and its mean and variance are

$$\text{mean} = m\bar{x} + k$$

$$\text{variance} = m^2 \sigma^2$$

The above discussion was limited to a description of changes occurring in the mean and variance when constants are added to or multiplied by all the samples of a distribution. It was stated earlier, that these operations could be used to simplify the calculations required in obtaining the mean and variance. How these operations simplify the calculation of the mean and variance will not be described here. For those who must perform these calculations manually, it is recommended that the references listed at the end of the chapter be consulted. Otherwise, wherever possible, these calculations should be performed on existing time-sharing or batch-processing systems to save time and reduce computational errors.

Example 4.16 (Transformed Mean and Variance). Using the information provided in Example 4.9 on percentage return on stockholders' equity, determine the new mean and variance if tax reforms, price and wage controls, and sales allowed each industry to increase its return on stockholders' equity by 10% over existing profit levels ($m = 1.1$), while each industry gained an additional flat 2% ($k = 2$) return on investment. ■■

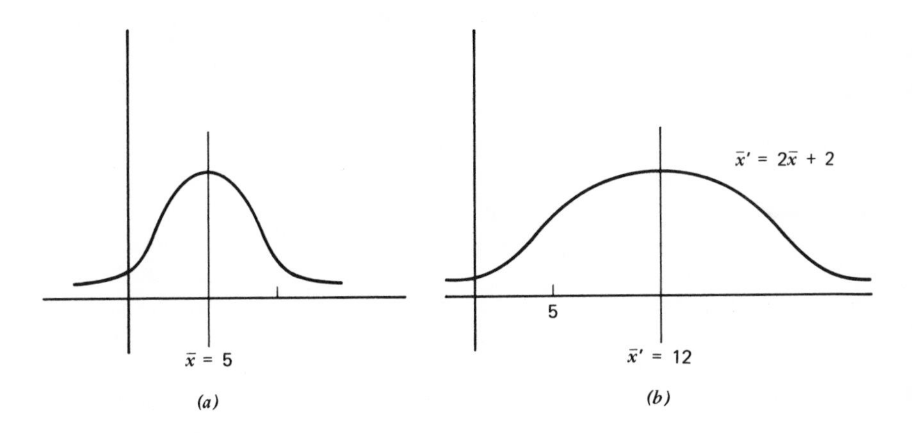

Figure 4.23 Expansion and translation of distribution with $k = 2$, $m = 2$.

Solution: There is no need to recompute the mean and variance which were previously given as

$$\bar{x} = 9.35 \qquad \sigma^2 = 7.60$$

The new mean $(\bar{x}')$ and variance $(\sigma')^2$ are given by

$$\bar{x}' = m\bar{x} + k \qquad (\sigma')^2 = m^2\,\sigma^2$$

Hence

$$\bar{x}' = (1.1)(9.35) + 2 \qquad (\sigma')^2 = (1.1)^2(7.6)$$
$$\bar{x}' = 10.285 + 2 \qquad (\sigma')^2 = (1.21)(7.6)$$
$$x' = 12.285 \qquad (\sigma')^2 = 9.196$$

Exercise 4.10. Compute the new mean and variance for the data of Example 4.15, if $m = 1.4$, and $k = 2.5$.

Exercise 4.11. Compute the mean and variance for all even numbers from 2 to 40. What is the mean and the variance if each number is divided by 2 and a constant equal to 0.5 is added to each resulting number?

Standard Deviation

Up to this point we have restricted our attention to the calculation of the mean and variance of a distribution. The mean is useful in itself as a measure of central tendency. The variance is an intermediate computational requirement to obtain a more meaningful and descriptive measure of dispersion, the standard deviation. The standard deviation is simply the square root of the variance, i.e.,

$$\text{variance} = \sigma^2$$
$$\text{standard deviation} = \sigma$$

One reason why the standard deviation is more meaningful than the variance is that it is expressed in the same dimensional units as the original samples. If the samples are measured in units of percent or feet, the standard deviation is in percent or feet also. The variance is in square percent or square feet. Thus the standard deviation can be represented on the same scale and in the units as the original samples. A second reason that the standard deviation is preferred over the variance as a measure of dispersion is that for many types of distributions, the standard deviation can be related to specific percentages of samples which lie within a distance of 1 standard deviation from the mean. This is a significant descriptor which is used widely in comparing the dispersion of one distribution with another. The geometrical and arithmetical properties of the standard deviation and their uses in the study of dispersion will now be discussed.

The standard deviation of a distribution is computed by the formula,

$$\sigma = \sqrt{\sigma^2}$$

$$= \sqrt{\frac{1}{N} \Sigma f_i (x_i - \bar{x})^2}$$

Once the variance is obtained, it is a simple matter to obtain the standard deviation by completing the square root manually or by checking the table.

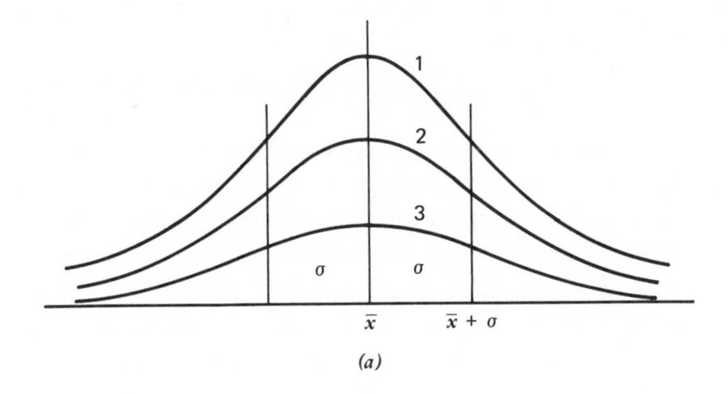

(*a*)

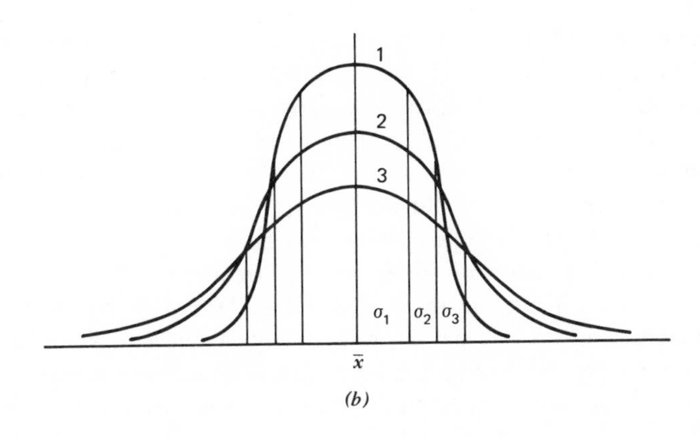

(*b*)

Figure 4.24 (*a*) Different sample, same mean and variance, and (*b*) different variance, same mean and sample size.

The standard deviation is also called the root-mean-square deviation (RMS deviation). The name may assist you in recalling the sequence in which you perform the computational operations required to find the standard deviation of a distribution. The name contains the operations to be performed in reverse order. The first operation is to compute the deviations $(x_i - \bar{x})$, then square them $(x_i - \bar{x})^2$ before finding the mean of the squared deviations $(1/N) \Sigma (x_i - \bar{x})^2$. The final operation is finding the square root of the mean-square deviation.

The standard deviation is most meaningful when used to describe the dispersion of the *normal* distribution. The normal distribution is one of the best known distributions and its properties are reference marks against which the properties of other distributions are compared. The normal distribution has a bell-shaped appearance. It is symmetrical about its mean and extends indefinitely in either direction from the mean. It possesses some basic properties which are illustrated in Figure 4.24 (*a*) and (*b*). Note that the frequency distributions are drawn as smooth curves rather than the bar graphs used previously. This approximation simplifies the drawing and explanation, and will be consistent with the usage of smooth frequency curves in later discussions.

Recalling that when *percent* or *proportional* frequencies are used to draw a frequency distribution for intervals of length 1, the area under the frequency curve is equal to 1. Otherwise the area is equal to the total frequency N times the length of interval L, i.e., area = $N \cdot L$. Figure 4.24 (*a*) illustrates three normal frequency distributions having the same mean $\bar{x}$ and standard deviation σ, but having different sample sizes. The area under each frequency curve between the values $\bar{x}$ and $\bar{x} + \sigma$ is the same percent area of the total area for each of the three distributions. This area is approximately 34.13% of the total area. If the interval length $I = 1$ is used in constructing the frequency curves, then the area under each curve is equal to the number N_i of samples used to construct the curve. Thus the number of samples which lie between the mean $\bar{x}$ and $\bar{x} + \sigma$ of a normal distribution is equal to 34.13% of the total number of samples used to construct the distribution. Knowing the values N_i, $i = 1, 2$, and 3, we know that $0.34 N_i$ samples lie between $\bar{x}$ and $\bar{x} + \sigma$.

Figure 4.24 (*b*) illustrates three normal frequency distributions labeled 1, 2, and 3 with the same mean $\bar{x}$ and sample size N but with different standard deviations, σ_1, σ_2, and σ_3, respectively. Again, 34.13% of the samples which make up the distribution lies between the mean and 1 standard deviation measure from the mean for the distribution. For a normal distribution it is also true that the mean $\bar{x}$ and the standard deviation σ determines the percentage of samples which have values between any two values x_1 and x_2. The number of samples to the left or right of a value x_1 or between two values x_1 and x_2

are frequently required in statistical analysis. Special tables to facilitate obtaining these values are available and are included where needed in the sections which follow.

Example 4.17 (Normal Distribution). Consider a normal distribution made up from 1000 measurements. Suppose that the distribution has a mean of 5 and a standard deviation of 2 (variance = 4). Figure 4.25 illustrates the distribution which was constructed from proportional frequencies over intervals of unit length. The frequency curve is drawn as a smooth curve.

For this distribution, 34.13% of the measurements which make up the distribution have values between $\bar{x} = 5$ and $\bar{x} + \sigma = 5 + 2 = 7$. The same percentage of measurements have values between $\bar{x} = 5$ and $\bar{x} - \sigma = 5 - 2 = 3$. Thus 68.26% of the measurements have values between 3 and 7. By symmetry of the distribution about the mean, half of the measurements have values less than 5 and the remaining half have values greater than 5. From this symmetry it is observed that 15.87% of the measurements have values below 3 and 15.87% above 7.

If one proceeds 2 standard deviations (2σ) from the mean in either direction, then an additional 13.59% of the measurements have values between $x = \bar{x} + \sigma$ and $x = \bar{x} + 2\sigma$, and $x = \bar{x} - \sigma$ and $x = \bar{x} - 2\sigma$. The regions between the 2 and 3 standard deviation points contain another 2.15% each of the samples. Thus, 99.74% of the measurements have values between the 3 standard deviation points from the mean. These locations and approximate percentages are illustrated in Figure 4.26.

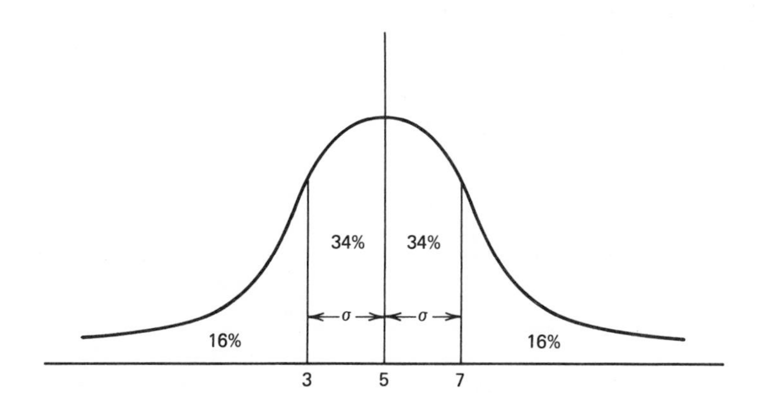

Figure 4.25 Normal distribution where $\bar{x} = 5$, $\sigma = 2$.

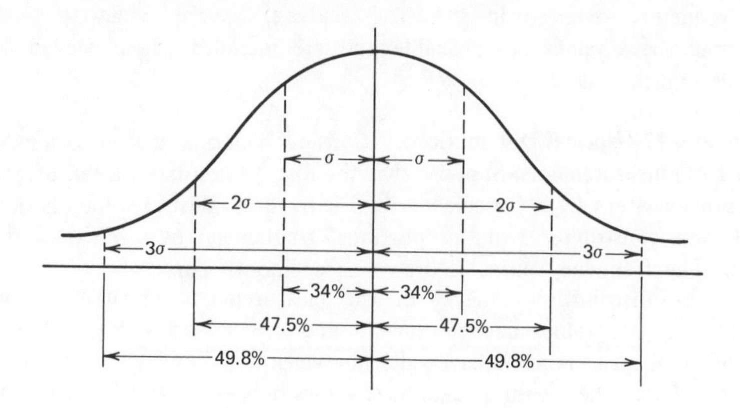

Figure 4.26 Normal distribution–dispersion.

One should note at this point that knowing the standard deviation of a *normal* distribution provides one with a considerable amount of information on the dispersement of the measurement values about the mean of the distribution. If the distribution has a mean of $\bar{x} = 5$ and a standard deviation $\sigma = 1$, then 68% of the sample values lie between 4 and 6, 95% lie between 3 and 7, and 99.7% lie between 2 and 8. Since we know that the frequency curve is bell-shaped, it is now possible to approximate the frequency curve from our knowledge of the mean $\bar{x}$, the significance of σ and the area relationships to σ, 2σ, and 3σ.

Dispersion in Nonnormal Distribution

We have shown that for normal distributions the mean and standard deviation serve as good descriptors of central tendency and dispersion, respectively, among the sample values. We know that when the standard deviation is small, the measurement values tend to cluster around the mean. A large standard deviation indicates that the measurement values are more

Table 4.6 Dispersion for normal distribution.

Interval about mean	Measurements contained in interval, %
$\bar{x} \pm \sigma$	68
$\bar{x} \pm 2\sigma$	95
$\bar{x} \pm 3\sigma$	99.7

dispersed from the mean. Further, as indicated in Table 4.6 we can estimate the percentage of measurements which lie within 1, 2, or 3 standard deviations from the mean.

Analysts appreciate the fact that many of the measurements or observations they make are seldom normally distributed. The key questions for analysts to answer are: If I want to be sure that in a nonnormal distribution $x\%$ of the measurements are included in an interval about the mean, how large should this interval be? If I take any interval centered about the mean, what percentage or measurements will lie in that interval?

If we are able to answer these questions, then the value of the mean and standard deviation as statistical descriptors will be enhanced considerably. The fact is that these questions can be answered and hence it is possible, through a very simple relationship, to approximate the percentage of measurements which fall in an interval centered at the mean for nonnormal distributions. The relationship is stated as follows: For *any* frequency distribution an interval of h standard deviation units centered on the mean $\bar{x}$ will contain at least

$$\left(1 - \frac{1}{h^2}\right) \times 100$$

percent of the measurements. Table 4.7 illustrates the percentages of measurements obtained for some key values of h.

Table 4.7

Number of standard deviation units		Measurements within interval x, %	
$h = 1$	$\bar{x} \pm \sigma$	0	(68)
$h = 1.5$	$\bar{x} \pm 1.5\sigma$	45	
$h = 2$	$\bar{x} \pm 2\sigma$	75	(95)
$h = 2.5$	$\bar{x} \pm 2.5\sigma$	84	
$h = 3$	$\bar{x} \pm 3\sigma$	88	(99.7)

Table 4.7 indicates that for *any* distribution one can assert what percentage of measurements can be expected to be within certain intervals centered about the mean. For intervals of 1 standard deviation or less–i.e., $h \leqq 1$–the relationship is useless. For intervals greater than 1 standard deviation, $h > 1$, the relationship is weaker than that obtained for normal distributions (values contained in parentheses in the table). One should keep in mind that these values are lower bounds; at least this percentage of measurements will fall in

the designated interval. As the distribution approaches the normal distribution in shape, the percentages will approach those achieved by the normal distribution. With experience, one should be able to accurately estimate how close to the normal one may estimate the percent measurements in any interval centered about the mean.

Example 4.18 (Nonnormal Distribution). To illustrate the use of the mean and standard deviation with nonnormal distributions, we consider the estimated birth rates (births per 1000 population) for the year 1965 for 59 national groups. The frequency distribution is given in Figure 4.27. The frequency table and the computations required to calculate the mean and variance are given in Table 4.8.

Table 4.8 Birth-rate computations.

Interval midpoint, x_i	Frequency, f_i	Product, $f_i x_i$	Product, $f_i x_i^2$
13	7	91	1883
19	21	399	7581
25	8	200	5000
31	4	124	3844
37	5	185	6845
43	9	387	16641
49	4	196	9604
55	0	0	0
61	1	61	3721
Totals	59	1,643	55,119

Computations:

$$\bar{x} = \frac{1643}{59} = 27.8$$

$$\frac{1}{N} \Sigma f_i x_i^2 = \frac{55,119}{59} = 934.2$$

$$\sigma^2 = \frac{1}{N} \Sigma f_i x_i^2 - (\bar{x})^2$$

$$\sigma^2 = 934.2 - (27.8)^2$$

$$= 934.2 - 772.8$$

$$\sigma^2 = 161.4$$

$$\sigma = 12.7$$

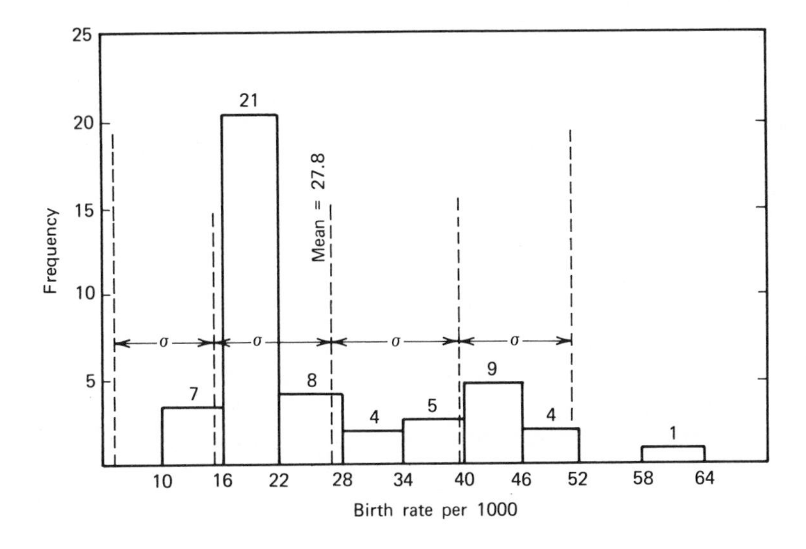

Figure 4.27 National group birth rates, 1965.

Figure 4.28 summarizes the relationships between the mean, various length intervals centered about the mean, and the percent of measurements falling within the interval for this distribution and for a normal distribution as a comparison.

Figure 4.28 illustrates the conservatism contained in the formula which relates the standard deviation to the percentage of measurements within intervals of prespecified lengths with centers at the mean of the distribution. This relationship is a useful guide for nonnormal distributions but its accuracy is suspect when interval lengths below 2σ are considered. The only expedient to better estimates is experience and the actual construction and measurement of the distribution.

Interval	Birth-rate distribution measurements, %	Normal distribution measurements, %
27.8 ± 12.7	29/59 = 49.1	68
27.8 ± 25.4	57/59 = 96.6	95
27.8 ± 38.1	59/59 = 100	99.7

Figure 4.28 Distribution Comparison.

Exercise 4.12. Table 4.9 is the frequency table for the heights of major world dams. The measurements are in 100 ft units.

(a) Find the mean and standard deviation and estimate the percentage of samples falling within 1, 2, and 3 standard deviations from the mean.

(b) Construct the frequency distribution and check the estimates obtained in (a) graphically.

(c) Note that the distribution appears to be composed of two normal distributions, one with its modal class at interval 1-2 and the other at 5-6. Assign half (five) of the measurements falling in interval 3-4 to each "normal distribution" and divide the distribution into two distributions. Check the two distributions for normalcy–i.e., how close to the correct percentages 68, 95, and 99.7 do the measurements of the two distributions fall within σ, 2σ, and 3σ of their means? Do you consider analyzing the distribution as a composite of two distributions more descriptive and amenable to greater accuracy in description?

Table 4.9 Height of major world dams in units of 100 ft.

Interval	Midpoint	Frequency
0-1	0.5	7
1-2	1.5	17
2-3	2.5	12
3-4	3.5	10
4-5	4.5	12
5-6	5.5	30
6-7	6.5	12
7-8	7.5	12
8-9	8.5	3
9-10	9.5	1
10-11	10.5	1

Properties of the Standard Deviation

In computing the mean, each measurement value x_i is given its own weight in the formula, $\bar{x} = (1/N) \Sigma f_i x_i$. In computing the variance, each measurement value x_i is given a weight equal to the square of its distance from the mean.

$$\sigma^2 = \frac{1}{N} \Sigma f_i (x_i - \bar{x})^2$$

Thus if the mean is 6, we can add two measurements –94 and 106 to the distribution and not disturb the location of its mean; the mean is insensitive to changes which are symmetric about the mean. But in adding these same samples to the distribution, we would be adding $2(100^2 - \sigma^2)/(N + 2)$ to the variance. Hence the variance and the standard deviation are very sensitive to changes in values which are greatly removed from the mean. Further, any sample value added to the distribution, except values equal to the mean, will change the standard deviation of the distribution.

Coefficient of Variation

The standard deviation is an absolute, not relative, measure of dispersion. The standard deviation measures dispersion about the mean regardless of the value or size of mean. For example, a standard deviation of 10 lb when weighing tons of a product is relatively insignificant. But a standard deviation of 10 lb when measuring 100 lb of the same product may be significant. A measure of dispersion which accounts for the standard deviation (error in measurement) relative to the average measurement size is the *coefficient of variation.* This measure is defined as the standard deviation divided by the mean, i.e.,

$$\text{Coefficient of variation} = \frac{\sigma}{\bar{x}} = \frac{\text{standard deviation}}{\text{mean}}$$

Note that the standard deviation and the mean have the same dimensional units, i.e., pounds, feet, birthrate, etc. Thus the coefficient of variation is dimensionless and hence can be considered as a ratio or percentage $(\sigma/\bar{x}) \cdot 100$.

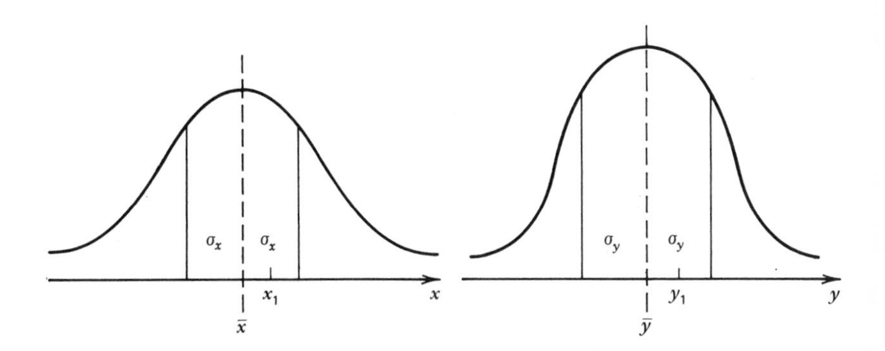

Figure 4.29 Distribution of production.

It gives a measure of variation in measurements relative to the average measurement. The coefficient of variation for $\sigma = 2, \bar{x} = 20$ is $2/20 \times 100 = 10\%$. If $\bar{x} = 200$, then the coefficient of variation is $2/200 \times 100 = 1\%$.

Standardized Variables

There are situations where an analyst wants to know how well one company, say company C, is doing relative to the industry in producing x_1 units of resource A and y_1 units of resource B. Other corporations in the industry produce resources A and B. Of all corporations producing resource A, the mean production is $\bar{x}$ units with a standard deviation of σ_x units. For resource B, the mean production is $\bar{y}$ units with a standard deviation of σ_y units. The two distributions are illustrated in Figure 4.29. Note that near normal distributions are assumed. The question is this: Is company C's production rate x_1 relative to the industry's mean rate $\bar{x}$ and the variance σ_x^2 better than its production rate y_1 relative to the industry's mean rate $\bar{y}$ and variance of σ_y^2? The answer to this question is obtained by *standardizing* the variables by means of the following ratios and comparing the results:

$$\frac{x_1 - \bar{x}}{\sigma_x} \qquad \frac{y_1 - \bar{y}}{\sigma_y}$$

where x_1 and y_1 are production rates of company C; $\bar{x}$ and $\bar{y}$ are mean production rates for the industries; σ_x and σ_y are standard deviations in production rates for the industries. Note that the magnitudes of the numbers x_1, y_1 are standardized or normalized and hence do not play a role in the ordering. The important quantity is the size of the deviation $x_1 - \bar{x}$, $y_1 - \bar{y}$ relative to the standard deviations σ_x, σ_y, respectively. When variable values are compared through the expression

$$\frac{x_i - \bar{x}}{\sigma_x}$$

the variable is said to be *standardized* and the values obtained are called *standard scores.* Higher standard scores indicate better performance.

To judge the production performance of company C relative to the industry, one compares the two quantities

$$\frac{x_1 - \bar{x}}{\sigma_x} \qquad \text{and} \qquad \frac{y_1 - \bar{y}}{\sigma_y}$$

The quantities indicate how many standard units the production rates x_1, y_1 deviate from the industry's mean production rates $\bar{x}$, $\bar{y}$, respectively. Performance in the industry is measured by the number of standard units the company's production rate exceeds the industry's average production rate. If the production rate x_1 is such that

$$\frac{x_1 - \bar{x}}{\sigma_x} = 3$$

then the production rate x_1 is better than the production rate of 99.7% of the companies which make up the industry. If the production rate y_1 is such that

$$\frac{y_1 - \bar{y}}{\sigma_y} = 2$$

then the production rate y_1 is better than 95% of the companies which make up the industry. Company C is doing better, relative to the industry, in producing resource A at a production rate x_1 than producing resource B at a production rate of y_1. Note that the relative sizes of x_1 and y_1 never entered into the determination.

Example 4.19 (Standard Variables). Is the production rate of 33 units with a mean of 25 and standard deviation of 4 better than the production rate of 120 units with a mean of 110 and a standard deviation of 7?

Table 4.10 Ordinates and areas for normal distribution.

Standard variate	Ordinate y	Area
0.0	.399	.000
0.2	.391	.079
0.4	.368	.155
0.6	.333	.226
0.8	.290	.288
1.0	.242	.341
1.2	.194	.385
1.4	.150	.419
1.6	.111	.445
1.8	.079	.464
2.0	.054	.477
2.2	.036	.486
2.4	.022	.492
2.6	.014	.495
2.8	.008	.497
3.0	.004	.499

To answer this question we standardize each variable and compare the results

$$\frac{33-25}{4}=\frac{8}{4} \qquad \frac{120-110}{7}=\frac{10}{7}$$

Since $8/4 > 10/7$, the production rate of 33 is better relative to the industry's rate of 25 than the production rate of 120 relative to the industry's rate of 110.

Standardized variables find application in other areas of statistics such as educational statistics, testing of hypothesis, and statistical estimating procedures. The reader is advised to spend some time in understanding and using standardized variables.

Table 4.10 contains the information required to construct an accurate normal distribution. Column 1 contains values for the standard variate Z. Column 2 gives the height of the normal curve at the value given for Z. Column 3 gives the area under normal curve from the mean $(x = \bar{x})$ $Z = 0$ to the value of the standard variate Z. Note the values for $Z = 0, 1, 2,$ and 3.

Exercise 4.13 A student received a grade of 75 on test A and 85 on test B. The remainder of the class had an average grade and standard deviation of 80 and 12, respectively, on test A, and 88 and 6 on test B. In which test did the student rank higher relative to the class? ■■

Exercise 4.14 Company A, in industry I_1, returns 6.2% on stockholders' equity while the industry averages 5.6% with a standard deviation of 1.2 units. Company B, in industry I_2, returns 7.4% on stockholders' equity while the industry averages 6.8% with a standard deviation of 1.8 units. Which company is doing better relative to its industry? ■■

Exercise 4.15 If industry I_1 has a higher growth rate than industry I_2, how much higher would it have to be before you would invest in company A looking for equity appreciation over 5 yr? Assume the return on stockholders' equity will remain constant for both companies in both industries. ■■

REFERENCES

Afifi, A. A., and Azen, S. P. *Statistical Analysis.* New York: Academic Press, 1972.

Dixon, Wilfrid J., and Massey, Frank J. Jr. *Introduction to Statistical Analysis.* New York: McGraw-Hill, 1969.

Hoel, Paul G. *Introduction to Mathematical Statistics.* New York: Wiley, 1966.

Longley-Cook, L. H. *Statistical Problems.* New York: Barnes & Noble, 1970.

McCullough, Celeste, and Van Atta, Loche. *Introduction to Descriptive Statistics and Correlation.* New York: McGraw-Hill, 1965.

Siegel, Sidney. *Nonparametric Statistics for the Behavioral Sciences.* New York: McGraw-Hill, 1956.

Chapter 5 / Correlation and Regression

5.1 INTRODUCTION

Chapter 4 reviewed the meaning and significance of some elementary descriptors used in statistics to describe sets of data. For the central tendency of data sets, the median, mode, and mean are the measures employed. For the spread or variability of a data set, the range and standard deviation respectively, are used. Several graphical techniques were employed to obtain simple and concise representations for data sets. The frequency table, frequency distribution, and cumulative frequency distribution provide a quick way of summarizing and recording information contained in a data set.

The descriptors and graphical techniques used in Chapter 4 dealt with single variables. The variables assumed nominal, ordinal, or interval values. Whatever the variable and the type of values it assumed, the objective of Chapter 4 was to describe the data set associated with the variable in a simple and concise manner. No attempt was made to compare or relate data sets from two or more variables. The objective of the present chapter is to introduce two additional statistical descriptors which find use as measures in assessing whether two variables are related. The determination of the two descriptors, called the *correlation coefficient* and the *line of regression* requires that the variables assume ordinal or interval values. Hence, these two descriptors are not applicable to variables assuming nominal values. Other statistical descriptors and measures of association between variables which assume nominal values are applicable and will be discussed in Chapter 7. For the present chapter, we shall restrict our attention to the statistical analysis of variables which assume ordinal or interval values.

Some main concerns of substantive analysis are:

1. Determination of whether relationships exist between variables, events, conditions, values, etc.

2. Identification of the type of relationship which exists.
3. Measurement of the "strength" of the relationship.

For example, an analyst may be concerned with analyzing the increased use of drugs in high school age children over the past 5 yr. In attempting to find the conditions or events which might be a cause of, or be related to, this incidence of increased drug usage, the analyst may turn to analyzing all changes which have occurred in high schools over the past 5 yr. He would find, as one social analyst recently reported, that in Ohio the number of teachers employed in high schools has increased over the past 5 yr. So have the teachers' salaries. Is the increase in the number of teachers or their salaries the cause for the increase in drug usage? Perhaps, but very unlikely. Yet, when the drug usage and the number of teachers employed statistics are checked for the 50 states over the 5 yr period, one finds an increase in both in almost all of the states. A coincidence? Of course not. A relationship exists. It is not a proximate or direct causal relationship, but it is a relationship. The two occurrences are strongly related, but in this instance they are related through other variables or factors such as higher incomes and the increases in population which are more direct causes for the increase in student enrollment and hence the incidence of drug usage and the number of teachers employed.

In the above example, drug usage in high schools (or in the general population) is said to *correlate* with the increase in teacher population in high schools. From what has been said, it should be clear that although the two events correlate, there is no guarantee that a direct or proximate causal relationship exists between them. The events can be completely independent of each other.

5.2 RELATIONSHIPS BETWEEN VARIABLES

There are many types of relationships which exist between variables which may or may not be causally connected. A type of relationship we are interested in is the *functional* relationship wherein it is specified how one variable is related, or dependent on, one or more other variables. Physical laws of nature and economic principles are normally expressed through functional relationships. For example, the physical relationship

$$\text{distance} = (\text{velocity}) \cdot (\text{time})$$

expresses the fact that if one travels at a velocity of v ft/sec for t sec, one will have travelled the distance

$$S = vt \text{ ft}$$

The following table illustrates this relationship for a 10 sec travel time, $t = 10$, and velocities of from 1 to 10 ft/sec:

Velocity y	1	2	3	4	5	6	7	8	9	10
Distance S	10	20	30	40	50	60	70	80	90	100

The next table illustrates the same relationship for a velocity of 10 ft/sec, $v = 10$, and for travel times of from 1 to 10 sec:

Time t	1	2	3	4	5	6	7	8	9	10
Distance S	10	20	30	40	50	60	70	80	90	100

The next table illustrates the same relationship with the distance S restricted to 50 ft. It gives values for time t and velocity v which yield the value $S = vt = 50$.

Time t	1	2	3	4	5	10	12.5	16.6	25	50
Velocity v	50	25	16.6	12.5	10	5	4	3	2	1

The graph of the relationship $S = vt = 50$ follows:

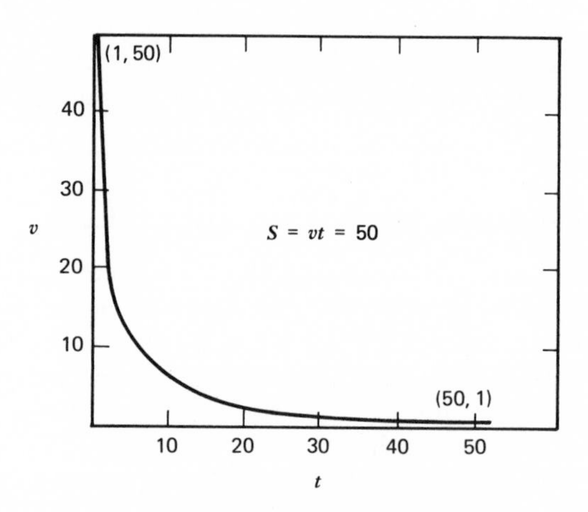

Each point on the graph determines a velocity and time required to travel 50 ft. Hence the graph gives a complete picture of the functional relationship $S = vt = 50$.

Functional relationships are normally expressed by means of a formula, a table of values, a graph of the formula, or a combination of the three. Equations, graphs, or tables are quantitative interpretations of the real world. For example, if one wishes to travel a distance of 400 mi at 40 mph, the relationship $S = vt$ tells him that it will take exactly 10 hr to do so.

Most physical laws are expressed as *essentially deterministic* functional relationships among the variables involved. However, the relationships one wishes to consider in economics, international affairs, and politics, social systems, and international trade are not all readily expressed in terms of functional relationships which are essentially deterministic. We say essentially deterministic because most functional relationships for physical laws such as $S = vt$ and $S = 1/2\ gt^2$ are operational in the sense that measurements are involved. Operationally (or practically) it is difficult, if not impossible, to measure or maintain a velocity of exactly 40 mph for a period of 10 hr. We say that each time someone travels at approximately 40 mph for approximately 10 hr, he will travel a distance of approximately 400 mi. The quantities involved are physical and can be measured or maintained to a high degree of accuracy so that the distance traveled can be predicted or determined to a very high degree of accuracy. In this sense the functional relationships of the physical sciences are essentially deterministic or probabilistic with a high degree of accuracy.

Returning to the variables one considers in the social, political, and economic sciences, it is difficult to express laws or relationships among these types of variables which are essentially deterministic. Uncertainty and chance are almost always involved and play a major role in the observation, measurement, and interpretation of the variables one considers. As an example consider the "quantity theory of money." It is expressed by one of the most famous functional relationships (equations) in economics.

$$MV = PT$$

where

M = quantity of money in circulation
V = velocity of circulation (rate of changing hands)
P = general level of prices (price index)
T = number of transactions as a measure of physical output

The quantity theory of money says that the expenditures in the economy MV (quantity available M times the number of turnovers V) are equal to the receipts PT (average price P per transaction times the number of transactions or sales T).

The equation looks quite deterministic. However, a survey of economic schools of thought indicates that not all professional economists have the same degree of faith in the fidelity with which the theory models economic "facts." Also, the variables involved in the equation are not easy to measure to a reasonable degree of accuracy. Thus the functional relationship expressed by $MV = PT$ has a lesser degree of determinism than the physical relationship $S = vt$. This example illustrates an important difference between physical relationships and economic, social, or political relationships. The physical relationship has a high degree of determinism yet retains a small probability that errors will be present in measurement or observations. The economic, social, or political relationship has a lesser degree of determinism and possesses a larger probability that errors are present in the relationship itself as well as the measurements and observations of the quantities involved.

Because functional relationships utilized in the political, social, and economic sciences are more prone to contain some uncertainties is not a good reason to avoid using them in the analysis of political, economic, or social events. Rather, it is the good analyst who recognizes the limitations involved and makes the best use of a situation over which he has some but not complete control. He learns how to obtain relationships among variables, evaluate their fidelity and reliability, and use them to make rational conclusions or to guide him to seek other information and relationships if the need for them is so indicated. Using a relationship of known capability and limitations is better than using no relationship.

Example 5.1 (Implication Relationship). In Chapter 3, much was said about the evaluation of the credibility of A and B and the plausibility of the argument $A \to B$. We wrote the functional relationship

$$b(B) = b(A)\, b(A \to B) + b(\bar{A})\, b(\bar{A} \to B)$$

and noted that when A is true and the argument $A \to B$ is valid, then the relationship is deterministic and states that B must be true. When either A is only credible or the argument $A \to B$ is only plausible, then the relationship gives an approximate value to the credibility of B as a function of the credibility of A and $\bar{A}$ and the plausibility of $A \to B$ and $\bar{A} \to B$. This particular relationship contains the deterministic case as a special instance. The special instances arise mainly in logic and mathematics. The nondeterministic (probabilistic or plausible) relationship applies in most instances in economic, social, political, and other pseudosciences wherein uncertainties exist in the credibilities of statements or in the plausibilities of the arguments.

The analyst may ask, "Why all the emphasis on functional relationships?" The answer is simply that there is much to be gained in expressing physical, social, economic, or political laws, theories, or hypotheses, through quantitative functional relationships. Having the relationship expressed quantitatively (or only partly so) allows the analyst to check the fidelity (faithful reproduction of reality) and the reliability of the relationship against many experiences or experiments and to use them as a basis to extend and adjust existing knowledge. With good quantitative relationships the analyst can make better predictions or prescriptions for future events in the real world. These relationships provide the analyst with a stronger, more valid basis to make decisions and hence lead to a more rational control over the environment and future events occurring in it. Hence, an objective of this chapter is to provide the analyst with some methods on how to derive functional relationships from information which is essentially political, social, or economic in substance and how to examine these relationships for fidelity and for their predictive ability.■■

Before we proceed to describe and apply the techniques of regression and correlation, we shall discuss linear functional relations. These relations are quite important and useful in developing and explaining the concepts of correlation and regression.

5.3 LINEAR FUNCTIONAL RELATIONSHIPS

In this section we review a few algebraic and geometric properties of linear relationships–i.e., those which can be represented by means of a straight line. Variables so related will be said to be linearly related. Thus we shall restrict our discussion to linear correlation and regression in which a principal objective is to find the best fit of a straight line to a set of data points (x_1, y_1), (x_2, y_2), ..., (x_n, y_n) for the variables x and y. For example, what is the relationship between Russia's (x) and U.S.'s (y) aircraft production since 1950? Has inflation in Russia (x) kept pace with inflation in the U.S. (y) since 1960? The relationships sought in these two questions can be approximated by a straight line.

5.4 TERMS AND NOTATION

The important concept to be described here is that of a function or functional relation. A functional relationship is an expression or formula which specifies how variables are related to each other. The functional expression can be verbal such as, "price will increase as supply decreases" or "Russia's defense budget increases at 1.5 times that of the U.S. defense budget." The respective functional expressions can be written as formulas such as

$$P = \frac{C}{S} \qquad \text{and} \qquad (DB)_R = 1.5\,(DB)_{US}$$

The functions contain both constants and variables. Variables can assume different values but constants only a fixed value. If P is a price level and S is a supply level, then P varies inversely as S varies, and the amount of variation is determined by the constant C. In the formula $DB_R = 1.5\, DB_{US}$, the constant is specified as 1.5 and fixes how DB_R varies with DB_{US}. Thus if DB_{US} changes from \$50 to 100 billion, DB_R will change from \$75 to 150 billion.

Functional relations between two variables (there can be more variables related, but we will restrict our attention to functional relations involving but two variables) are expressed notationally as

$$y = f(x)$$

This says that the variable y is a function of, i.e., depends on, the value of the variable x. The function is single-valued in the sense that for each value of x there is one and only one value for y. However, for each value of y there can be more than one value of x. For example, in the functional relation,

$$y = x^2$$

when $y = 1$, the value of x can be 1 or -1.

In practice, the substantive analyst is always seeking to express functional relationships among social, political, or economic variables. Some are difficult if not impossible to express. Others are simple yet illuminating. For example, a difficult functional relationship exists in defining the actions Red China must take in order for U.S. to consider favorably the resumption of trade with Red China. Another way of saying this is, "what are the important factors (variables) affecting the resumption of trade between Red China and United States, and what attitudes (values) must Red China assign to these in order for U.S. to react favorably?" The values assumed by the variables here are mostly nominal. The functional values (favorable, unfavorable) are also nominal. This example illustrates the generality of the application of functional relationships to nominal-valued variables. However, for ease in explanation, our discussions will be concerned principally with variables assuming either ordinal or interval values.

When we write the equation $y = f(x)$, we are saying that y is a function of x. We are not specifying the functional relationship between x and y. When we write $y = 2x + 5$, we are specifying the functional relationship as "multiply the value of x by 2 and add 5 to the result in order to obtain y." When we assign a value to the variable x, the value of y is completely determined. Changing the constants 2 or 5 in the functional relation $y = 2x + 5$ changes the functional relation between x and y. Thus

$$y = f(x) = 2x + 5$$
$$y = g(x) = 3x - 2$$

are different functional relations. Assigning a value to x, say $x = 2$, yields different values for y,

$$y = f(2) = 2 \cdot 2 + 5 = 9$$
$$y = g(2) = 3 \cdot 2 - 2 = 4$$

The functional relation $y = f(x) = ax + b$ is determined by the constants (also called parameters) a and b. We shall now show how to graph the function $y = f(x)$ and what roles a and b play in determining the graph of $f(x)$.

5.5 GRAPHS OF FUNCTIONS

Associated with every function of the form $y = ax + b$ is a graph. The graph is a pictorial representation of the function which relates values of y to values of x. The graph is plotted on the rectangular $x - y$ coordinate system wherein the variable x is normally scaled on a horizontal line and the variable y is normally

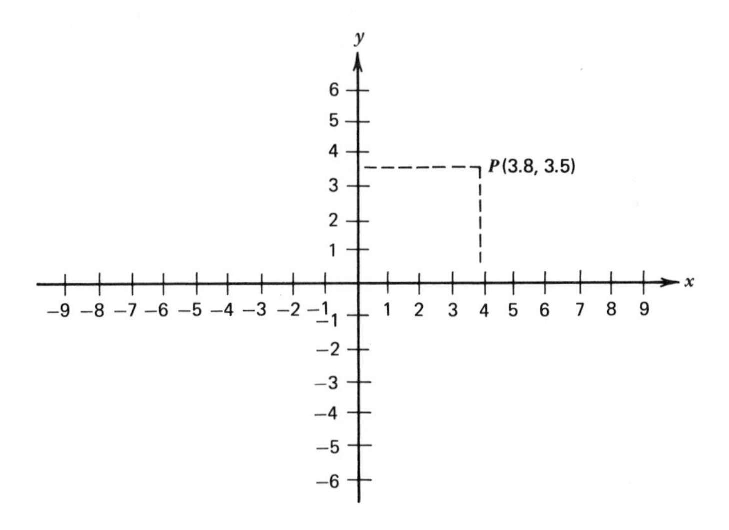

Figure 5.1 Rectangular coordinate system.

scaled on a vertical line. The scaled lines are called the x *and* y *axes*. The point where they intersect is called the *origin* and it is the point where x and y both have the value 0. The rectangular x-y coordinate system for x and y is depicted in Figure 5.1.

Values of x which are positive (> 0) are located to the right of the origin, negative values to the left. Positive values of y are located above the origin, negative values below. To locate a point P in the x-y coordinate system, a pair of values, one for x and one for y, is needed. The two values x, y are normally enclosed in a parenthesis and are called the coordinates of the point P. The coordinates of the point P will always be given in the order x, y, i.e., x followed by y, and hence are called an ordered pair. To each point P there is associated an ordered pair (x,y) and the values in the ordered pair (x,y) are called the coordinates of the point P. The point P in the coordinate system is labeled $P(x,y)$ and is determined by locating the value of x on the x axis and then moving y units above (or below if y is negative) the x axis. In Figure 5.1 the point $P(x,y)$ has the value $P(3.8,3.5)$. This point is located in the coordinate system by locating 3.8, the value of x, on the x axis and then moving vertically upward to the point at which $y = 3.5$.

Exercise 5.1. Plot the points (2,4) and (–2,–3) and draw a straight line through them.■■

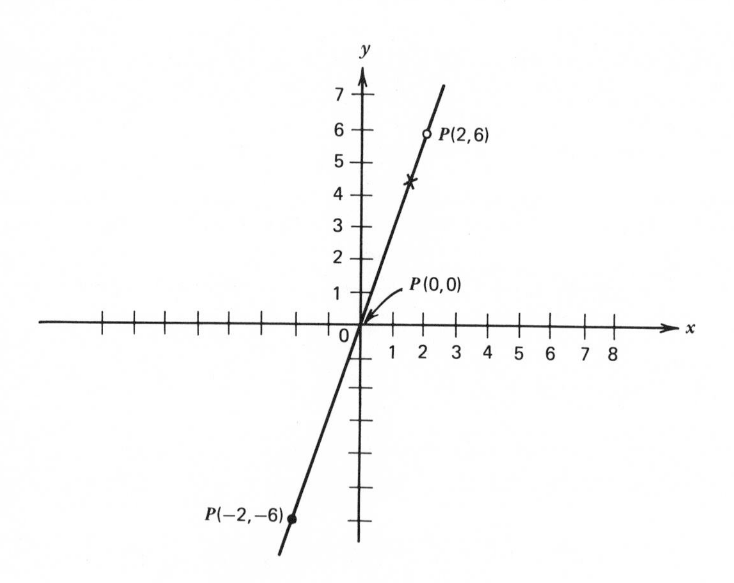

Figure 5.2 Graph of $y = 3x$.

Exercise 5.2. Plot the points (1,2), (2,4), and (3,6). Do these three points lie on the same straight line? ■■

Exercise 5.3. Draw any nonvertical line in the x-y plane. Determine where the line crosses the y axis. Starting with a point on the line, if you move 3 units in the positive x direction remaining on the line, how many units did you move in the y direction? ■■

Now let us find the values for y in the functional relationship $y = f(x) = 3x$. It is normal in this case to construct the following table which lists values of y for each value assigned to x:

x	$y = 3x$
−2	−6
0	0
2	6

There are only three entries in the table. The first column lists three values for x and the second column lists the associated values for y. Thus the table contains three ordered pairs or coordinates (−2,−6), (0,0), and (2,6). If these three points are plotted (or graphed) in the x-y coordinate system, they will represent points which are associated with the function $y = 3x$. These points are illustrated in Figure 5.2 wherein they are connected by a straight line.

Recalling from geometry that any two distinct points in a plane determines a unique straight line, we can see that the graph of any point which is obtained through the relation $y = 3x$ will lie on this straight line. For example, for $x = 1.5$, we have $y = 4.5$ and the point $(x,y) = (1.5, 4.5)$. This point, when graphed, lies on the line and is located on the line at the asterisk mark. Equations such as $y = 3x$ are called linear functions because their graphs are straight lines. We shall see that the graph of every straight line determines a unique linear function. Conversely, every linear function determines a unique straight line when graphed in the x-y coordinate plane.

5.6 LINEAR FUNCTIONS AND THEIR GRAPHS

The general linear function is given by

$$y = f(x) = mx + b$$

The literals m and b are called *parameters* or *constants of the equation* and their values uniquely determine the linear function. For example, in the following six linear functions

$y = f(x) = 2x + 3$	$m = 2, b = 3$
$y = f(x) = -x - 6$	$m = -1, b = -6$
$y = f(x) = -2x + 4$	$m = -2, b = 4$
$y = f(x) = x - 3$	$m = 1, b = -3$
$y = f(x) = 2x$	$m = 2, b = 0$
$y = f(x) = 7$	$m = 0, b = 7$

it is illustrated that m and b can assume values positive, negative, or zero. All of the six functions have graphs which are straight lines. The last linear function $y = f(x) = 7$ is a function which is independent of x. Its graph is a straight line located 7 units above and parallel to the x axis. We will now see how one can construct the graph from the function or the function from its graph.

In the equation for the general linear function, the parameter m is called the *slope* of the graph of the straight line representing the equation, and b is called the *y intercept* of the graph. We shall show that from knowledge of the slope and *y intercept* of a line, or any two distinct points on the line, one can readily construct the line. Also, given the graph of an equation for a straight line, one can readily estimate the slope and *y intercept* of the graph of the line and hence write the equation for the line. First, we will start with the equation

$$y = mx + b$$

and illustrate two ways one can construct (or graph) the straight line which is associated with the equation when values for m and b are specified.

Two-Point Form

One method of constructing the line is called the *two-point form.* In this method two distinct values of x are chosen and corresponding values for y are computed. For example, in the equation

$$y = 2x - 3$$

y has the values -3 and 1 when x is 0 and 2, respectively. The values of x and y are listed as two ordered pairs (points)

$$(x_1, y_1) = (0, -3) \qquad \text{and} \qquad (x_2, y_2) = (2, 1)$$

The ordered pairs state that the variable y has a value y_1 when x has the value x_1, etc. These points are said to *satisfy the equation* $y = 2x - 3$ and hence lie on the graph of the equation. Since two points determine a straight line, it suffices to draw (or plot) the two points in the x-y coordinate system and draw a line through the two points by means of a straight edge. The line determined by the two points

$$(x_1,y_1) = (0,-3) \qquad (x_2,y_2) = (2,1)$$

is illustrated in Figure 5.3.

Slope-Intercept Form

There are several properties of the equation of a straight line which help in constructing its graph. First, the equation has two parameters, the y intercept b and the slope m. Once values for b and m are selected, the ordered pairs (x,y) which satisfy the equation $y = mx + b$ are uniquely determined. Just as two points determines a line uniquely so does the y intercept and the slope of the line. This leads us to the second method of constructing the graph of an equation. The method is called the *slope-intercept* form of the line. To use this method we first note that in the equation

$$y = mx + b$$

that $y = b$ when $x = 0$. This states that the graph of the equation $y = mx + b$ passes through the point $(0,b)$. The point $(0,b)$ lies on the y axis. Hence the graph of the line passes through the y axis at the point $y = b$. This line is illustrated in Figure 5.4. If the equation $y = 2x - 3$ is written as $y = mx - 3$, then there are an infinity of lines passing through the y intercept $(0,-3)$. These lines are illustrated in Figure 5.5. Unique lines are determined by specifying

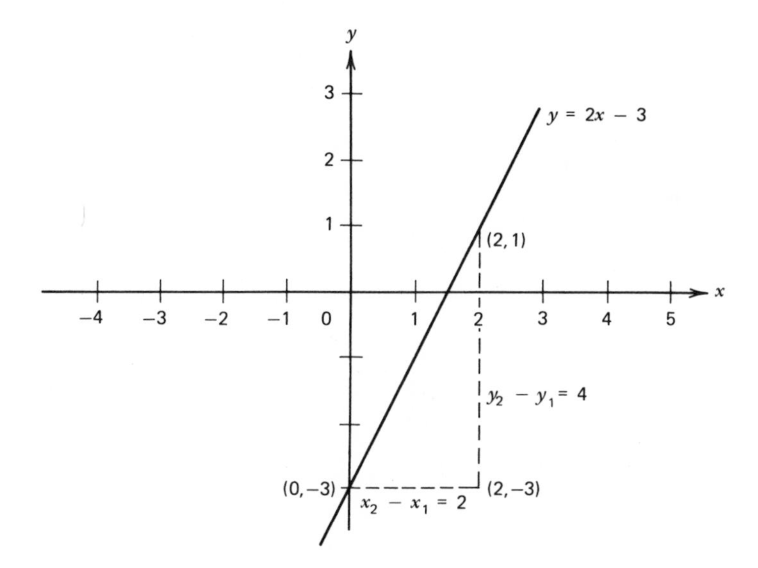

Figure 5.3 Line through two points.

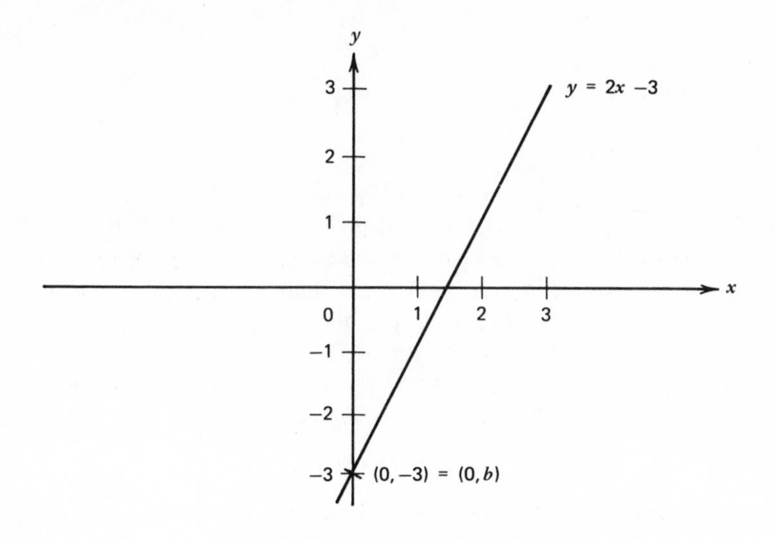

Figure 5.4 Slope-intercept form.

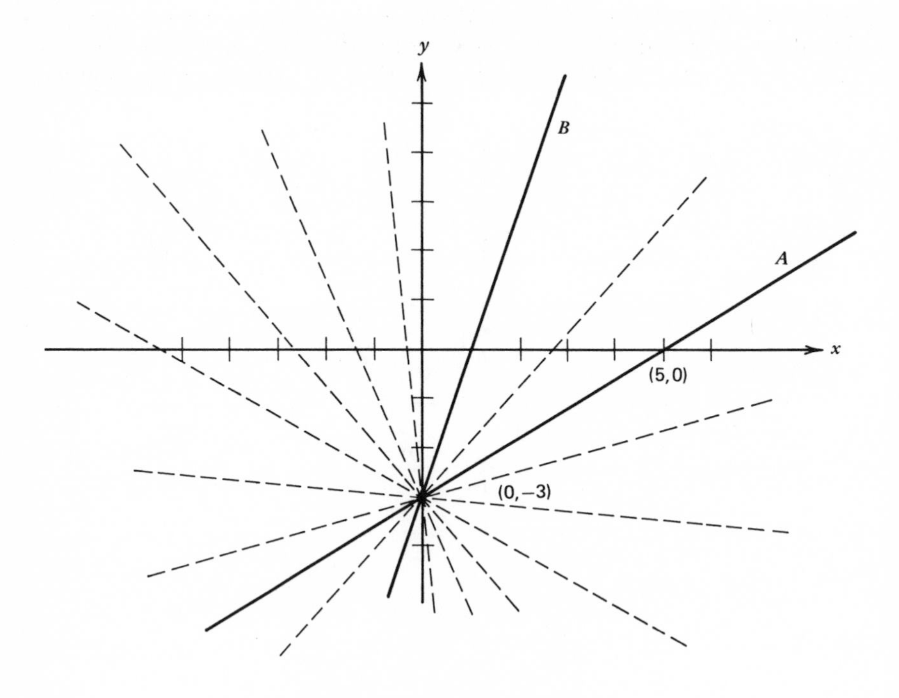

Figure 5.5 Family of lines.

the slope m or selecting another point in the plane. Two such lines are illustrated in Figure 5.5. The line A was selected by specifying the point (5,0) and the line B by specifying the slope $m = 3$. Hence, to construct the graph of the equation

$$y = mx + b$$

one need only locate one point of the line at the y intercept $(0,b)$ and draw a line through the y intercept having the slope $(y - b)/x = m$. For example, if we construct the graph of the line

$$y = -2x + 1$$

the y intercept is $b = +1$ and the slope is $m = -2$. Thus the line passes through the intercept $(0,b) = (0,1)$ with a slope equal to $m = -2$. To determine the line with slope m, we use the intercept point as the starting point and move x units horizontally and $y = mx$ units vertically to find a second point on the line. For example, from the point (0,1) we use the value $x = 1$ to find $y = mx = (-2) \cdot 1 = -2$. This gives us the new point $x = 1$ and $y = -1$, i.e., from (0,1) we move 1 unit in the x direction and -2 units in the y direction. These two points are used to graph the line illustrated in Figure 5.6

The y intercept is an easy concept to remember and it is easy to determine and locate in the x-y coordinate system. The slope is an easy concept to

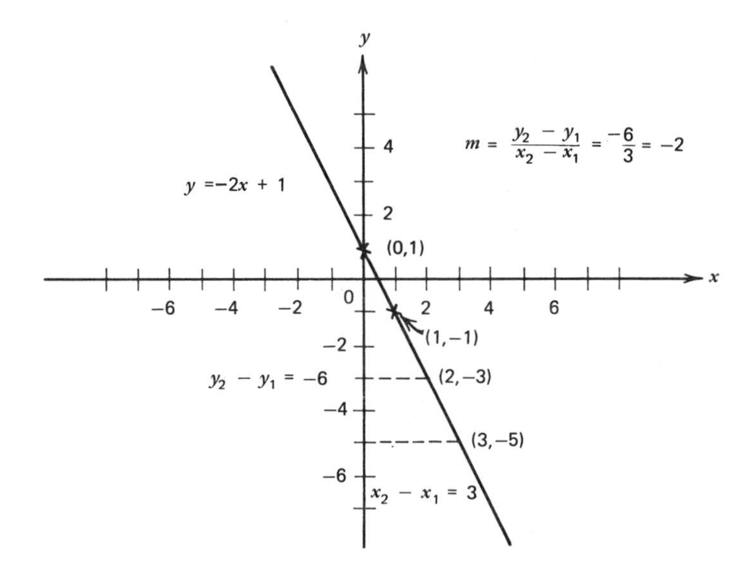

Figure 5.6 Slope-intercept form.

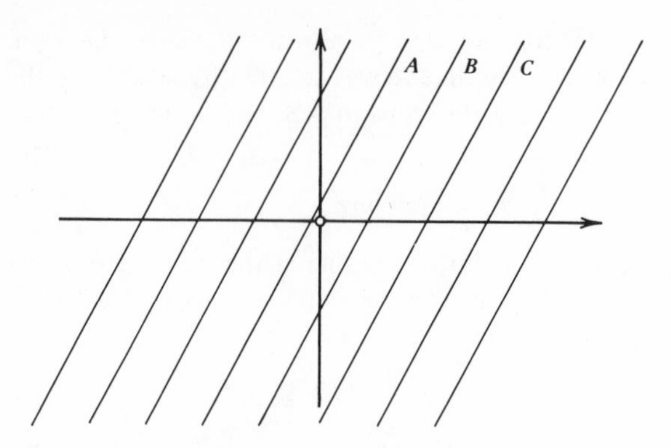

Figure 5.7 Family of lines.

remember in that all are familiar with the slope or incline of a hill. We speak of steep slopes and mild or small slopes. Skiers are especially familiar with slopes and their magnitudes since so much of their art of skiing depends on handling the slopes. However, to draw a line with a slope of a specified value requires a little figuring. If one remembers that the slope of a line is defined as the ratio of the vertical change (y) to the horizontal change (x) as a point moves along the line, then one should have little trouble in constructing lines with a specified slope. For example, in the equation

$$y = -2x + 1$$

the slope m is given as -2. This means that as x is varied k units, y will vary $-2k$ units along the line. Thus starting from any point on the line (preferably

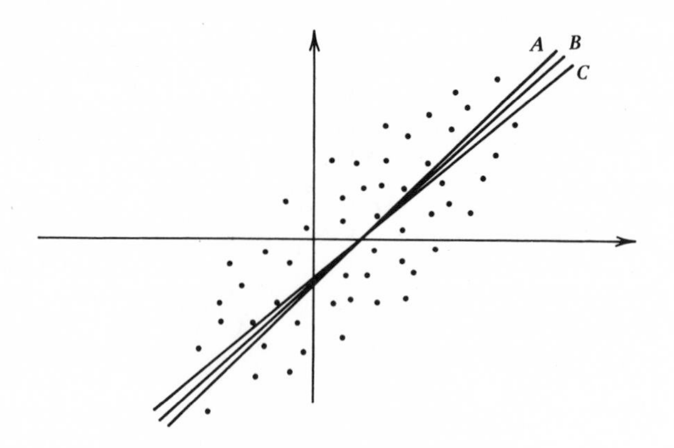

Figure 5.8 Curve fitting.

the y intercept), one will move 1 unit in x and -2 units in y to locate the second point. The two points determine the line with slope -2 and intercept (0,1).

One can construct a second family of lines with a prespecified slope. In this family of lines, the intercept b is not specified. One such family of lines is illustrated in Figure 5.7. This family of lines has slope $m = 2$. The y intercept for line A is $b = 0$, for line B is $b = -3$, and for line C is $b = -6$.

There are certain features about linear equations and their graphs that should be highlighted here. First, equations with positive slopes have lines which rise (values for y increase) as values of x increase. Values of y decrease for increasing values of x when the slope is negative ($m < 0$). Equations with positive intercepts ($b > 0$) have graphs which cross the y axis above the x axis. If the intercept is 0 ($b = 0$), the graph passes through the origin (0,0). Otherwise, the graph crosses the y axis below the x axis when the intercept is negative ($b < 0$).

5.7 ESTIMATING EQUATIONS FROM GRAPHS

In correlation and regression analysis an objective is to find the equation of the line (or graph) which "best" fits a set of points that are scattered in the plane.

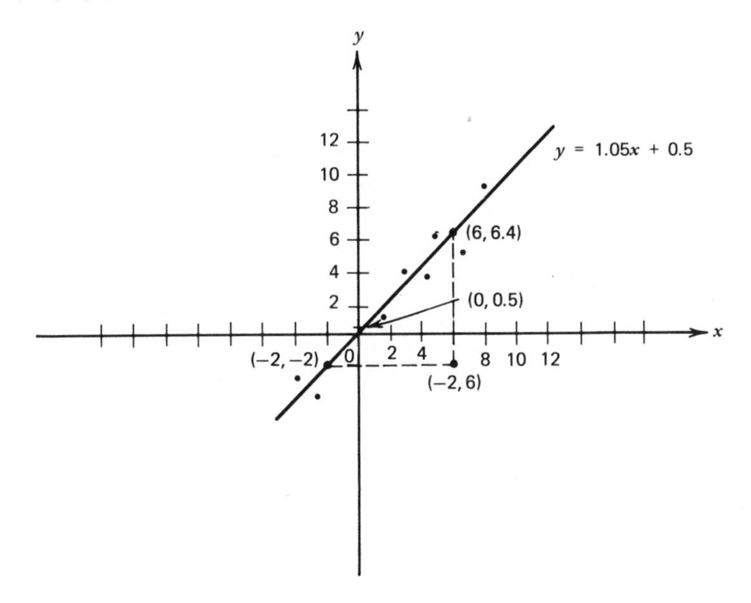

Figure 5.9 Curve fitting.

For example, Figure 5.8 shows the plot of a set of points. Three lines *A*, *B*, and *C* which appear to "fit" the set of points are superimposed on the set of points. The problem is to position a line which "best fits" the set of points and then estimate parameters for the equation of that line. We now show how one can "eye ball" the position of the line to best fit the set of points. From this line we shall show how one can estimate the y intercept and slope. We illustrate the estimating process by means of Example 5.2.

Example 5.2 (Line Fitting). We shall attempt to estimate the equation of a line which best fits the set of points (–4,–3), (–3,–4), (–2,–2), (–1,–2), (1,1), (2,2), (3,4), (4,4), (5,4), (5,5), (5,6), (7,5), (8,8), and (8,9). We use a transparent straight edge in attempting to fit a line to the set of points. We note that no straight line passes through all the points. The straight line that best fits the line is one that comes "closest" to all the points in the set. Using a transparent straight edge we try to eye ball the line through the points. Once the line is drawn, the problem reduces to estimating its slope and y intercept. The y intercept is easy to estimate. One need only determine where the line crosses the y axis. This value determines the value for b in the equation for the line,

$$y = mx + b$$

From the graph of the "estimating" line drawn in Figure 5.9, our estimate for b is $b = 0.5$. The next step is to estimate the value for the slope m of the line. To do this we start with a point on the line for which the coordinates (x_1, y_1) are known. If the line passes through one of the given points, that point is a prime candidate for selection. We have selected the point (–2,–2) which happens to be a point in the original set of points. We call this point $P_1(x_1, y_1)$. From this point we move horizontally to a point $P_2(x_2, y_1)$, where x_2 is selected so that $x_2 - x_1$ is a whole number (integer). Note that the point P_2 is selected to have the same y coordinate as the original point $P_1(x_1, y_1)$. We now move from the second point $P_2(x_2, y_1)$ vertically to the point $P_3(x_2, y_2)$ which lies on the line. The distance from P_3 to P_2 is $y_2 - y_1$. Recalling the definition of the slope of a line which previously was given as

$$m = \frac{y_2 - y_1}{x_2 - x_1}$$

we now have two points P_1 and P_3 on the line from which the slope m can be determined. From the graph we estimate

$$m = \frac{6.4 - (-2)}{6 - (-2)} = \frac{8.4}{8} = 1.05$$

Having estimated the value of the slope m and the y intercept b, we can now write the equation of the line which best fits the set of points. The equation is,

$$y = mx + b$$
$$y = 1.05x + 0.5$$

■■

The estimating procedures described above result in a simple and reasonably accurate estimate of the equation of the line which best fits the set of points. The slope of this line, as we shall see in the sequel, has a special significance in correlation and regression analysis.

Example 5.3 (Equations for Lines)

1. What is the equation for the line which passes through the points (1,5) and (2,8)?

In this example, two points on the line are given. The slope m is found by the formula

$$m = \frac{y_2 - y_1}{x_2 - x_1} = \frac{8 - 5}{2 - 1} = 3$$

Thus the equation of the line is

$$y = 3x + b$$

We need to determine b. To do this, substitute the coordinate values of either point in the equation. For the point (1,5)

$$5 = 3 \cdot 1 + b \qquad \text{or} \qquad b = 2$$

For the point (2,8)

$$8 = 3 \cdot 2 + b \qquad \text{or} \qquad b = 2$$

For any point on the line we would obtain the value $b = 2$. Thus the equation of the line passing through the two points (1,5) and (2,8) is

$$y = 3x + 2$$

2. What is the equation of the line with y intercept $b = 3$ and passes through the point (–2,6)?

In this example one point on the line and the y intercept are given. Our knowledge about the line allows us to write

$$y = mx + 3$$

If we substitute the coordinates of the point (–2,6) into the equation, we can find the value for m. Doing this yields

$$6 = m(-2) + 3$$

or

$$m = \frac{6-3}{-2} = -\frac{3}{2}$$

Thus the equation for the line is,

$$y = -\frac{3}{2}x + 3$$ ■■

Exercise 5.4. Estimate the slope and y intercept for the line which best fits the set of points (–2,–6), (–1,–1), (0,2), (1,6), and (2,8). Write the equation for the line.■■

Exercise 5.5. Write the equation for the line which has a slope of 3 and a y intercept $b = 2$.■■

Exercise 5.6. Write the equation for the line passing through the two points (–2,3) and (3,1).■■

Exercise 5.7. Write the equation for the line passing through the point (–2,3) and having a slope equal to –3.■■

5.8 SCATTER DIAGRAMS–UNGROUPED DATA

Regression and correlation analysis is a method of finding functional relationships among two or more variables. When more than two variables are involved the analysis is called multiregression or multivariate analysis. In the discussions to follow we shall restrict our analysis to two variables. This is called bivariate analysis. In regression and correlation analysis we seek to identify relations between lists or tables of values for two variables. The variables may or may not be causally related. The objective is to determine whether a relation between the variables exists and to use it to determine how variations in one of the variables affect the value (or distribution of values) in the remaining variable.

The starting point in a correlation or regression analysis is the assemblage of a table or list of values assumed by the variables. For example, one can consider U.S.S.R. iron production for 10 yr as one variable (x) and U.S.S.R. automobile production (passenger cars) as the second variable (y). The values for the two variables are matched on a year-by-year basis for 10 yr and listed as follows:

Iron production	31	47	54	60	66	71	76	81	85	90
Automobile production	77	124	139	148	165	173	185	201	230	251

The lists (or table) contains ten ordered pairs which can be thought of as coordinates of points in the x-y coordinate plane. By plotting the points we obtain a graphic assist to determining quickly whether there is any relationship between the two variables. We would like to know whether increases in iron production are accompanied by comparable increases in automobile production. Naturally, the converse relationship is also of interest. On plotting the ten points we obtain what is called a *scatter* diagram for the two variables. The scatter diagram for the two variables is given in Figure 5.10.

Figure 5.10 is but a simple example of a scatter diagram. For each value of x there is only one value for y and the listing contains distinct values for x. In a more general situation there may be several values of y for each value of x and vice versa. For example, consider the 20 data points (obtained by an industry association) which relate inventory levels against price levels for a product. The data points were taken at the end of 20 consecutive fiscal quarters. The inventory and price level are normalized to the levels which are considered normal for the industry (where 100 = normal level).

Inventory	Price	Inventory	Price
100	100	101	97
104	102	95	95
106	104	95	98
107	100	100	101
105	98	101	98
101	98	98	100
96	101	101	104
98	106	103	101
103	106	100	97
103	100	97	99

Note that in this listing the same inventory level at different instances assumes different price levels. Also, the same price levels, over time, take on different inventory levels. The scatter diagram for these data are given in Figure 5.11.

It is not easy to determine visually, with accuracy, the functional relation between inventory level and price level. There appears to be a slight increase in price level with inventory level. If this is the case, we appear to have an economic situation for a product which does not follow the economic theory which states

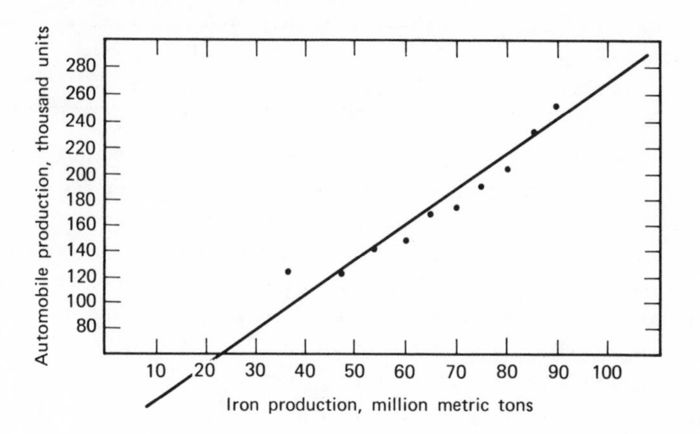

Figure 5.10 Iron versus automobile production.

that the price should fall as inventories for that product increase. However, other factors such as a time lag between inventory level and price change and critical inventory levels enter into the picture and must be considered before making any final conclusions. As an example and exercise, the reader should plot the scatter diagram for the inventory-price levels wherein the price level for the second quarter is paired with the inventory level of the first period. The one fiscal quarter time differential is retained in obtaining the 18 additional points for the scatter diagram. The result will be a scatter diagram from which

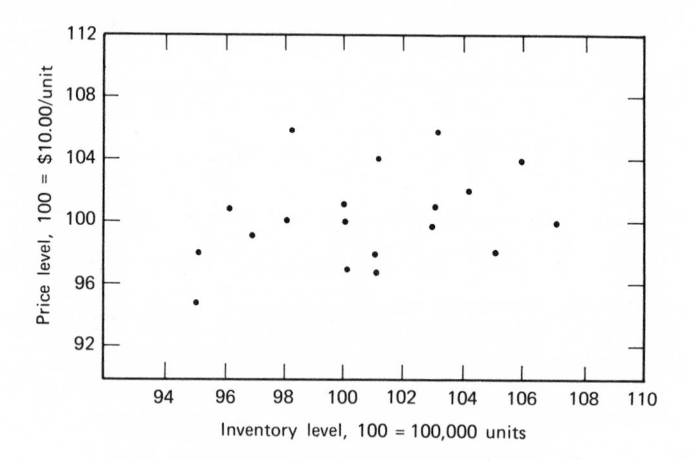

Figure 5.11 Price versus inventory levels.

one can estimate whether a time delay of one quarter results in a decrease in price levels with previous increases in inventories (supply). One might also try a two quarter delay between inventories and price. And if one is really interested, one should try a one fiscal quarter delay in inventory changes relative to price. All yield some interesting scatter diagrams when compared to Figure 5.11.

5.9 INFORMATION FROM SCATTER DIAGRAMS

One can ask "What information is contained in a scatter diagram?" The answer depends on the positional relationships among the points of the scatter diagram. If the points are well scattered and have little order, then one can say little about the relationship (if any exists) between the two variables x and y. Figure 5.12 illustrates one scatter diagram in which the variables have little or no relationship.

When there is a semblence of order in the position of the points to each other in the scatter diagram, then information on one variable x contains some information on the variable y (and vice versa). For example, in Figure 5.13, it is indicated that for increasing values of x, the variable y also increases. The amount of increase differs in the three diagrams. The degree of orderliness between x and y, as exhibited in the scatter diagram, is a measure of the co-relation (or correlation) of x and y. When the correlation between x and y is high, then information on x yields a considerable amount of information on y. Correlation and regression analysis is concerned with measuring how much information about y can be obtained from information on x, and the confidence one can place in this derived information.

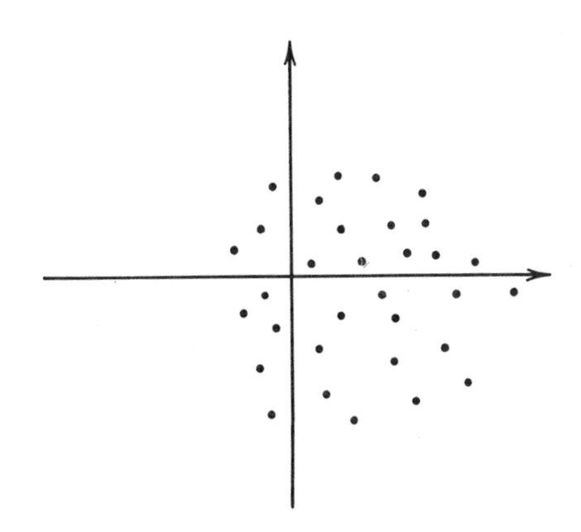

Figure 5.12 Random scatter.

5.10 SCATTER DIAGRAMS–GROUPED DATA

When dealing with a listing of paired variables (x,y) we can consider each variable independent of the other and construct frequency tables and distributions for each using the methods described previously. However, we can combine the two frequency tables into a single two-dimensional frequency table which obviates the construction of the individual frequency distributions and serves as the scatter diagram for the two variables. We now illustrate how the scatter diagram for grouped data is assembled.

The intervals selected for recording the data in the frequency tables for x and y are plotted as intervals on the x and y axis, respectively, on the x-y coordinate plane. Vertical lines are constructed at x-interval end points and horizontal lines are constructed from the y-interval end points. The result is a network of cells formed at the intersection of x-intervals and y-intervals.

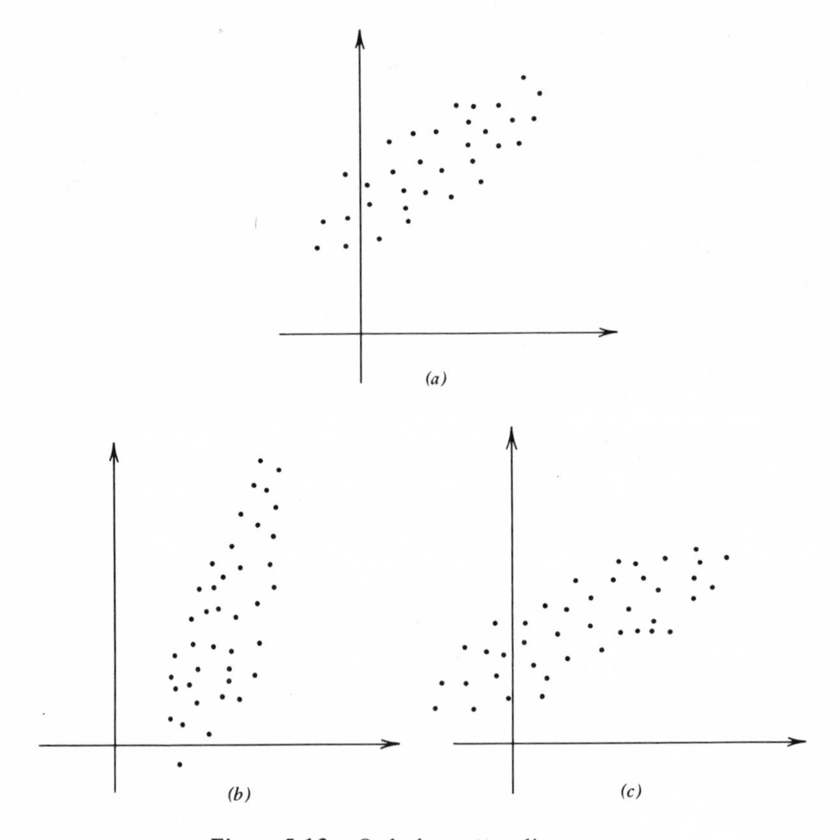

Figure 5.13 Orderly scatter diagrams.

Figure 5.14 depicts one set of cells wherein the x and y data are to be assembled into classes or intervals of two units in length. The intervals for each variable are selected independently and hence can be of different lengths. Each interval on the y axis identifies a *row.* Each interval on the x axis identifies a *column.* The intersection of each row and column forms a *cell.* Each cell is identified by the column interval (x axis) and the row interval (y axis) from which it is formed.

Figure 5.14 contains the scatter diagram for the price-inventory levels given in Section 5.8. The first point (100,100) is tabulated in the cell whose x interval is 100-101, and whose y interval is 100-101. The cell is four over and four up from the left-hand corner of the diagram. The second point (104,102) is located in the cell six over and five up. Locating points in the scatter diagram is very much like locating points in the x-y coordinate plane. In the scatter diagram, intervals instead of points are used to locate the x and y values of the ordered pairs.

At the top of the scatter diagram is a row which contains the sum of the frequency counts in the columns below it. This row and the bottom row is the

fx	2	2	2	7	3	2	2	0	20
108-109									0
106-107			1		1				2
104-105				1			1		2
102-103						1			1
100-101		1	1	11	11		1		7
98-99	1	1		11		1			5
96-97				11					2
94-95	1								1
	94-95	96-97	98-99	100-101	102-103	104-105	106-107	108-109	*fy*

Figure 5.14 Scatter diagram, grouped data, of x intervals (inventory)

frequency table for the x variable with the intervals defined in the bottom row. To the right of the scatter diagram is a column which contains the sum of frequency counts in the rows to the left of it. This column and the leftmost column is the frequency table for the y variable with intervals defined in the leftmost column. The scatter diagram, as depicted in Figure 5.14, contains the frequency tables for the two variables x and y, the total frequency count (upper right-hand corner), and the scatter diagram (two-dimensional distribution) for the two variables x and y. The clustering and orderliness of the distribution of the points in the scatter diagram is a measure of the correlation between the two variables x and y.

5.11 LINES OF REGRESSION

In Section 5.7 it was shown how to eye ball a straight line to fit the points of a scatter diagram. From this line an estimate of the y intercept and slope for the line could be readily made. The line selected to best fit the points of the scatter diagram is called a *line of regression* for the data set. We will now describe an analytical method for determining the line of regression for any data set on two variables x, y, and their associated scatter diagram.

The method used to fit a line to the set of points (x_1,y_1), (x_2,y_2), ..., (x_n,y_n) is called the *method of least squares.* We will not show how the method is derived. We will describe only what it accomplishes and how we can use the results to determine lines of regression. First, the least squares method is the analytic procedure used to find the parameters for the line which best fits the set of points of the scatter diagram in a least square sense. If all the points on the scatter diagram are collinear (lie on a straight line), then that line is the regression line and there is no need for the method of least squares. In most

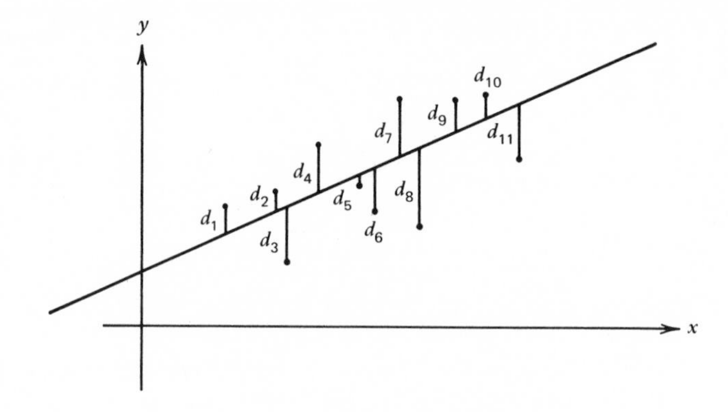

Figure 5.15 Vertical deviations.

instances, most points of the scatter diagram will not lie on any line drawn on the scatter diagram. One could draw many different lines and all could look like good fits to the scatter diagram. What the method of least squares does is determine that line for which the sum of the squares of vertical distances from all points to the line is the least. These vertical distances (or deviations) are depicted in Figure 5.15. The line shown there minimized the sum $D = d_1^2 + d_2^2 + \cdots + d_{11}^2$. Any other line drawn through the points would result in a higher value for this sum. For pedogogical reasons we will next discuss correlation and then return to complete our discussion on regression analysis.

5.12 MEASURES OF CORRELATION

In the sections to follow we shall illustrate two methods for determining the degree of correlation between two variables. First, we will illustrate a graphic technique which uses some simple counting formulas and a partitioning of the scatter diagram. This method is quick, but it is not as accurate (efficient) as the quantitative method that is most widely used. However, for substantive analysis, where the data lacks a high degree of accuracy or credibility, the graphic method will suffice for most purposes. Next, we will illustrate the quantitative method which is systematic and easily performed, using existing data-processing techniques. It is suggested that substantive analysts become proficient in the graphic technique in order to arrive at "quick and dirty" estimates and then, if the situation calls for it, employ the quantitative method to obtain more efficient estimates for the correlation between the variables under analysis.

Graphic Estimate for Correlation

In attempting to determine the relationship (if any) which exists between two variables x, y, we normally have available, or are able to obtain, a set of measurements or observations on the two variables. The measurement set (also called a population or sample) is labeled (x_1,y_1), (x_2,y_2), (x_3,y_3), ..., (x_n,y_n). The first step in determining the correlation between x and y is to construct a scatter diagram. A typical scatter diagram is given in Figure 5.16. The second step is to compute the means $\bar{x}$ and $\bar{y}$ for the samples plotted in the scatter diagram.

$$\bar{x} = \frac{1}{n} \sum_{i=1}^{n} x_i$$

$$\bar{y} = \frac{1}{n} \sum_{i=1}^{n} y_i$$

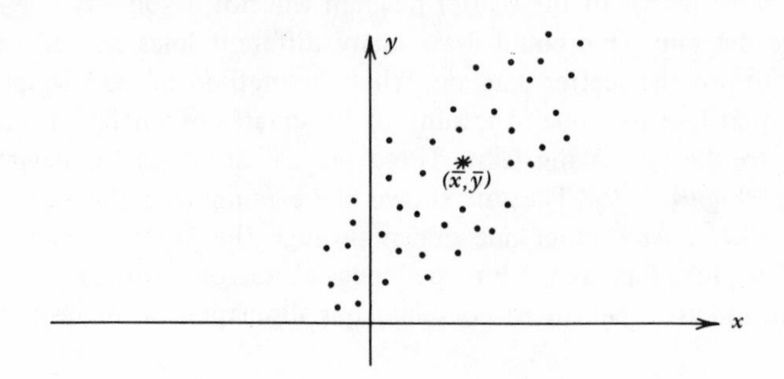

Figure 5.16 Scatter diagram.

The point $(\bar{x}, \bar{y})$ whose coordinates are the means of the x and y data is located on the scatter diagram as in Figure 5.16.

The third step is to draw a new set of coordinate axes with origin at the point $(\bar{x}, \bar{y})$. The scatter diagram with the new coordinate system is shown in Figure 5.17

The fourth step is to label the four quadrants of the new coordinate plane I, II, III, and IV, starting from the upper right-hand quadrant and proceeding counterclockwise around the new origin $(\bar{x}, \bar{y})$.

The fifth step is to count the number of points falling in each quadrant. Label these counts n_1, n_2, n_3, and n_4 as shown in Figure 5.17. If a point falls on an axis, count one-half for each of the adjoining quadrants.

The sixth step is to make two simple computations given by the formulas

$$r(+) = \frac{n_1 + n_3}{n}$$

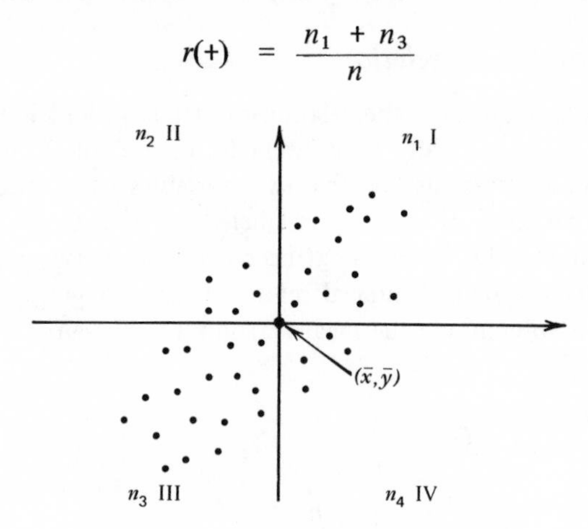

Figure 5.17 Scatter diagram counts.

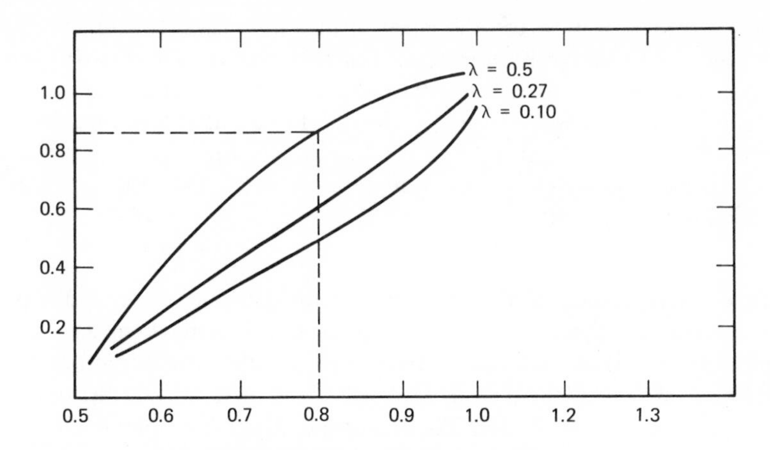

Figure 5.18 Estimating chart.

$$r(-) = \frac{n_2 + n_4}{n}$$

The quantity $r(+)$ is the ratio of the number of points falling in first and third quadrants to the total number of points $n_1 + n_2 + n_3 + n_4 = n$. The quantity $r(-)$ gives the ratio for the second and fourth quadrants. Intuitively and geometrically, if the quantity $r(+)$ is larger than $r(-)$, then the correlation is positive; otherwise it is negative.

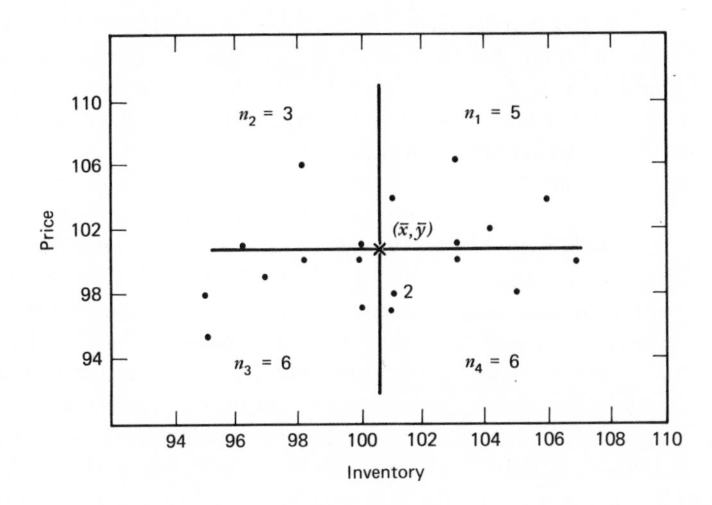

Figure 5.19

The seventh step is to use the larger of the two quantities $r(+)$ and $r(-)$ to enter Figure 5.18 at the bottom scale. For this example say that $r(+) > r(-)$. Say also that $r(+) = 0.8$. One enters the bottom scale at 0.8 and proceeds vertically to the curve labeled $\lambda = 0.5$. From the curve one proceeds horizontally to the vertical scale at the left to read the degree of correlation between the two variables. In this example $r = 0.83$. The value of r is called the *correlation coefficient.*

Example 5.4 (Graphical Method). Consider the data assembled for the price-inventory level example of Section 5.8. The scatter diagram is repeated here in Figure 5.19 for ready reference. The means of the inventory-price (x, y) variables are $\bar{x} = 100.7$ and $\bar{y} = 100.25$. A new set of axes is drawn through the point (100.7, 100.25). A numeric 2 is placed besides the point (101,98) since it occurred twice in the listing and is to be counted twice in the quadrant counts. The quadrant counts are

$$n_1 = 5, \quad n_2 = 3, \quad n_3 = 6, \quad \text{and} \quad n_4 = 6$$

Hence

$$r(+) = \frac{n_1 + n_3}{n} = \frac{5 + 6}{20} = \frac{11}{20} = 0.55$$

$$r(-) = \frac{n_2 + n_4}{n} = \frac{3 + 6}{20} = \frac{9}{20} = 0.45$$

Since $r(+) > r(-)$, the correlation is positive (as previously estimated). We use $r(+) = 0.55$ to enter the bottom scale, proceed to the graph labeled $\lambda = 0.5$, and then to the vertical scale to read the correlation coefficient as 0.2. While this number is positive, it is not large enough to say with confidence that the two quantities are linearly related.

One can obtain an improved estimate on the correlation if one uses the following method for obtaining the values of n_1, n_2, n_3, and n_4: First determine the number $m = 0.27n$ where n is the number of points in the scatter diagram. Next, starting from the leftmost point of the scatter diagram and using a straight edge held vertically on the scatter diagram, move to the right and expose points until the count includes m points (= 27% of n). Draw a *vertical* line separating these points from the rest. Next, starting from the rightmost point of the scatter diagram count over m(= 27% of n) points to the left and draw another *vertical* line separating these points from the remainder of the scatter diagram. Discard all points *between* the two vertical lines. Draw a *horizontal* line which separates the remaining $2m = 54\%$ of n points into two equal halves. Label the remaining sectors as in Figure 5.20. Estimate $r(+)$ and

$r(-)$ as before. In using the data in Figure 5.18, the line $\lambda = 0.27$ will replace the line $\lambda = 0.50$. The value read from the chart for the correlation is a better estimate than that obtained in using all the points of this scatter diagram and $\lambda = 0.50$.■■

Exercise 5.8 Use the data on iron production versus automobile production of Section 5.8 and determine the correlation between the two variables by the graphic method explained above using the $\lambda = 0.5$ and the $\lambda = 0.27$ curves.■■

Quantitative Measure of Correlation

The graphic method for measuring the correlation between two variables illustrates those characteristics of a scatter diagram which enter into determining the degree of correlation between two variables. Briefly, these are

1. There must be an upward or downward trend (but not both) in the position of the points relative to their center $\bar{x}, \bar{y}$.
2. The points should remain in a channel along the direction of the trend. The smaller the channel, the higher the correlation. Hence correlation, in a sense, measures how close the points of the scatter diagram come to lying on a straight line.

The graphic method should be employed where the points of the scatter diagram are distributed reasonably uniformly along the line which appears to fit them. Specifically, on using the graphic method for determining the correlation coefficient with scatter diagrams which are skew symmetric and have few points near the center $(\bar{x}, \bar{y})$, as illustrated in Figure 5.21, the graphic

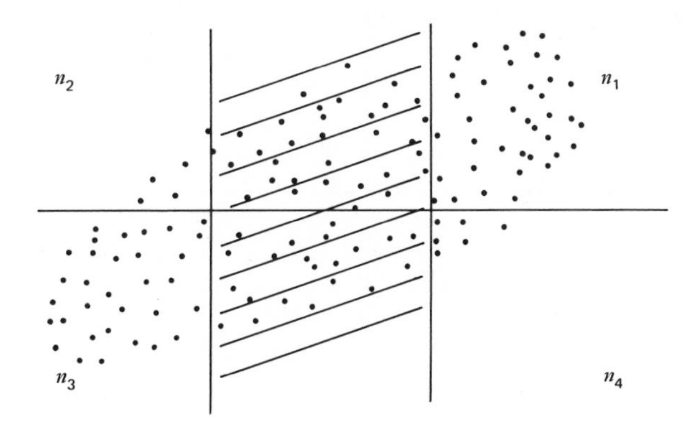

Figure 5.20

methods will generally yield higher coefficients of correlation than the analytic method. With well-distributed scatter diagrams, the opposite is usually true.

The graphic method for estimating correlations should be used with the knowledge that errors of 10% or more are possible. If the results are to be used in further important analyses or decisions, then it is strongly recommended that the analytic method for measuring correlation to be described in the next section, be used to verify any results obtained graphically.

Analytic Estimate for Correlation–Method I

Many of the steps used in the graphic method of estimating correlation are also used in the analytic method. First, the coordinate axes are centered at the point $(\bar{x}, \bar{y})$ which identifies the means of the sets $x_1, x_2, \ldots, x_n$ and $y_1, y_2, \ldots, y_n$, respectively. In the analytic method one additional operation is performed. The variables x and y are scaled to standard units which are, as previously explained, obtained by the formulas

$$u_i = \frac{x_i - \bar{x}}{\sigma_x} \qquad v_i = \frac{y_i - \bar{y}}{\sigma_y}$$

Thus the standard deviations for the measurements x_i and y_i must be computed in order to express the measurements x_i, y_i in standard units. On plotting the new variables u_i and v_i we obtain a scatter diagram which has its center at the point $\bar{x}$, $\bar{y}$ and has its coordinate axis rescaled to standard units. As in the graphic measure of correlation, points of the scatter diagram are now centered at the origin $(\bar{x}, \bar{y})$ and distributed in the four quadrants in the same manner as in the graphic method. If one so choses, one can apply the graphic method at

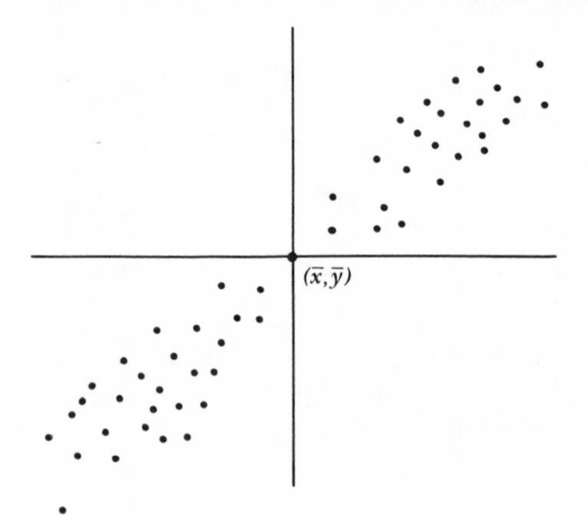

Figure 5.21 Skew symmetric scatter.

this point to obtain an estimate for the correlation between the two variables. Before proceeding with the remainder of the analytic method we will use the example on inventory versus price to illustrate the steps taken so far and show that graphically the two scatter diagrams are equivalent except for scale. To compute the standard deviation for the x and y variables and to convert each measurement to standard units, we construct and complete Table 5.1.

The scatter diagram for the 20 points expressed in standard units is given in Figure 5.22. Note that all points lie within 3σ units of the origin. This is due to dividing the deviations $x_i - \bar{x}$ and $y_i - \bar{y}$ by σ_x and σ_y, respectively, to express the variables in standard units and thus have more than 99% of the points within 3 standard deviations of their means $\bar{x}$ and $\bar{y}$.

Figure 5.22 contains the quadrant point counts n_1, n_2, n_3, and n_4. Note that these counts are identical with those obtained with the scatter diagram for x, y. Thus the translation of the origin to $(\bar{x}, \bar{y})$ and changing the scale to standard units had no effect on the results of the computation for the correlation coefficient using the graphic technique. This is true, in general, for both the graphic and analytic method. Invariance under transforms is desirable for any computation, and in particular, for the calculation of the correlation coefficient.

We have noted previously that the correlation coefficient for this scatter diagram was positive and equal to 0.2. This states that the majority of the points are located in the first and third quadrants. It should be noted, although it is not

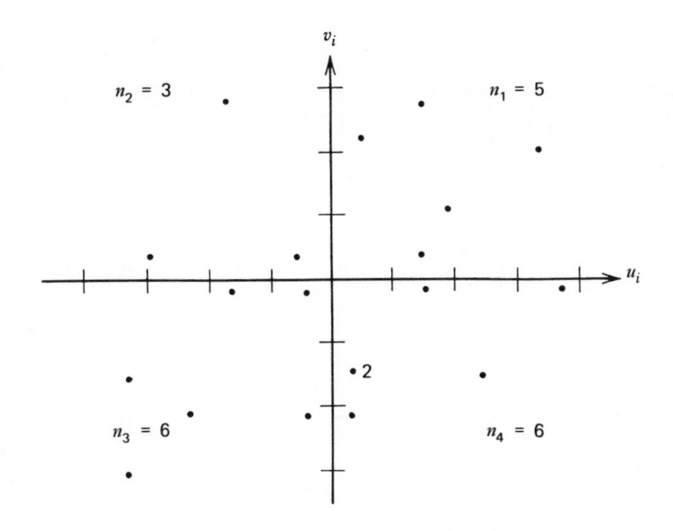

Figure 5.22 Scatter diagram in standard units.

Table 5.1 Converting to standard units.

Inventory, x_i	Price, y_i	$x_i - \bar{x}$	$y_i - \bar{y}$	$(x_i - \bar{x})^2$	$(y_i - \bar{y})^2$	$u_i = \frac{x_i - \bar{x}}{\sigma_x}$	$v_i = \frac{y_i - \bar{y}}{\sigma_y}$	$(x_i - \bar{x})(y_i - \bar{y})$
100	100	−0.7	−0.25	0.49	0.06	0.206	−0.0833	0.175
104	102	3.3	1.75	10.89	3.06	0.973	0.583	5.755
106	104	5.3	3.75	28.09	14.06	1.56	1.25	19.875
107	100	6.3	−0.25	39.69	0.06	1.86	−0.0833	−1.575
105	98	4.3	−2.25	18.49	5.06	1.27	−0.750	−9.675
101	98	0.3	−2.25	0.09	5.06	0.0884	−0.750	−0.675
96	101	−4.7	0.75	22.09	0.56	−1.39	0.250	−3.525
98	106	−2.7	5.75	7.29	33.06	−0.796	1.92	−15.525
103	106	2.3	5.75	5.29	33.06	+0.678	1.92	13.225
103	100	2.3	−0.25	5.29	0.06	+0.678	−0.0833	−0.575
101	97	0.3	−3.25	0.09	10.56	0.0884	−1.08	−0.975
95	95	−5.7	−5.25	32.49	27.56	−1.68	−1.75	29.925
95	98	−5.7	−2.25	32.49	5.06	−1.68	−0.750	12.825
100	101	−0.7	0.75	0.49	0.56	−0.206	0.250	−0.525
101	98	0.3	−2.25	0.09	5.06	0.0884	−0.750	−0.675
98	100	−2.7	−0.25	7.29	0.06	−0.796	−0.0833	0.675
101	104	0.3	3.75	0.09	14.06	0.0884	1.25	1.125
103	101	2.3	0.75	5.29	0.56	0.678	0.250	1.725
100	97	−0.7	−3.25	0.49	10.56	−0.206	−1.08	2.275
97	97	−3.7	−3.25	13.69	10.56	−1.09	−1.08	12.025

Computations for Table 5.1

$\bar{x} = \frac{1}{20} \Sigma x_i = 100.7$	$\bar{y} = \frac{1}{20} \Sigma y_i = 100.25$
$\sigma_x^2 = \frac{1}{20} \sum_{i=1}^{20} (x_i - \bar{x})^2 = \frac{230.20}{20}$	$\sigma_y^2 = \frac{1}{20} \sum_{i=1}^{20} (y_i - \bar{y})^2 = \frac{178.70}{20}$
$\sigma_x^2 = 11.51 \qquad \sigma_x = 3.39$	$\sigma_y^2 = 8.93 \qquad \sigma_y = 2.99$

obvious from Figure 5.22, that the points in the first and third quadrants will also have coordinates u_i, v_i which generally are larger than those in the second and fourth quadrants. The coordinates u_i, v_i are both positive in the first quadrant and negative in the third quadrant. Thus their product $u_i \cdot v_i$ is positive in the first and third quadrants. The coordinates u_i, v_i are of opposite sign in the second and fourth quadrants. Thus their product $u_i \cdot v_i$ is negative in the second and fourth quadrants. Hence, if the points (u_1, v_1), (u_2, v_2), ..., (u_n, v_n) are predominantly in the first and third quadrants and have relatively larger coordinate values there than those in the second and fourth quadrants, the sum of the coordinate products $u_i \cdot v_i$ will have more positive terms of larger values than the negative terms from the second and fourth quadrants resulting in the sum

$$\sum_{i=1}^{n} u_i v_i$$

being positive. Hence large positive values will tend to indicate a lower left to upper right trend in the scatter diagram. However, the value of the sum increases with an increasing number of points having essentially the same relative scatter. To account for the dependence of the sum on the number of terms in the sum, the sum is divided by n, the number of terms in the sum. This yields an average value for the products $u_i \cdot v_i$ and results in a useful measure of the correlation existing between the two variables x and y. The sum

$$r = \frac{1}{n} \sum_{i=1}^{n} u_i v_i$$

is called the *correlation coefficient* and the symbol r is used to denote its value. When u_i and v_i are expressed in terms of the original values x_i, y_i, their means $\bar{x}$, $\bar{y}$, and their standard deviations σ_x, σ_y, we have the following formula:

$$r = \frac{\sum_{i=1}^{n} (x_i - \bar{x})(y_i - \bar{y})}{n\, \sigma_x\, \sigma_y}$$

We can now use the quantities listed in Table 5.1 to compute the correlation coefficient r. The first step is to multiply the quantities $(x_i - \bar{x})$ and $(y_i - \bar{y})$ in columns 3 and 4 to obtain a new column headed $(x_i - \bar{x})(y_i - \bar{y})$. The new column is summed to obtain the numerator for the correlation coefficient r. This value is divided by the product $n\, \sigma_x\, \sigma_y$ where the quantities n, σ_x, and σ_y are given at the bottom of Table 5.1.

In Table 5.1 the sum of the new column yields 66. The denominator is $20 \times 3.39 \times 2.99 = 202.80$. Thus, $r = 66/202.80 = 0.325$. This value for r differs from the one obtained by the graphic method and is the more accurate of the two estimates.

Exercise 5.9. For the data on iron production v. automobile production of Section 5.8, determine the correlation coefficient using the analytic method. Compare the results with those obtained in Exercise 5.8.■■

If one is interested in calculating the correlation coefficient directly from the ungrouped original data (x,y), the formula for the correlation coefficient

$$r = \frac{1}{n} \Sigma\, u_i\, v_i$$

can be rewritten in terms of the original data. The formula reduces to

$$r = \frac{n \sum_{i=1}^{n} x_i y_i - \sum_{i=1}^{n} x_i \sum_{i=1}^{n} y_i}{\sqrt{\left[n \sum_{i=1}^{n} x_i^2 - \left(\sum_{i=1}^{n} x_i\right)^2\right] \left[n \sum_{i=1}^{n} y_i^2 - \left(\sum_{i=1}^{n} y_i\right)^2\right]}}$$

While this formula looks complicated, it is well-suited for calculation by manual means or a desk calculator. The calculations are readily performed with less effort and fewer errors if the computations indicated in Table 5.1 are performed in the following order:

1. Complete products for columns 3, 4, and 5.
2. Sum columns 1, 2, 3, 4, and 5.
3. Square sums of columns 1 and 2.
4. Multiply sums of columns 3, 4, and 5 by n.
5. Product the sums of columns 1 and 2.
6. Substitute results in formula for r and perform the indicated operations.

The result is the correlation coefficient r. The computational procedure is illustrated in Example 5.5.

Example 5.5 (Analytic Method). Determine the correlation coefficient for the following pairs of values for the two variables x, y. Use the original data to plot the scatter diagram, and estimate the correlation coefficient using the graphic-analytic technique. (Data points in columns 2, 3 of Table 5.2.)■■

Solution: Table 5.2 is constructed to order and record the necessary computations. The data are entered into columns headed x_i and y_i. The columns headed $x_i \cdot y_i$, x_i^2, and y_i^2 are then completed by performing the indicated operations. The row titled sum is completed next by summing the entries in each of the five columns. The row titled products is completed next by squaring the sums of the first two columns and multiplying the sums of the next 3 columns by n = 14. The next row labels the results of the above calculations ④, ⑥, ①, ③, and ⑤. It also includes another product involving the sums of the first two columns. This product is labeled ②. The six quantities are then substituted into the formula in the row titled computations and the indicated operations are performed. The result is a value r = 0.68 for the correlation coefficient.

In performing the calculation manually, a table of squares and square roots provide a worthwhile assist both in reducing errors and effort. Naturally, if the number of data points is large, every effort should be expended to obtain access to a desk calculator or data-processing facilities, especially if the calculations are to be repeated for several data sets over a period of time.

We now use the graphic method to calculate the correlation coefficient for the data. Using the scatter diagram of Figure 5.23 and finding the new origin $\bar{x}$ = 2.64, $\bar{y}$ = 9.0, we construct the new coordinate axes and determine that

$$n_1 = 5.5 \qquad n_2 = 1 \qquad n_3 = 5 \qquad n_4 = 2.5$$

$$r(+) = \frac{n_1 + n_3}{n} = \frac{5.5 + 5}{14} = 0.75$$

$$r(-) = \frac{n_2 + n_4}{n} = \frac{1 + 2.5}{14} = 0.25$$

Using $r(+)$ = 0.75 to enter Figure 5.18 we obtain the value r = 0.72 for the correlation coefficient. This does not differ significantly from the value r = 0.68 obtained using the analytic method.

Table 5.2

i	x_i	y_i	$x_i y_i$	x_i^2	y_i^2
1	3	10	30	9	100
2	2	8	16	4	64
3	1	8	8	1	64
4	2	11	22	4	121
5	4	12	48	16	144
6	3	12	36	9	144
7	3	10	30	9	100
8	5	13	65	25	169
9	1	6	6	1	36
10	2	5	10	4	25
11	3	8	24	9	64
12	1	7	7	1	49
13	3	9	27	9	81
$N = 14$	4	8	32	16	64
Sum	37	127	361	117	1,225
Products	$37^2 =$ 1,369	$127^2 =$ 16,129	$14 \cdot 361 =$ 5,054	$14 \cdot 117 =$ 1,638	$14 \cdot 1{,}225 =$ 17,150
Formula terms	④	⑥	①	③	⑤

② $(\Sigma x_i)(\Sigma y_i) = 4{,}699$

Computations:

$$r = \frac{①-②}{\sqrt{[③-④][⑤-⑥]}}$$

$$= \frac{5054 - 4699}{\sqrt{(1638 - 1369)(17150 - 16129)}}$$

$$= \frac{355}{(16.3)(32.6)}$$

$$= 0.68$$

Table 5.3 GNP-expenditures-manpower correlation.

Country	GNP, billions	Defense expenditures, millions	Armed forces, thousands
Belgium	16.7	520	107
France	93.5	4,215	535
Germany (W)	112.2	4,607	440
Italy	56.7	1,908	376
Netherlands	19.0	746	129
Denmark	10.0	283	50
Norway	7.0	270	34
Portugal	3.7	230	148
Sweden	19.3	800	82
United Kingdom	99.2	5,937	437
Switzerland	13.9	352	31

Example 5.6 (Military-Economic Correlation). Military analysts and foreign affairs specialists are interested in determining relationships which might exist between economic and military strengths of nations. In this example we investigate whether defense expenditures correlate with the GNP for the 11 nations listed in Table 5.3. On the surface, one would expect that these two variables would be highly correlated. We also investigate whether the size of

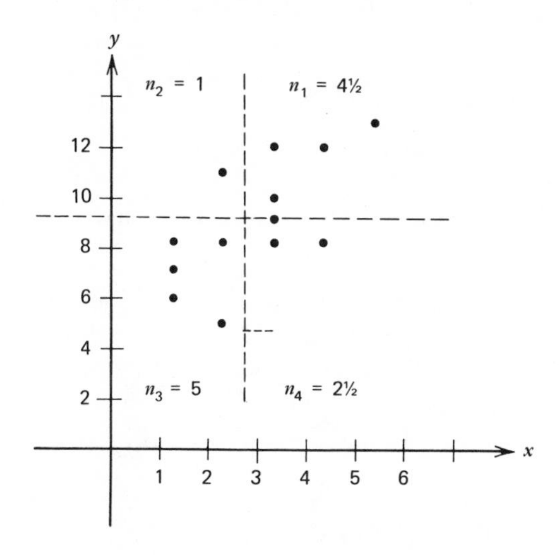

Figure 5.23 Scatter diagram.

the armed forces of these nations correlates with defense expenditures. A high correlation might indicate that military expenditures are concentrated in maintaining and building up manpower levels. A low correlation might indicate that military expenditures are concentrated in equipment and weapons. In either case, additional analysis on manpower and equipment structures of the armed forces would be indicated.

The computation for the correlation coefficient follows the procedure used in computation Table 5.2. The results are entered in Table 5.4 which contains all quantities needed in the computation except the products of the sums given at the bottom of columns 2 and 3, and 3 and 4. These are

$$(\Sigma G) \cdot (\Sigma E) = 8{,}765{,}761$$
$$(\Sigma E) \cdot (\Sigma M) = 47{,}067{,}292$$

To simplify the computation, we divide (scale) all quantities by 10^5 and round off the results. Substituting in the formula, we obtain for the GNP and military expenditure correlation:

$$\begin{aligned} r &= \frac{①-②}{\sqrt{(③-④)(⑤-⑥)}} \\ &= \frac{182-88}{\sqrt{(7.0-1.9)(8764-3947)}} \\ &= \frac{94}{\sqrt{(5.1)(4817)}} \\ &= 0.60 \end{aligned}$$

For the correlation between military expenditures and manpower levels, we obtain,

$$\begin{aligned} r &= \frac{①-②}{\sqrt{(③-④)(⑤-⑥)}} \\ &= \frac{644-471}{\sqrt{(8764-3947)(96-56)}} \\ &= \frac{173}{\sqrt{(4817)(40)}} \\ &= 0.39 \end{aligned}$$

Table 5.4 Correlation computations.

Country	GNP, G 109	Expenditure, millions, E 106	Manpower, M 103	$G \cdot E$	$E \cdot M$	G^2	E^2	M^2
Belgium	16.7	520	107	8,684	55,640	27,889	270,400	11,449
France	93.5	4,215	535	394,102	225,502	8,742	17,766,225	286,225
Germany	112.2	4,607	440	516,905	2,027,080	12,589	21,224,449	193,600
Italy	56.7	1,908	376	108,186	717,408	3,215	3,640,464	141,376
Netherlands	19.0	746	129	14,174	96,234	361	556,516	16,641
Denmark	10.0	283	50	2,830	14,150	100	80,089	2,500
Norway	7.0	270	34	1,890	9,180	49	72,900	1,156
Portugal	3.7	230	148	851	34,040	14	52,900	21,904
Sweden	19.3	800	82	15,440	65,600	372	640,000	6,724
United Kingdom	99.2	5,937	437	588,950	2,594,469	9,841	35,247,969	190,969
Switzerland	13.9	352	31	4,892	10,912	193	123,904	961
Sums	441.2	19,868	2,369	1,656,904	5,850,215	63,365	79,675,816	873,505
Products	194,657	394,737,420	5,612,161	18,225,944	64,352,365	697,015	876,433,970	9,608,555

The results show a high level of correlation ($r = 0.60$) between GNP and military expenditures. However, the correlation between military expenditures and military manpower levels, while significant, is not as high as that between military expenditures and GNP. This appears to support the thesis that high technology and sophisticated weaponry accounts for the major part of military expenditures.

5.13 CORRELATION SUMMARY

In correlation analysis we observe two variables simultaneously–i.e., each is a free variable. We seek to determine whether the two variables are linearly related, and to what extent they are so related. The correlation relationship (or coefficient) has a purely mathematical (or statistical) connotation and does not imply or confirm any causal connections between the two variables.

The correlation coefficient assumes values from -1 to $+1$. If $r = 0$ or is small ($r < 0.25$), the data are "normally" said to be uncorrelated–i.e., a linear relationship does not exist between the variables. However, other relationships might exist. Correlation, other than linear, will be necessary to determine these.

There are two methods for determining the correlation coefficient. The graphic-analytic method is easier to use than the purely analytic method. The value of the correlation coefficient r determined by the graphic-analytic method is not as accurate or efficient an estimator for correlation as that value of r obtained by the purely analytic method.

5.14 LINES OF REGRESSION

Previously, in Section 5.11, we described the least squares method of fitting a straight line to a scatter diagram. The line was called a *line of regression.* We now discuss the concepts of regression analysis and illustrate a simple method for obtaining the equation for the line of regression. We will also show how the constants of the regression line are related to the correlation coefficient and how the calculations for the correlation coefficient can be utilized in determining the constants (parameters) of the regression line.

Regression analysis is employed in situations where one wishes to investigate the variations of one variable when another variable, to which a linear relationship is suspected (as determined by the correlation coefficient or other measure), is held fixed at each of a set of values in a predetermined interval. For example, we could be interested in the yield of corn for differing levels of irrigation. The experiment could be run simultaneously over many plots of ground. Thus to each irrigation level there will be a set of yield values, one for each plot held to the specified irrigation level. Thus, in regression analysis one variable (x)

is controlled while the values of the second variable (y) are measured. The pairs of values (x,y) are plotted and yield a scatter diagram. A regression line can be fitted to the scatter diagram and used in making estimates or predictions on yield as a function of irrigation level.

Regression analysis establishes the nature of association between variables while correlation analysis is concerned with the degree of association between variables. In this sense regression analysis attempts to gain greater understanding of causal relationships while correlation analysis attempts to determine whether the relationship is strong enough to warrant further study to uncover causal relationships between the variables.

As stated previously, the line of regression is a straight line fitted in a least square sense to the scatter diagram of the two variables. The line of regression is given by the equation

$$y' = mx' + b$$

where m is the slope of the line and b is the y' intercept. The y' intercept can be calculated from the means of the two variables $\bar{x}$, $\bar{y}$ and the slope m of the regression line. We first show how the slope m of the regression line is calculated. The equation for m is obtained by using the least square principle to "fit" the straight line to the scatter diagram. The result is,

$$m = \frac{n \, \Sigma \, x_i y_i - \Sigma \, x_i \, \Sigma \, y_i}{n \, \Sigma \, x_i^2 - (\Sigma \, x_i)^2}$$

This formula is obtained from the formula for the correlation coefficient by substituting x_i for y_i in the denominator of that equation. The numerator remains unchanged. Hence, in calculating the slope of the regression line we can use precisely the same procedure as that used in calculating the correlation coefficient. Except in this computation the terms $n \, \Sigma \, y_i^2$ and $(\Sigma \, y_i)^2$ need not be computed, thus simplifying the calculations for the denominator.

Example 5.7 (Regression Line). Find the slope of the regression line for the data given in Example 5.6, Table 5.2.■■

Solution: Referring to the computations and notations in Table 5.2, the formula for the slope m of the regression line is

$$m = \frac{①-②}{③-④}$$

$$= \frac{5054 - 4699}{1638 - 1369}$$

$$= \frac{355}{269}$$

$$= 1.32$$

The equation for the y' intercept, as derived from a least square fitting of the line, is

$$b = \bar{y} - m\bar{x}$$

Thus to calculate b, we need calculate the values for the means $\bar{x}$ and $\bar{y}$ of the scatter diagram. Continuing the example started above, we have previously determined that

$$\bar{x} = 2.64 \qquad \bar{y} = 9.0$$

With $m = 1.32$, we have

$$\begin{aligned} b &= 9.0 - 1.32 \times 2.64 \\ &= 9.0 - 3.49 \\ &= 5.51 \end{aligned}$$

Thus the equation for the regression line in the example is

$$y' = 1.32x' + 5.51$$

The calculations for the slope and intercept parameters for the line of regression can be accomplished in the following four steps:

Step 1. Calculate the slope m from the equation for the correlation coefficient r replacing x_i for y_i in the denominator for that equation.■■

Step 2. Find the means $\bar{x}$, $\bar{y}$ for the data set (x_1,y_1), (x_2,y_2), ..., (x_n,y_n).

$$\bar{x} = \frac{1}{n}\Sigma x_i \qquad \bar{y} = \frac{1}{n}\Sigma y_i$$ ■■

Step 3. Compute the intercept b for the regression line by substituting the value of m obtained in Step 1 and $\bar{x}$, $\bar{y}$ in Step 2, into the equation for the intercept

$$b = \bar{y} - m\bar{x}$$ ■■

Step 4. Substitute the values for m and b in the slope intercept form for the regression line

$$y' = mx' + b$$ ■■

5.15 RELATIONS BETWEEN CORRELATION AND REGRESSION CONSTANTS

Up to this point we have restricted our attention to regression lines of y on x. In doing this, we have been describing y as a function of x. If the variables x and y are interchanged, we obtain the regression line of x on y. The two regression lines, generally, are not the same. If all the points of the scatter diagram lie on the line of regression, then the line will be the line of regression of y on x and of x on y. The two lines of regression will pass through the point $(\bar{x},\bar{y})$ since every line of regression passes through the means of the variable samples. The slopes of the lines of regression of y on x and x on y will be equal if the standard deviations σ_x of variable x and σ_y of variable y are equal. In this case, the slope m of the regression line is equal to the correlation coefficient. In general

$$m\,(y \text{ on } x) = \frac{\sigma_y}{\sigma_x} \cdot r$$

and

$$m\,(x \text{ on } y) = \frac{\sigma_x}{\sigma_y} \cdot r$$

where r is the correlation coefficient and σ_x, σ_y are the standard deviations of the x data set and y data set, respectively.

5.16 APPLICATIONS OF CORRELATION AND REGRESSION

Below are listed pairs of variables which have been previously reported as being correlated. However, most reports do not state whether the relationship is causal or by chance. Consider each pair of variables in turn and satisfy yourself whether the relationship is causal. The ambitious or interested few should collect data on each pair of variables and calculate the correlation coefficient and the line of regression.

1. Income level against education level.
2. Ratio of urban to rural population against industrialization (by countries).
3. Trade levels against distance between trading centers.
4. High school grades against college level grades.
5. Crop output against levels of
   (a) Rainfall or irrigation.
   (b) Fertilization.
   (c) Insecticides and pesticides.

6. Mortality rates against number of doctors (medical facilities).
7. Birthrates against education level.
8. Stock prices against corporate earnings.
9. College aptitude scores against college grades.

One must keep in mind the distinctions which exist between the correlation coefficient r and the slope m of the regression line. These two quantities are equal only under two conditions: (a) when the variable values are ordinal-valued (ranks), and (b) when the distributions for x and y have the same standard deviation, i.e., $\sigma_x = \sigma_y$. Otherwise, the two quantities are related as previously stated.

To illustrate how a regression analysis is conducted, we will determine the line of regression for income level against education level. The example indicates the necessity to differentiate between the independent variable (education level) and the variable one suspects is dependent on it (income level). If we reverse the role of the two variables, the data have to be collected accordingly, i.e., determine income levels to be investigated and then sampling at each income level to determine the education level for those sampled.

Example 5.8 (Regression Analysis). To convince a class of 40 college students that education really pays off in lifetime income, the statistics professor had the class conduct the following statistical experiment:

He divided the class into ten groups of four students each and assigned the groups the responsibility of obtaining the income level of four married males twenty five years old in each of ten education levels as given in Table 5.5. The incomes obtained were rounded to the nearest $100 and summarized in Table 5.5.

Education level, yr	Income, thousands
8	3.1, 4.7, 4.2, 3.8
9	6.2, 3.7, 5.1, 5.0
10	5.6, 6.3, 6.7, 6.2
11	5.4, 7.5, 8.0, 7.2
12	9.1, 6.4, 6.9, 7.5
13	5.7, 9.1, 7.5, 6.8
14	7.5, 5.9, 8.6, 8.4
15	10.1, 9.3, 7.9, 8.2
16	9.6, 13.7, 12.2, 10.1
Over 16	10.7, 12.7, 12.1, 10.7

The students were asked to determine the line of regression for the date set using the education level as the independent variable and the income level as the dependent variable. The objective of the analysis was to determine: (a) whether income level could be predicted with reasonable accuracy from a person's education level and (b) whether the income level would increase with the education level.

The regression line slope m was computed for the 40 data points. Each education level had associated with it four incomes, e.g., (9,3.1), (8,4.7), (9,4.2), and (8,3.8) are the four data points for education level equal to 8. Table 5.5 contains all the data points and the computations to determine the slope of the line of regression and the y intercept.

The calculation for the y intercept was obtained from the means $\bar{x}$, $\bar{y}$, and the slope m as follows:

$$\begin{aligned} b &= \bar{y} - m\bar{x} \\ &= 7.63 - 0.78 \times 12.5 \\ &= 7.63 - 9.75 \\ &\;\; -2.12 \end{aligned}$$

Thus, the regression line was written as

$$y^1 = 0.78x^1 - 2.12$$

From this equation the students estimate that a high school graduate at age 25 could expect a yearly income of

$$\begin{aligned} y^1 &= 0.78 \times 12 - 2.12 \\ &= 9.36 - 2.12 \\ &= 7.24 \qquad \text{or} \qquad \$7{,}240/\text{yr} \end{aligned}$$

A college graduate at age 25 could expect a yearly income of

$$\begin{aligned} y^1 &= 0.78 \times 16 - 2.12 \\ &= 12.48 - 2.12 \\ &= 10.36 \qquad \text{or} \qquad \$10{,}360/\text{yr} \end{aligned}$$

■■

Rank Correlation

Rank correlation is a statistical measure used to assess whether an action or event has any effect on the order or rank of some attribute as it applies to the participants in the action or event. The attribute could be power, export level, industrial capacity, etc. For example, questions such as the following can be answered partially through the use of rank correlation: Did World War II have a significant effect on the *relative* ability or capacity of the participants

Table 5.5

x_i	y_i	$x_i y_i$	x_i^2	x_i	y_i	$x_i y_i$	x_i^2
8	3.1	24.8	64	13	5.7	74.1	169
8	4.7	37.6	64	13	9.1	1,183	169
8	4.2	33.6	64	13	7.5	975	169
8	3.8	30.4	64	13	6.8	884	169
9	6.2	55.8	81	14	7.5	1,050	196
9	3.7	33.3	81	14	5.9	826	196
9	5.1	45.9	81	14	8.6	1,204	196
9	5.0	45.0	81	14	8.4	1,176	196
10	5.6	56.0	100	15	10.1	1,515	225
10	6.3	63.0	100	15	9.3	1,395	225
10	6.7	67.0	100	15	7.9	1,185	225
10	6.2	62.0	100	15	8.2	1,230	225
11	5.4	59.4	121	16	9.6	1,536	256
11	7.5	82.5	121	16	13.7	2,192	256
11	8.0	88.0	121	16	12.2	1,952	256
11	7.2	79.2	121	16	10.1	1,616	256
12	9.1	109.2	144	17	10.7	1,819	289
12	6.4	76.8	144	17	12.7	2,159	289
12	6.9	82.8	144	17	12.1	2,057	289
12	7.5	90.0	144	17	10.7	1,819	289
Totals				500	305.3	4,073.7	6,580
$r = 40$						162,948	263,200

Computations:

$$m = \frac{n \, \Sigma \, x_i \, y_i - \Sigma \, x_i \, \Sigma \, y_i}{n \, \Sigma \, x_i^2 - (\Sigma \, x_i)^2}$$

$$= \frac{162{,}948 - 152{,}650}{263{,}200 - 250{,}000}$$

$$= \frac{10{,}928}{13{,}200}$$

$$= 0.78$$

$$\bar{x} = 12.5$$

$$\bar{y} = 7.63$$

Table 5.6 Export rank.

Nation	Export rank, 1938	Export rank, 1948	D	D^2
A	1	1	0	0
B	3	3	0	0
C	7	5	2	4
D	4	8	4	16
E	10	10	0	0
F	6	2	4	16
G	5	4	1	1
H	8	6	2	4
I	2	7	5	25
J	7	9	2	4

in the war to export and market their raw materials and furnished products? If World War II had a significant effect, was this effect more pronounced for the losers than the winners? Does the rank ordering, as measured by test grades, of the same set of students taking two tests differ significantly? Did the second test results differ sufficiently from the first test results to warrant further investigation on the reasons for the difference?

To apply rank correlation requires that the rankings of variable values be known at two different times for the same variable, or at anytime for two different variables. The ranks could be obtained before and after an event (like a war) or after two events (like two tests). When two variables are involved, the rankings can be done simultaneously (like two judges ranking contestants in a beauty contest) or the rankings can be spaced in time (like the ranking of students in high school and college).

As an example, consider the export level of nations involved in World War II. We can rank the nations in accordance with their export level in 1938 and then again in 1948. In this way we can investigate whether World War II (or other events which occurred between the years 1939 and 1948) had any effect on the ability of the participants of the war to maintain their rankings as exporters. To do this, we require information on exports in the 2 yr sufficient to complete a table of ranks similar to that given in Table 5.6. Also included in the table are columns headed by the quantities D and D^2. The meaning and use of these quantities in rank correlation will be described in the discussion to follow.

With the information contained in the table, it is possible to assess whether the differences in export ranking of nation participants in World War II are significant enough to warrant additional investigation. The measure of rank correlation is given by Spearman's formula, which is

$$\varphi = 1 - \frac{6 \sum_{i=1}^{n} D_i^2}{n(n^2 - 1)}$$

In the formula (and in Table 5.6) D_i is the difference between ranks of participant i. The quantity n is the number of participants and φ (phi) is called the rank-correlation coefficient. The quantities D_i and D_i^2 are given in columns 4 and 5 of Table 5.6. To compute the rank-correlation coefficient for the ranks in the table we have $n = 10$ and $\sum_{i=1}^{10} D_i^2 = 70$. Thus

$$\varphi = 1 - \frac{6 \times 70}{10\,(100 - 1)}$$

$$= 1 - \frac{420}{990}$$

$$= 0.576$$

This value indicates that the differences in rank are significant and warrant additional investigation to determine whether World War II was a direct and important cause for the differences noted.

Example 5.9 (Rank-Correlation Index). Rank correlation can be useful in determining whether there are national attributes or factors which can be used to predict how one nation will act toward another nation. In this example the level of internal citizen discontent with country A's leaders is considered as the leading cause for country A's hostility toward other countries. To test this assertion required a considerable amount of analysis of country A's citizen unrest and hostility to other countries to be analyzed and scored on a 0-1 numbered scale. Twelve periods of citizen unrest were analyzed and scored. Using these scores, the citizen unrest and national hostility levels were ranked for the 12 periods. The results were as follows:

Period	Citizen Unrest	National Hostility	D_i	D_i^2
1	5	6	1	1
2	7	4	3	9
3	1	1	0	0
4	6	3	3	9
5	12	9	3	9
6	11	10	1	1
7	10	8	2	4

Period	Citizen Unrest	National Hostility	D_i	D_i^2
8	8	12	4	16
9	9	11	2	4
10	3	5	2	4
11	4	2	2	4
12	2	7	5	25

To determine whether the assertion (a hypothesis) had some substance, we apply Spearman's formula to the data and obtain that the rank-correlation coefficient $\varphi = 0.7$. This value of φ is high enough to assert that the level of hostility country A displays toward others is partly caused by the citizen unrest within the country. Such hostility toward other countries can be deliberate to transfer public attention from internal to external problems, or it can reflect hostility to all due to internal strife. One can be reasonably sure that country A will display hostility toward other nations whenever there is a high level of unrest within country A.

REFERENCES

Afifi, A. A., and Azen, S. F. *Statistical Analysis.* New York: Academic Press, 1972.

Dixon, Wilferd J., and Massey, Frank J. Jr. *Introduction to Statistical Analysis.* New York: McGraw-Hill, 1969.

Gerbner, George, Holsti, Ole R. et al. *The Analysis of Communication Content.* New York: Wiley, 1969.

Hoel, Paul G. *Introduction to Mathematical Statistics.* New York: Wiley, 1966.

Longley-Cook, L. H. *Statistical Problems.* New York: Barnes & Noble, 1970.

McCollough, Celeste, and Van Atta, Loche. *Introduction to Descriptive Statistics and Correlation.* New York: McGraw-Hill, 1965.

Siegel, Sidney. *Nonparametric Statistics for the Behavioral Sciences.* New York: McGraw-Hill, 1956.

Chapter 6 / Probability and Bayesian Analysis

6.1 INTRODUCTION

In this chapter we shall see that the theory of probability deals with classes, sets, and events and the assignment of numbers (called probabilities) to these classes, sets, and events in such a manner that three conditions are satisfied. First, if E is a set, class or event, then the probability of E, denoted $p(E)$ is a number greater than or equal to 0 (nonnegative). Second, if E is an event which must occur, then the probability of E, $p(E)$, is equal to 1, i.e. $p(E) = 1$. And third, if E and F are two events such that the occurence of one precludes the occurrence of the other, i.e., E and F are mutually exclusive events, then the probability of either (or both) occurring, denoted $p(E + F)$, is the sum of the separate probabilities of each occurring, i.e.

$$p(E + F) = p(E) + p(F)$$

Probability theory starts with these simple assumptions and develops how to find the probabilities of more complex sets or events occurring, provided of course, that numbers are assigned to the events so that the three assumptions are satisfied.

Probability formulas for complex events are developed along lines similar to the development of complex sets (Chapter 1) and formal logic (Chapter 2). If A and B are simple events (sets, statements) and $p(A)$, $p(B)$ are probabilities associated with A and B which satisfy the three conditions stated previously, then probabilities for complex events (sets, statements) such as $A + B$, $A \cdot B$, $\bar{A} + AB$, etc., can be expressed in terms of the probabilities $p(A)$ and $p(B)$.

Some results of this development are that we will know the methods, formulas, and techniques to use to obtain the probabilities of complex events. However, the formulas do not tell us how to assign numbers (probabilities) to the events or even how to combine the simple events to obtain the complex event to be investigated. In Chapter 3 we indicated several approaches to assigning credibilities (equals subjective probabilities or beliefs) to events and how complex events can be structured. The credibility and plausibility charts, relative frequency, analogical arguments, and causal relationships were all discussed from the point of view of assigning credibility, plausibility, or probability numbers to events, sets, classes, etc. The set operations of union, intersection, and complementation, which are dual to the logical connectives "or," "and," and "not," were used to structure complex sets, events, and statements from simple ones. Thus much of the work covered in the first three and also the fourth chapters purposefully developed a background for the better understanding of probability theory and its applications. Before we proceed with the objective at hand, which is to provide requisite skills in concepts and methods so that substantive analysts can think "probabilistically and statistically" without being "expert" in the theory, we shall distinguish between three types of probability statements or judgements. The first type of probability statement depends on prior knowledge of the universe under observation or measurement. For example, in the probability statement, "what are the chances (or probability) of throwing three sevens in ten tosses of the dice?" it is assumed (a priori) that the dice are fair dice, i.e., each of the six faces has an equal chance of turning up top side. In the probability statement, "what are the chances of obtaining five tails in five tosses of a coin?," it is assumed that the coin used is a fair coin. The results of most games of chance like poker, dice, or roulette rely on a priori probabilities of events and a calculus for determining the probabilities of possible outcomes. The assignment of a priori probabilities usually results from experiments which verify that the physical objects employed (dice, coins, wheels) behave as their manufacturers designed them to behave.

The second type of probability statement is statistical in nature and depends on the ratio of the frequency of occurrence of some event (or property) to the number of times the event (or property) could possibly have occurred. For example, the probability that the next 1971 car you see will be a G.M. product is 0.52, is based on the fact that G.M. produced 52% of all cars sold in the United States in 1971. Observations or measurements of past occurrences are required in statistical probability and actual frequencies are estimated. These frequencies determine the stated probabilities.

The third type of probability statement is the subjective probability or credibility statement. These statements depend on opinions and judgements that do not involve statistical or a priori probabilities. Statements are made without a history of occurrences from which statistical probabilities can be estimated. And the events or occurrences lack properties, either physical or logical, that admit a reasonable estimate of a priori probabilities to be assigned. For example, if we ask, "what is the probability that peace will come to Viet Nam in 2 yrs?," we are seeking learned opinions and judgements which are not and cannot be based on statistical or a priori probabilities. For the event "peace in Viet Nam in 2 yrs" has not had a history of occurrence to yield a statistical probability, and nor has it properties which can be observed or measured to yield on a priori probability. Most probability statements of interest to substantive analysts are of the subjective probability type. All analysts, and mainly the good experimentalists, develop a competency in estimating subjective probabilities. Most of the statements contain the words "possibly," "probably," "likely," "some," "many," "few." The resulting statements are imprecise and they indicate a level of uncertainty. Yet such statements have a value because they admit forming limited judgments where precise judgments are out of the question. Yet, we need not, as good substantive analysts, be straddled by this indefiniteness in judgments. As indicated in Chapter 3, it is often possible to assign more precise ranges of values to the terms probably and likely which denote subjective probabilities and to the terms some, many, and often which denote indefinite quantities.

The results of probability applies equally as well to the three types of probability statements. Basically, if we assume that certain a priori, statistical (ratio), or subjective (credibility or inductive) probabilities (numbers) hold on the occurrence of some event or the possession of some property, then we must accept the conclusion that certain other probabilities (numbers) are determined. With a priori probabilities, these other probabilities are determined precisely, with statistical probabilities they are normally placed within determined levels of confidence, but with subjective probabilities the results are estimates which must be tempered with caution and good judgment.

6.2 PROBABILITY CONCEPTS

Before we discuss probability concepts, we will introduce some notations and definitions which are required in the development of probability theory.

Normally we will be making probability statements about statements (true or false), occurrence of events, or the possession of some property. We shall use capital letters A, B, C, ..., to denote a statement, an event or a property. The notation $p(A)$ will stand for the probability that,

A is true, if A is a statement
A will occur, if A is an event
A is satisfied, if A is a property

For example, if A is the statement, "Russia will recall her French ambassador within 30 days," then $p(A)$ is the probability (subjective) that the statement is true or, equivalently, that the event will occur as stated. If A is the statement, "the economy of country x is agrarian," then $p(A)$ is the probability that country x's economy satisfies the definition for being called agrarian.

As stated in the introduction to this chapter, when dealing with probabilities, we make assumptions about the fairness or randomness of the materials or processes which are employed. These assumptions are in the form of a priori probabilities (dice are fair, wheel is random), statistical ratios (successes to failures, occurrence to nonoccurrence), or subjective probabilities (Kent chart, credibility scales, plausibility scales). We require a way of making known exactly what our assumptions are in each probability statement we make. We do this with the notation $p(A/B)$ which denotes the probability of A (as a statement, event, or property), assuming that B is true (as a statement), occurred (as an event), or is satisfied (as a property). The slanted line reads "given that" or "assuming that." Thus the notation $p(A/B)$ reads as follows:

$$p(A/B) \equiv \text{the probability that } A \text{ is true given that } B \text{ is true}$$

Note, the above is read as if A and B are statements. Keep in mind that A or B could just as well be events or properties.

The probability $p(A/B)$ is called the *conditional* probability of A since it is conditioned on the assumption that B is true. For example, consider the two statements,

A: A Republican will be our next president.

B: Polls show that 52% of voters sampled will vote Republican in the next presidential election.

If we write only $p(A)$, then different values for $p(A)$ are possible based on the opinions, wishes, values, etc., of those evaluating $p(A)$. If we write $p(A/B)$ then the opinions of those evaluating $p(A)$ are tempered by the results of the poll. This example involves subjective probabilities so in this example it is not possible to show exactly how $p(A)$ and $p(A/B)$ differ. When the probability values can be assigned deterministically, then the values for $p(A)$ and $p(A/B)$ can be expressed quantitatively and their differences noted.

The notation for conditional probability and the notation used in sets and logic can be employed in simplifying probability statements involving complex statements and conditions. For example, consider the statements or events

A: Military bases are being constructed in country X.

B: Agents report seeing construction activity in country X.

The conditional probability statement is,

$p(A/B) \equiv$ probability that military bases are being constructed in country X given that agents reported seeing construction activity there.

Three other simple conditional probability statements are possible. Using the negation or complementation notation from logic or sets respectively, the three are written as,

$$p(A/\bar{B}),\quad p(\bar{A}/B),\quad \text{and}\quad p(\bar{A}/\bar{B})$$

where $\bar{A}$ is the negation of the statement A; or if A is an event, $\bar{A}$ is its nonoccurrence. As an example, consider A and B as the statements listed previously. The conditional probability $p(\bar{A}/\bar{B})$ is the statement, " the probability that military bases are *not* being constructed in country X given that agents *have not* observed construction there."

Exercise 6.1 Using the statements A and B above, write out in full the conditional probability statements $p(\bar{A}/B)$ and $p(A/\bar{B})$.

6.3 COMPLEX PROBABILITY STATEMENTS

The conditional probability statement $p(A/B)$ contains the simple statements A and B. One of the principal findings of probability is the calculus for computing probability values for complex statements in terms of the probability values of its simple statements. Hence we need a notation for expressing complex (or compound) statements. This was developed for us in Chapter 2 on logic. We review briefly the section on logical connectives and illustrate their use in probability. As an example of this usage, consider the three simple statements (or events)

A: virus of type 1 is detected

B: virus of type 2 is detected

C: illness of type x diagnosed

Many probability statements can be formed from these three simple statements. Two such statements are,

$p(C/A \wedge B) \equiv$ probability that illness of type x will be diagnosed if viruses of type 1 and type 2 are detected

$p(\bar{C}/A \wedge \bar{B}) \equiv$ probability that illness of type x will not be diagnosed if viruses of type 1 but not type 2 are detected

Exercise 6.2 Complete the probability statements for $p(C/\bar{A} \wedge B), p(C/A \wedge \bar{B})$. List the remaining four combinations of statements C, A, and B.

In the above, the conjunction "and" was used to form complex statements. The disjunction "or" is also used to form compound statements from simple ones. As in logic, the disjunction is used in a nonexclusive sense, i.e., A or B $(A + B)$ means either A or B or both. Hence from the statements A, B, and C we can form compound probability statements such as

$$p(A \vee B/C) \qquad p(C/A \vee B)$$
$$p(\bar{A} \vee B/C) \qquad p(C/\bar{A} \vee B)$$

and all other combinations of A, B, C, their negations, and the connectives and, or, and not. For example, the probability statement

$$p(C/A\bar{B} + \bar{A}B)$$

is read, "the probability of C given that A or B, but not both, occurs."

Note that the notations for the conjunction ($\wedge$) and disjunction ($\vee$) are replaced by the simpler symbols $\cdot$ and +. We shall continue to use $A \cdot B$ or simply AB in place of $A \wedge B$, and $A + B$ in place of $A \vee B$ where the context makes it clear whether A and B are numbers, events, sets, properties, or statements.

One of the difficult things for a substantive analyst to do is to express complex probability statements in terms of the simpler probability notation. For example, analysts find it difficult to take a probability statement such as "what is the probability that illness x will be diagnosed given that neither virus of type 1 nor virus of type 2 are detected" and write it as a probability formula. For example, the above statement is written in probability notation as

$$p(C/\bar{A} \cdot \bar{B})$$

The main problem here normally is that substantive analysts are not familiar with identifying the logical structure (simple statements and connectives) of complex statements because they never were impressed with the need to do it. In probability, the need exists and the analyst who lacks this ability should review Chapters 1 and 2 and refamiliarize himself with the notational equivalences there.

6.4 BASIC PRINCIPLES OF PROBABILITY THEORY

As stated in the introduction, there are three basic principles of probability theory. For any event, set, class, or property A, the probability of A, denoted $p(A)$ is a nonnegative number. If the event A is certain to occur, then the probability of A is 1, $p(A) = 1$. If A and B are two mutually exclusive events, i.e., the occurrence of one precludes the occurrence of the other, then the probability of the disjunction of A and B is equal to the sum of the separate probabilities.

$$p(A + B) = p(A) + p(B)$$

From these three assumptions it is possible to derive some other useful fundamental probability relationships. These are listed here for use in future applications.

1. If a statement B is irrelevant to the truth of statement C, then we may write

   $$p(C/AB) = p(C/A)$$

   This states that the imposition of conditions which do not affect the truth or occurrence of C, can be ignored in computing the probability of C. An example of such a condition is in our previous medical example where virus type 2 is known to be unrelated to illness type x. If this is the case then the formulas $p(C/AB)$, $p(C/A + B)$, $p(C/A\bar{B})$, etc., all reduce to $p(C/A)$. This states that the occurrence or nonoccurrence of B (virus type 2) does not enter into the diagnosis of the illness.
2. If it is impossible for the event A to occur, then its probability of occurrence, denoted $p(A)$, is 0, i.e., if A cannot occur, then $p(A) = 0$.
3. It is impossible for an event to occur and not occur simultaneously. Equivalently, statements cannot be true and false and properties possessed and not possessed. Thus if A is an event, statement, or property, then this property of probability is expressed as

   $$p(A/\bar{A}) = 0$$

4. If A is an event (statement or property), and if B is relevant to A, then the probability that A will occur or not occur is certainty. We express this by the equation

   $$p(A/B) + p(\bar{A}/B) = 1$$

5. If two events A and B are independent, i.e., the occurrence of one does not affect the occurrence or nonoccurrence of the other, then the probability that both occur simultaneously is equal to the product of their probabilities

   $$p(AB/C) = p(A/C) \cdot p(B/C)$$

These properties of probability are quite useful in the development of the applications considered in the remainder of this chapter. The examples which follow illustrate some of the applications of these principles to practical problems.

Example 6.1 (Notation Interpretation). Let the three statements (events) A, B, and C be as follows:

$$A \equiv \text{virus of type 1 is detected}$$
$$B \equiv \text{virus of type 2 is detected}$$
$$C \equiv \text{illness of type } x \text{ diagnosed}$$

1. When we write the formula $p(A/C) = 1$, we are saying that whenever illness of type x is diagnosed, we are certain to find viruses of type 1. This is another way of writing the implication (valid in this case), if C, then A or $C \rightarrow A$.

2. When we write the formula $p(A/\bar{C}) = 0$, we are saying that whenever illness of type x is not diagnosed, we are certain not to find viruses of type 1. This is another way of writing the implication (valid in this case), if $\bar{C}$, then $\bar{A}$ or $\bar{C} \rightarrow \bar{A}$.

3. When we write the formula $p(C/AB) = 1$, we are saying that whenever viruses of type 1 and type 2 are detected, the illness of type x will be diagnosed. Written as an implication, this statement is $(A \wedge B) \rightarrow C$.

4. There are eight combinations of conditional probability statements containing two statements and having probability values 0 or 1. If U and V are any events, statements, or properties, the eight probability statements are,

$$p(U/V) = 0 \qquad p(U/V) = 1$$
$$p(U/\bar{V}) = 0 \qquad p(U/\bar{V}) = 1$$
$$p(\bar{U}/V) = 0 \qquad p(\bar{U}/V) = 1$$
$$p(\bar{U}/\bar{V}) = 0 \qquad p(\bar{U}/\bar{V}) = 1$$

There are many interesting combinations of these eight statements. For example, $p(U/V) = 1$ states that V is a sufficient condition for U, and $p(U/\bar{V}) = 0$ states that V is a necessary condition for U. Hence if both $p(U/V) = 1$ and $p(U/\bar{V}) = 0$ hold, then V is a necessary and sufficient condition for U. ■■

Exercise 6.3 Identify three other pairs of statements from which you can conclude that U and V will be necessary and sufficient conditions. ■■

6.5 EVALUATING COMPLEX PROBABILITY STATEMENTS

Up to this point we have considered conditional probability statements having probabilities equal to 0 or 1. We now show how to evaluate complex conditional probabilities with values other than 0 or 1. Suppose that a series of tests performed on many patients having disease x indicated the presence of viruses of type 1 10% of the time and viruses of type 2 15% of the time. This information can be expressed as conditional probabilities as follows:

$$p(A/C) = 0.10$$
$$p(B/C) = 0.15$$

Note that the probability values assigned depend on the relative frequency of the presence of the viruses. Hence, the probabilities stated above are statistical probabilities.

From these statements and the relation $p(A/C) + p(\bar{A}/C) = 1$, we can also write that

$$p(\bar{A}/C) = 0.90$$
$$p(\bar{B}/C) = 0.85$$

Now suppose that the virus of type 1 was detected in the last ten diagnoses of disease x. *What is the probability that on the eleventh diagnosis of disease x, the virus of type 1 will be detected?* Let us abstract and denote the statements involved as follows:

$A \equiv$ diagnosed disease x on patients $P_1, P_2, \ldots, P_{10}$

$C \equiv$ viruses of type 1 detected on patients $P_1, P_2, \ldots, P_{10}$

Let us also assume that the historical data collected previously,

$$p(A/C) = 0.10 \text{ (for type 1 virus)}$$

remains relevant to this situation and that the diagnoses of different patients are independent of each other. With these assumptions, if we are to avoid the Monte Carlo fallacy, we must evaluate $p(A/C) = 0.1$, i.e., the probability of detecting viruses of type 1 on the eleventh patient with disease x after a run of ten remains equal to the original value of 0.1. The reason for this is that the run of detections of viruses of type 1 was highly unlikely [$= (0.1)^{10}$] based on the 0.1 probability, and its occurrence should not change the statistical probabilities previously established. Naturally, if long runs occur more frequently, we should question the credibility of the assumption and look for physical or medical reasons which justify a change in the probabilities.

Example 6.2 (Monte Carlo Fallacy). At this point we need a word of caution on the assignments of statistical probabilities on the basis of samples of limited size. We caution against falling into the *Monte Carlo fallacy* which is best explained by means of an example. Gamblers, including amateurs and experts, on seeing a run of reds (or blacks) in roulette, or a run of sevens in dice, have a tendency to bet against the run on the basis that the probabilities should eventually equal out. Thus gamblers concoct systems of betting which state that if n blacks (or reds) occur in a row, they will bet red (or black) on the $n + 1$, $n + 2$, and $n + 3$ spin of the wheel. Their assumption is that since red and black should appear with equal probability, then a long run of one color should eventually be covered by a sequence in which the other color is the winner a large majority of times. There are several things wrong with this reasoning. First, the spins of the wheel are independent of each other. What happens on spin n has no effect on what happens on spin $n + 1$, $n + 2$, etc. Second, to say that the probabilities will start to equalize after a run of n is equivalent to saying that you know that over a sequence of predetermined length, the probability of occurrence for red and black will be equal. This statement cannot be justified within the realm of probability theory even if it is true that the wheel (or coin) is biased. To illustrate this point in another way, consider an urn filled with equal numbers of red and black balls. If balls are drawn and not replaced, and if at some stage considerably more red balls than black have been drawn, then it is rational to assume that the next draw will favor black being drawn. However, if the balls are replaced after each draw, and if a run of red balls of length n are drawn, then on the $n + 1$ draw it is as if a new urn (after mixing) is to be used. Thus, no matter how much red was favored in previous draws, the next draw will select red or black with equal probabilities.■■

Now let us show how we can translate verbal probability statements to formulas through the notation conventions previously established. For example, let us write an expression for *the probability that aerial photographs taken over region x indicate that missile silos of type 1 or type 2, but not both, are under construction in region x.* To do this we extract and label the simple statements of the complex statement as follows:

$C \equiv$ interpret aerial photograph
$A \equiv$ photograph indicates construction of type 1 silos
$B \equiv$ photograph indicates construction of type 2 silos

The condition is that the probabilities are to be derived only from the interpretation of aerial photographs. This has been labeled C. The complex event is of the exclusive or form, i.e., type 1 or type 2 but not both. This is expressed (as in logic) as

$$A\bar{B} + \bar{A}B$$

Thus the final expression is

$$p[(A\bar{B} + \bar{A}B)/C]$$

As another example we shall write the expression for *the probability that the new photograph indicates silos of type 1 but not type 2 are being constructed.* Using the notation of the last example, C is the conditional statement, and $A\bar{B}$ is the complex probability statement. Hence the complete probability expression can be written as

$$p(A\bar{B}/C)$$

There are several other complex probability statements analysts should be able to recognize. If A and B are two events and C is a conditional statement, then the probability that either event A or B, but not both, occur under the condition C is expressed as

$$p[(A\bar{B} + \bar{A}B)/C]$$

The events $A\bar{B}$ and $\bar{A}B$ are mutually exclusive (from set theory and logic) so that the above expression can be rewritten as

$$p[(A\bar{B} + \bar{A}B)/C] = p(A\bar{B}/C) + p(\bar{A}B/C)$$

Another probability statement which finds wide usage in probability is the one expressing the joint occurrence of two events. Under the conditional statement C, this probability is expressed as

$$p(AB/C)$$

Associated with the joint probability of AB is the probability that only one event will occur ($A\bar{B}$ or $\bar{A}B$), neither event will occúr ($\bar{A}\bar{B}$) and both will not occur simultaneously ($\overline{AB}$). Note that $\overline{AB}$ is the complement of AB, $\overline{AB}$ is not to be confused with $\bar{A}\bar{B}$. Some numerical examples should illustrate the differences.

Example 6.3 (Evaluating Probability Statements). Let $p(A/C) = 0.1$, $p(B/C) = 0.15$. Then if A and B are independent, we have

$$\begin{aligned}
p(AB/C) &= p(A/C) \cdot p(B/C) = 0.015 \\
p(A\bar{B}/C) &= p(A/C) \cdot p(\bar{B}/C) = 0.1 \cdot 0.85 = 0.085 \\
p(\bar{A}B/C) &= p(\bar{A}/C) \cdot p(B/C) = 0.9 \cdot 0.15 = 0.135 \\
p(\bar{A}\bar{B}/C) &= p(\bar{A}/C) \cdot p(\bar{B}/C) = 0.9 \cdot 0.85 = 0.765 \\
p(\overline{AB}/C) &= 1 - p(AB/C) = 1 - 0.015 = 0.985
\end{aligned}$$

These examples illustrate the importance of not confusing the last two expressions. One should rely on the Venn diagram to ascertain the difference between $\bar{A}\,\bar{B}$ and $\overline{AB}$. Their Venn diagrams are repeated below in Figure 6.1 for comparison purposes.■■

6.6 AN APPLICATION OF PROBABILITY

One of the tasks every analyst performs is the searching or scanning of items to identify those with prespecified characteristics. The researcher scans texts for relevant information. The military analyst scans land masses to locate new construction or troop concentrations. Or he scans the sea to locate ships and submarines. The environmentalist scans forest for tree types, and the geologist looks for signs of mineralization. Most analysts start with some preconceived notion (a priori probabilities) of what to expect and the ratio in which to expect the different items sought. These expectations cannot be stated as precisely as those stated in the examples where balls of various colors are drawn randomly from a theoretical urn. However, the probability principles employed are the same and can be applied equally as well in all such situations. Rather than use the "clean" academic or mathematical example of an urn with colored balls in it, we shall use the "dirtier," but more realistic applications as they occur in military, medical, and political probability analysis. What we will attempt to explain is the calculation of probabilities for a sequence of events which in one instance are independent of each other, and in the second instance are dependent. Recall that two events are independent if the occurrence of one does not affect the occurrence or nonoccurrence of the other. Draws from a deck of cards are independent if the drawn cards are replaced and the deck reshuffled after each draw. If the cards are not replaced, then the draws are dependent on the cards drawn on previous draws.

Example 6.4 (Dependent and Independent Probabilities). Our first example is from military surveillance where a reconnaisance plane is asked to verify that 15

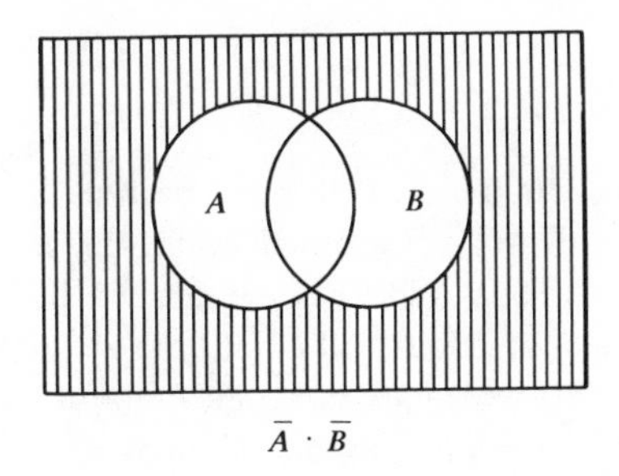

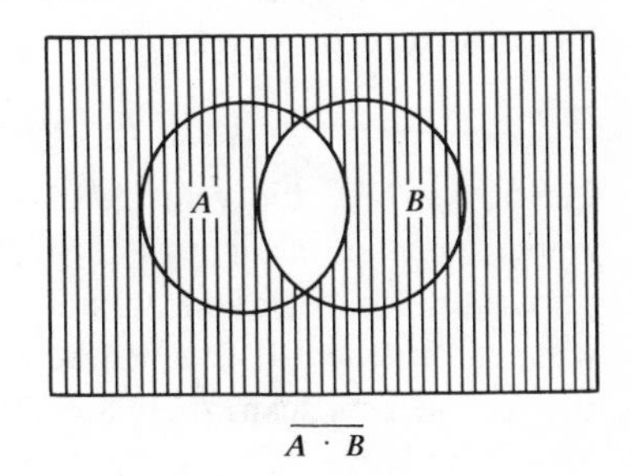

Figure 6.1 Venn diagrams for $\bar{A}\bar{B}$ and $\overline{AB}$.

armament emplacements of three different types (types 1, 2, and 3) are being constructed in an area of interest. Further information has it that five each of types 1, 2, and 3 are being constructed. The reconnaisance aircraft scans the area until an emplacement is located. Having located an emplacement, its type and position is marked so that the area will not be covered again, and the aircraft proceeds and scans another area. The probability statement then reduces to determining the probability that emplacements of type 1, 2, and 3 will be observed on successive scans. We can simplify the problem slightly by assuming that the reconnaisance aircraft will find one emplacement in each area scanned, but it will not know the type until observed. To write the probability statements for this type of problem, we need a notation which will indicate the order in which emplacements were observed; to do this let

A_n: Emplacement of type 1 is the nth one detected.
B_n: Emplacement of type 2 is the nth one detected.
C_n: Emplacement of type 3 is the nth one detected.

The conditions under which the observations are conducted are

X: Five each of emplacements type 1, 2, and 3.
Y: One emplacement identified on each scan.

Thus A_1 denotes the event, "emplacement of type 1 is the first one detected," B_3 denotes the event, "emplacement of type 2 is the third one detected," C_2 denotes the event, "emplacement of type 3 is the second one detected." With this notation convention we can now express the probability that an emplacement of type 1 will be detected on the first area scan as

$$p(A_1/XY) = 1/3$$

The value 1/3 is the frequency ratio of type 1 emplacements (5) to the total number of emplacements (15).

The probability that an emplacement of type 1 *will not* be detected on the second area scan after an emplacement of type 2 was identified on the first area scan can be expressed as

$$p(A_2/B_1XY) = 9/14$$

Let us see how the value 9/14 was obtained. On the first area scan a type 2 emplacement was identified. Since this area will not be scanned again, this is equivalent to selection without replacement. Thus 14 areas remain to be scanned. Of the 14, five will be of types 1 and 3, and four will be of type 2. On the second area scan, there are nine emplacements which are not type 1 (four of type 2 and five of type 3). The relative frequency ratio of *not* identifying a type 1 to the total number of possible identifications on the second scan is 9/14.

The next probability statement involves three successive scans. The probability sought is that of *not* identifying an emplacement of type 1 after emplacements of type 1 were *not* identified on the first and second scans. The notation for this probability is

$$p(\bar{A}_3/\bar{A}_2\bar{A}_1XY) = 8/13$$

The value 8/13 is obtained by noting that after the first scan five emplacements of type 1 and nine of type 2 or 3 remain to be identified. On the second scan another emplacement of type 2 or 3 is identified. Thus for the third scan eight emplacements of type 2 or 3 remain to be identified out of the 13 remaining emplacements. Hence the value 8/13.

As a final example, we shall express and compute the probability of identifying an emplacement of type 1 on the first and second scans, and one of type 3 on the third scan. The expression is as follows

$$p(A_1A_2C_3/XY)$$

On the first scan the probability of identifying a type one emplacement is 5/15. This leaves four of type 1 and ten of type 2 and 3 to be identified. On the second scan the probability is changed to 4/14 since four of type 1 and a total of 14 remain to be identified. On the third scan five emplacements of type 3 remain to be identified out of the 13 emplacements which remain unidentified. Hence the probability of identifying a type 3 emplacement on the third scan is 5/13. The final probability is the product of the separate probabilities or

$$p(A_1A_2C_3/XY) = 5/15 \cdot 4/14 \cdot 5/13 = 100/2730$$ ■■

Example 6.5 (Voting Probabilities). Let X, Y, A_n, B_n be the statements,

X: There are 100,000 registered voters in city x, 50,000 are Republicans and 50,000 are Democrats.

Y: 60% and 70% of registered Republicans and Democrats, respectively, vote on election day.

A_n: The nth registered voter voting on election day will vote the Republican ticket.

B_n: The nth registered voter voting on election day will vote the Democratic ticket.

Express the following formulas in words and determine their values.

1. $p(A_1/X), p(B_1/X)$
2. $p(A_1/XY), p(B_1/XY)$
3. $p(A_3A_2A_1/X)$
4. $p(B_3B_2B_1/X)$
5. $p(A_3A_2A_1/XY)$
6. $p(B_3B_2B_1/XY)$ ■■

Solution: For equation 1 the only information used is X which gives a relative frequency of 0.5 to the Republicans and Democrats. Hence,

$$p(A_1/X) = 1/2 \quad \text{and} \quad p(B_1/X) = 1/2$$

To answer 2 we need consider the statement that 60% of Republicans and 70% of Democrats vote on election day. This means that the number of actual voters is 30,000 and 35,000, respectively. Hence

$$p(A_1/XY) = 30{,}000/65{,}000 = 6/13$$
$$p(B_1/XY) = 35{,}000/65{,}000 = 7/13$$

Equations 3 and 4 are evaluated using only statement X and the fact that since voters only vote once, the probability of getting a sequence of voters is similar to the selection of balls without replacement from an urn. The probabilities reduce to

3. $p(A_3A_2A_1/X) = 1/2 \cdot 49{,}999/99{,}999 \cdot 49{,}998/99{,}998$
4. $p(B_3B_2B_1/X) = 1/2 \cdot 49{,}999/99{,}999 \cdot 49{,}998/99{,}998$

Equations 5 and 6 are evaluated by using the statements X and Y. This changes the statistical probabilities to

5. $p(A_3A_2A_1/XY) = 6/13 \cdot 29{,}999/64{,}999 \cdot 29{,}998/64{,}998$
6. $p(B_3B_2B_1/XY) = 7/13 \cdot 34{,}999/64{,}999 \cdot 34{,}998/64{,}998$

Exercise 6.4 Two political analysts are to interview voters for a political poll. Analyst A knows that 60% of the voters being polled are registered Republicans. Analyst B does not have this information. What value of probability will each attach to the event that each voter interviewed is a Republican? Select a, b, or c.

	Analyst A	Analyst B
(a)	0.5	0.6
(b)	0.6	0.5
(c)	0.6	can't say

■■

Exercise 6.5 Eight voters are to be interviewed. Three are Republicans and five are Democrats. The conditions are

X: The above information is known.

Y: Two Republicans will not disclose their voting preference.

Z: Two Democrats will not disclose their voting preference.

A: Voter interviewed identified as Republican.

B: Voter interviewed identified as Democrat.

Determine the probability for the following three events. Select (a), (b), or (c).

	$p(A/XY)$	$p(A/XZ)$	$p(A/XYZ)$
(a)	1/6	1/4	1/4
(b)	1/8	3/8	1/8
(c)	1/6	3/6	1/4

■■

Exercise 6.6 Of ten voters to be interviewed, three are known to be Democrats and seven are Republicans. On the basis of the above information, what is the probability that the first voter interviewed will be Republican? If you learn that one Republican and one Democrat are not available for interviewing, what is the probability that the first voter interviewed will be Republican?

Exercise 6.7 The conditional statements are:

X: Of ten voters, six are Democrats and four are Republicans.
A_n: The nth voter will be a Democrat.
B_n: The nth voter will be a Republican.

Determine the probabilities listed below. Select (a), (b) or (c).

	$p(A_2/A_1X)$	$p(A_2/X)$
(a)	5/9	5/9
(b)	5/9	6/10
(c)	6/10	indeterminant

■■

Exercise 6.8 Write out in full your understanding of the following expressions.

(a) $p(A/X) = 1$
(b) $p(\bar{A}/X) = 1$
(c) $p(A/X) = 0$
(d) $p(\bar{A}/X) = 0$
(e) $p(A/X) + p(\bar{A}/X) = 1$
(f) $p(A_1A_2/X)$
(g) $p(A_1 + A_2/X)$
(h) $p(A_1\bar{A}_2 + \bar{A}_1A_2/X)$
(i) $p(A_3/A_2A_1X)$
(j) $p(A_3A_2/A_1X)$ ■■

6.7 SAMPLE SPACE

Up to this point we have used the terms statements, events, sets, and properties to indicate or denote elements to which probabilities are assigned. We considered sets of events, statements, properties, subsets, without specifying the universe containing them. For example, we considered the event, "the dice total to seven" as one of eleven possible events, i.e., the dice totaling to 2, 3, 4, 5, 6, 7, 8, 9, 10, 11, and 12, without specifically identifying the ten other events in the event space. The set of all possible events or outcomes is called the *sample space* for the dice-tossing experiment. The *sample space* of any event or experiment is the exhaustive collection of all its mutually exclusive outcomes.

There are many ways of specifying or denoting a sample space. The ones most frequently used are

1. Listing or enumeration.
2. Outcome tree.
3. Matrix.
4. Venn diagram.

Examples for each of these means for denoting a sample space follow.

Example 6.6 (Single-Coin Sample Space). Let the event be the tossing of a coin. The outcomes are the tosses resulting in a head (H) or tail (T). The sample space is denoted as follows.

1. Enumeration or listing

H, T or T, H

2. Outcome tree

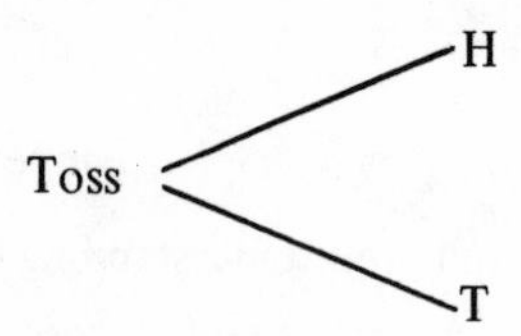

3. Matrix

(H, T) or (T, H)

4. Venn diagram

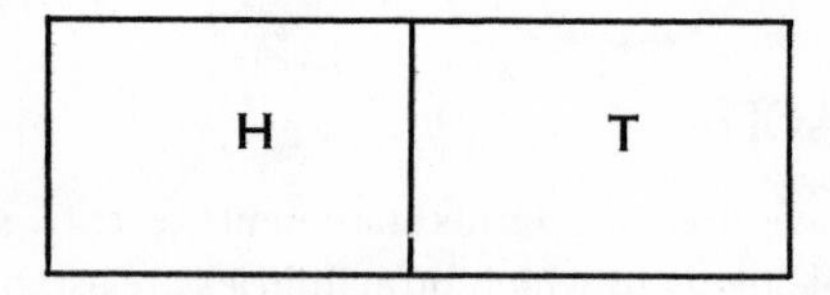

■■

Example 6.7 (Two-Coin Sample Space). Let the event be the tossing of two coins simultaneously. The outcomes are the tosses resulting in two heads, two tails, or one head and one tail. The sample space is denoted as follows:

1. Enumeration or listing

HH, HT, TH, TT (no order)

2. Outcome tree

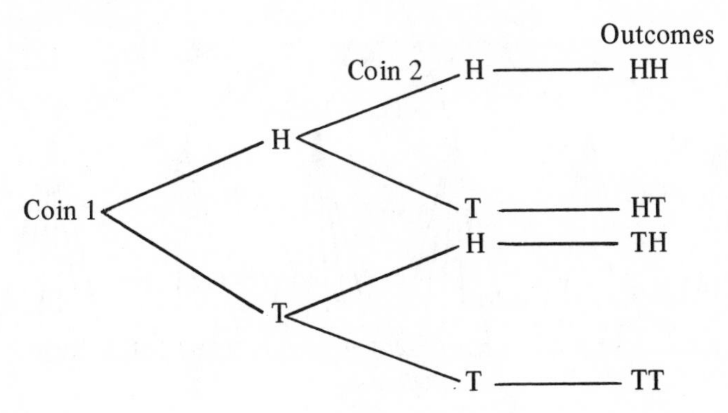

3. Matrix

		Coin 2	
		H	T
Coin 1	H	HH	HT
	T	TH	TT

(Order depends on how rows and columns are labeled.)

4. Venn diagram

(No order.)■■

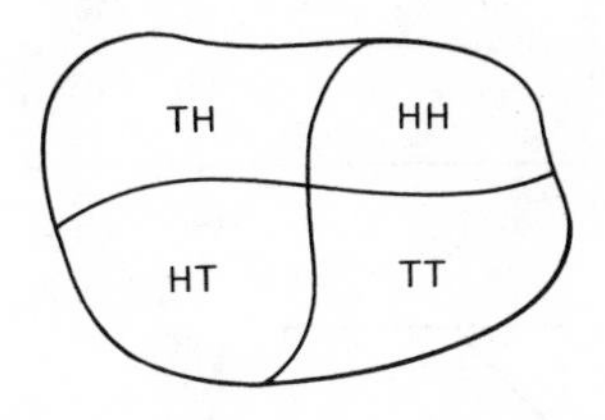

Example 6.8 (Two-Die Sample Space) Let the event be the tossing of two die. The outcome is the sum of dots showing on the top side of the die when they come to rest on a flat surface. The sample space is denoted as follows:

1. Enumeration or listing

2, 3, 4, 5, 6, 7, 8, 9, 10, 11, 12

2. Outcome tree

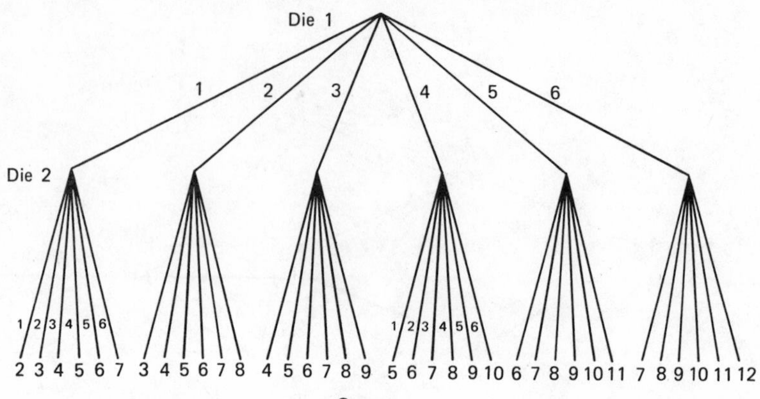

3. Matrix

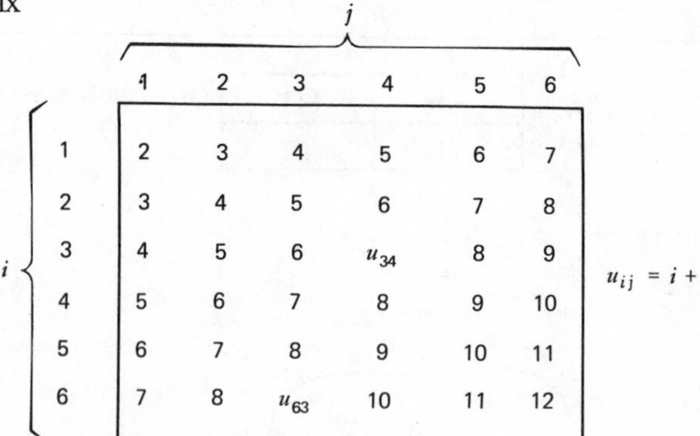

i \ j	1	2	3	4	5	6
1	2	3	4	5	6	7
2	3	4	5	6	7	8
3	4	5	6	u_{34}	8	9
4	5	6	7	8	9	10
5	6	7	8	9	10	11
6	7	8	u_{63}	10	11	12

$u_{ij} = i + j$

4. Venn diagram

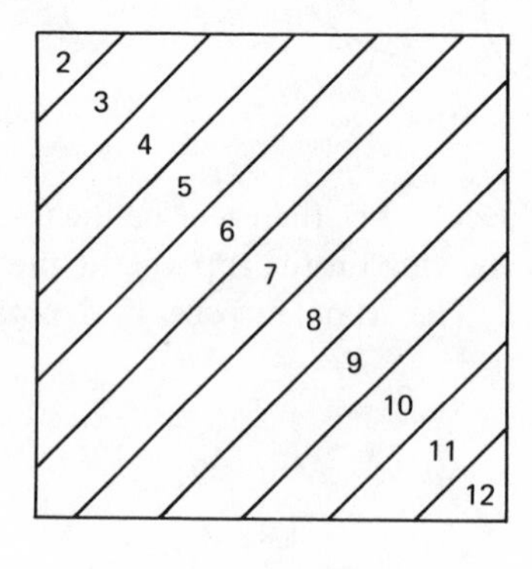

(One of many possible area arrangements)

■■

Example 6.9 (Subjective Sample Space). In the previous examples the event and all its possible outcomes were easy to identify. This is not the case where the definition of the outcomes of an event or experiment are subjective. Subjective outcomes are difficult to identify uniquely–let alone list them exhaustively. For example, let the event be specified as, "the candidate will increase taxes if elected." One analyst might consider only two alternatives as meaningful. He will label the event space simply as

E_1: Will increase taxes.

E_2: Will not increase taxes.

A second analyst might label the event space more complexly as,

E_1: Will increase taxes at least 50%.

E_2: Will increase taxes between 35 and 50%.

E_3: Will increase taxes between 20 and 35%.

E_4: Will increase taxes between 5 and 20%.

E_5: Will increase taxes between 1 and 5%.

E_6: Will not increase taxes.

Obviously, the event space constructed by the second analyst is of finer grain. It would be difficult to have the two analysts agree as to which event space should be used. Normally, the preferred event space for analysis purposes is the one which matches the accuracy with which the probabilities of the events can be determined. If information is available which allows one to assign the candidate's tax policies to one of the six outcomes, then certainly this event space is preferred to the simpler event space. However, if information is lacking and the candidate is noncommittal, then the added outcomes are window dressing and five of the six outcomes can be reduced to the one outcome–the candidate will increase taxes.■■

Example 6.10 (Sample Space Structure). In Example 6.9, it was suggested that the event space be structured no finer than that needed to match the information available. The events were capable of being stated in terms of the accuracy in which a quantitative measure (percent tax increase) could be achieved. In this example we illustrate a subjective event whose outcomes are not specifiable by a quantitative objective measure. Instead, the outcomes are events or courses of action which are ordered in accordance with subjective measures such as severity, appropriateness, i.e., the measure is itself subjective.

Let the event be identified as "country *X*'s reaction to country *Y*'s action." Some outcomes of the resulting event space are listed below in order of decreasing severity.

All out war
Large-scale war
Limited conventional war
Guerrilla warfare
Boycott, blockade, or seizure
Sever diplomatic relations
United Nations action
Diplomatic protest

If it is assumed that country *X* will initiate only one action in retaliation for country *Y*'s action, then the events can be considered as being mutually exclusive. Once initiated, the chosen action implies that all other actions of lesser severity were considered but were not chosen because they did not reflect the severity of the situation.

In this situation, the event space may be ordered or unordered. The outcomes may be listed without regard to order and the list may be larger or smaller than the one selected here. All these differences are due to the subjectivity involved in selection and ordering of outcomes. Most lists compiled by competent specialists will be representative. The more important properties the events in the event space should possess are

1. Mutually exclusive.
2. Well-defined.
3. Exhaustive.
4. Independent.

Achieving these properties for the events of a subjectively derived sample space will be one of the more difficult tasks to accomplish properly in the analyst's dealings with subjective event spaces. ■■

6.8 ASSIGNMENT OF PROBABILITIES

For finite event spaces an objective is to assign a probability of occurrence to each event in the space. If the events are independent, mutually exclusive, and exhaustive, then the sum of the probabilities assigned to the events must equal to 1. For example, in the dice game, the assignment of probabilities can be given with a matrix description of the event space.

	1/36	2/36	3/36	4/36	5/36		6/36
1	2	3	4	5	6	7	
2	3	4	5	6	7	8	5/36
3	4	5	6	7	8	9	4/36
4	5	6	7	8	9	10	3/36
5	6	7	8	9	10	11	2/36
6	7	8	9	10	11	12	1/36

We observe that there are 36 events in the event space but only 11 are distinct events. By proceeding diagonally from lower left to upper right, we can count the number of 2s, 3s, 4s, etc. These totals are noted, divided by 36 and entered above or to the right of the matrix as indicated by the lines. There are five possible ways of making 8 (2 + 6, 3 + 5, 4 + 4, 5 + 3, 6 + 2) with the dice. This number divided by 36 gives the probability of making an 8 on a single toss of the dice, i.e., $p(8) = 5/36$. The number 5/36 is entered to the right of the second row.

The probability given above depended on an a priori knowledge of the universe (sample space) under observation (or experimentation). We could have estimated the probability of each outcome by conducting a large number of tosses and tallying the number of times the dice totaled 2, 3, ..., 11, 12. We would then compute the probability of throwing any number i from 2 through 12 as follows:

$$p(i) = \frac{\text{number of times dice totalled } i}{\text{number of times dice thrown}}$$

This, of course, is the relative frequency of occurrence and hence is a way of assigning statistical probabilities to the events in the space.

There is also a third method of assigning probabilities to the events in the event space. We could just play with the dice for awhile, tossing them a few times and then make estimates of the probabilities. In this case care must be taken in assigning subjective probabilities to assure that they sum to a value of 1. Naturally, assigning subjective probabilities to events where the event space is known a priori is not recommended. Yet most people who gamble with dice play the game using experience alone. Very few crap shooters really know what the probabilities of the event space are. For example, from the event space it is easy to determine that a side bet against an 8 or 6 with even odds is a good bet, against a 10 or 4 with 2/1 odds is an even bet, and against a 9 or 5 with 3/2 odds, is an even bet. Also from the event space and event tree it is possible to determine whether it is advisable to roll the dice or bet against the roller. This last determination is left as an exercise.

In Example 6.5, where the event was identified as, "country *X*'s reaction to country *Y*'s action," a listing of outcomes and the assignment of probabilities subjectively may be the only recourse. The use of the term subjectively here means that all information is sought and utilized in making the assessments but the information is not sufficient to completely define the event space statistically or in an a priori fashion. Some human judgement enters into the assignment of probabilities. The assignment of subjective probabilities to events may vary from analyst to analyst although each is in possession of the same information. Once the probabilities are assigned, then the probabilities of combined events can be obtained using the formulas developed in the remainder of this chapter. The results will be correct only if the assignment of probabilities to the simple events is done properly and the computations performed correctly.

6.9 PROBABILITY OF COMPLEX EVENTS

Up to this point we have discussed assigning probabilities to simple, or fine grain, events in each of three ways: a priori, statistically, and subjectively. We are now interested in computing the probability for complex events, or events that are composed of a combination of simple events. As stated previously, complex events are combinations of unions, negations, and intersections of simple events from the event space. For example, consider an event space consisting of the three events E_1, E_2, and E_3 with assigned probabilities $P(E_1), P(E_2)$, and $P(E_3)$ where automatically

$$P(E_1) + P(E_2) + P(E_3) = 1$$

We can form many complex events. A list of several typical complex events follows.

$$E_1 \text{ and } E_2 = E_1 \cap E_2 = E_1 \cdot E_2$$
$$E_1 \text{ or } E_2 = E_1 \cup E_2 = E_1 + E_2$$
$$E_1 \cdot E_2 + E_3, \quad \bar{E}_1 \cdot E_2 \cdot \bar{E}_3$$
$$(\overline{E_1 + E_2}) \cdot \bar{E}_3, \quad \bar{E}_1 + \bar{E}_2 + \bar{E}_3$$

In the following sections we develop probability formulas which cover the calculation of complex event probabilities.

6.10 INTERSECTION (PRODUCT) OF EVENTS

Complex events can occur in either of two ways. In the first instance, one or more events can occur in a single experiment or trial. Examples are: (a) drawing two red and three black balls on a draw of five balls from an urn, (b) throwing a 3 and a 2 to make 5 on a throw of two die, and (c) obtaining two heads on the toss of two coins. In the second instance one or more events can occur as a result of a sequence of events and the sequence is considered as a single (complex) event. Examples are: (a) drawing two red balls and then three black balls on two successive draws of four balls from an urn, (b) throwing a single die twice and obtaining a 3 and 2 on the first and second throws, respectively, and (c) throwing a single coin twice and obtaining heads each time.

There appears to be no significant difference in the two ways in which complex events can be formed from simple events. They can occur sequentially or simultaneously. But we shall see that when the events are not mutually exclusive, i.e., the occurrence of one event does not preclude the occurrence of the second event, the order of occurrence of events becomes important in that the knowledge that one event has occurred alters the probability that the second event will occur.

Let us consider two events labeled A and B, and let us assume that the events A and B are mutually exclusive, i.e., as event sets in event space, their intersection is empty, $A \cdot B = \phi$. This states that the occurrence of one event precludes the occurrence of the other event. It also states that the two events cannot occur simultaneously. Thus the probability that both events A and B will occur is 0,

$$p(A \cdot B) = 0$$

When the events A and B are not mutually exclusive, then the event sets for A and B intersect in the event space E. We denote this by means of the following Venn diagram:

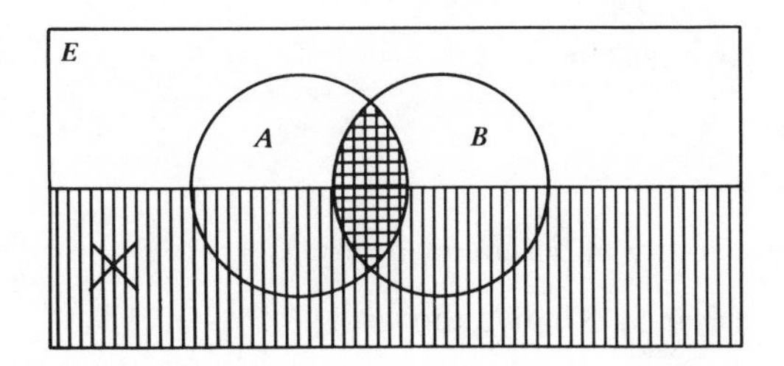

The intersection of A and B is denoted by the area in E marked with horizontal lines. The probability that event A or event B will occur in the event space E can be given by the ratio of the area of A or B, respectively, to the total area of E.

$$p(A) = \frac{\text{area of } A}{\text{area of } E}$$

$$p(B) = \frac{\text{area of } B}{\text{area of } E}$$

$$p(AB) = \frac{\text{area of intersection } A \cdot B}{\text{area of } E}$$

However, it is usually difficult to estimate or represent these areas accurately. Thus in most instances the relative areas are only schematic of the actual probabilities involved. For this reason Venn diagrams are utilized mainly in representing, describing, and visualizing the quantities involved in the formation of complex events.

Example 6.11 (Finite Sample Event–Space). The above areas designate event sets having an infinite number of points. When the event sets and spaces can be represented by a finite yet tractably small number of points in the Venn diagram, then the description and calculations can be simplified. For example, consider the throw of two die. The event space, which includes all possible outcomes, consists of 36 points given and labeled in the following Venn diagram:

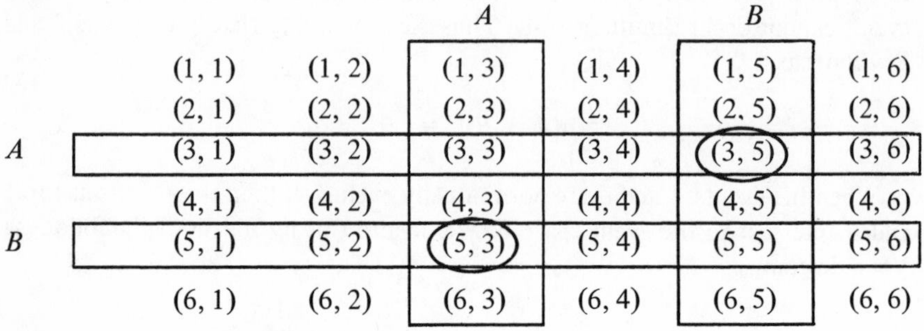

			A		B	
	(1, 1)	(1, 2)	(1, 3)	(1, 4)	(1, 5)	(1, 6)
	(2, 1)	(2, 2)	(2, 3)	(2, 4)	(2, 5)	(2, 6)
A	(3, 1)	(3, 2)	(3, 3)	(3, 4)	(3, 5)	(3, 6)
	(4, 1)	(4, 2)	(4, 3)	(4, 4)	(4, 5)	(4, 6)
B	(5, 1)	(5, 2)	(5, 3)	(5, 4)	(5, 5)	(5, 6)
	(6, 1)	(6, 2)	(6, 3)	(6, 4)	(6, 5)	(6, 6)

Let the two simple events be

A: Throw a 3.

B: Throw a 5.

Consider the probability of the following event:

C: Throw a 3 and a 5 in two throws.

From the diagram it is easy to see that there is one chance in six to obtain a 3 in a single toss. The same holds for the 5. The complex event, "throw a 3 and a 5," can be obtained in two ways: 3 then 5 or 5 then 3. Hence, there are only two favorable events out of a possible 36 events. From the event space we see that the probabilities are

$$
\begin{aligned}
p(3) &= 1/6 \\
p(3 \text{ and } 5) &= 1/36 \\
p(5 \text{ and } 3) &= 1/36 \\
p(3 \text{ and } 5, \text{ or } 5 \text{ and } 3) &= 1/18
\end{aligned}
$$

The events A and B or B and A are circled in the Venn diagram. ■■

Exercise 6.9 Using the Venn diagram for the throw of two die, calculate the following probabilities:

(a) Two die have the same number showing.
(b) Two die differ by 1.
(c) Number thrown (sum of the two die) is even. ■■

When two events are independent of each other, i.e., the outcome of one event does not affect the outcome of the other, several statements can be made. First, recall that the notation $p(A/B)$ signifies the conditional probability of event A given that event B has occurred. When A and B are independent, then we can write

$$p(A/B) = p(A), \qquad p(B/A) = p(B)$$

More important to the discussion is the fact that whenever A and B are independent,

$$p(AB) = p(A)\ p(B)$$

Also, if X is any condition, we can write

$$p(AB/X) = p(A/X) \cdot p(B/X)$$

Keep in mind that this formula holds only when A and B are independent. When A and B are not independent, we write

$$
\begin{aligned}
p(AB/X) &= p(A/BX) \cdot p(B/X) \\
&= p(B/AX) \cdot p(A/X)
\end{aligned}
$$

These latter forms are obviously the more general forms. By using them instead of the former, one need not check for independence of the two events. However, in calculating the quantities $p(A/BX)$ or $p(B/AX)$, one will have to determine what effect the occurrence of one event has on the other. If none, then the two events are independent. If there exists some effects, then one must determine these to complete the evaluation of the probabilities $p(A/BX)$ or $p(B/AX)$.

We will illustrate the use of the two formulas listed above and how to use Venn diagrams, event trees, etc., to assist in obtaining probabilities for the simple and complex events involved in the formulas.

Example 6.12 (Dependent Events). This example illustrates a method for determining probabilities of simple or complex events wherein the event probabilities can be assigned a priori. Consider a box containing four balls. Two of the balls are red and two are black. Balls will be drawn one at a time and will not be replaced after each draw. *We wish to determine the probability of drawing two red balls.* To determine this probability, we will use a procedure which will help in avoiding errors in assigning probabilities to the simpler events and in computing the complex probabilities.

First we will identify all the conditions under which the probabilities are to be determined and introduce an appropriate notation.

X: { Four balls in box, two red and two black. Balls drawn without replacement. }

We label these conditions X. The following notation will be employed: $p(R_i/X)$, $p(B_i/X)$ are the probabilities that a red and black will be drawn, respectively, on the ith draw, $i = 1, 2$. Hence, $p(R_2B_1/X)$ is the probability that a black ball will be withdrawn on the first draw (B_1) followed by a red ball (R_2) on the second draw, subject to not replacing the ball drawn on the first draw (X). Other combinations of draws are $p(R_2R_1/X)$, $p(B_2B_1/X)$. As we shall see in the event tree, there are six possible outcomes for the drawing of two balls without replacement. We wish to find the probability given by $p(R_2R_1/X)$.

Second, we use one or more of the four ways used to describe the event space. We can use the listing technique, an event tree, a matrix, or the Venn diagram. In this particular example, we will illustrate an event tree and a combination of the Venn diagram and listing methods. Using the Venn diagram and listing techniques, we have that the event space before the first draw is a box containing two red and two black balls. This space is denoted S_1 and is displayed as follows:

S_1

R	R
B	B

After the first draw (without replacement) the event space is modified to either of two spaces, depending on whether the first ball drawn was red (S_2) or black (S_2^1).

S_2

	R
B	B

or S_2^1

R	R
B	

The probability of drawing a red ball on the first draw in event space S_1 is

$$p(R_1/X) = \frac{\text{number of red balls in } S_1}{\text{total number of balls}} = 2/4 = 1/2$$

The probability of drawing a red ball on the second draw in event space S_2 is

$$p(R_2/R_1X) = \frac{\text{number of red balls in } S_2}{\text{total number of balls}} = 1/3$$

The probability of drawing a red ball on the second draw in event space S_2^1 is

$$p(R_2/B_1X) = \frac{\text{number of red balls in } S_2^1}{\text{total number of balls}} = 2/3$$

Now we use the event tree technique to represent the event space. The event space for draws without replacement is shown in Figure 6.2.

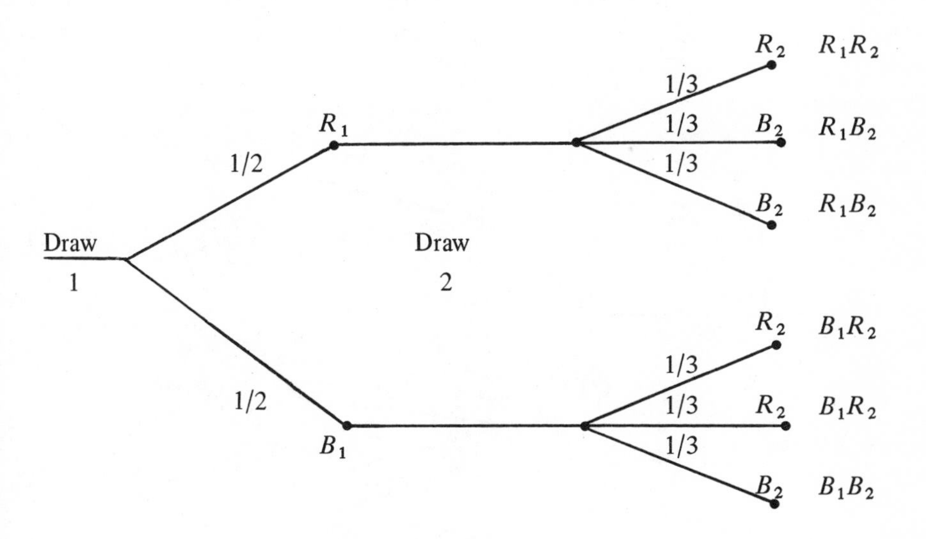

Figure 6.2 Event tree.

On the first draw, red or black will be drawn with probability equal to 1/2. On the second draw, the probability of drawing a red or black ball is dependent on whether a red or black ball was drawn and not replaced on the first draw. The two branches for the second draw represent these two alternatives. That these alternatives occur with equal probability is indicated by the values 1/2 assigned to the two branches on the tree. The three branches at the second draw indicate that each of the three remaining balls can be drawn on the second draw. The fact that they can be drawn with equal probability is indicated by the values 1/3 assigned to the three branches. The tree can be simplified at the second draw to indicate only two possibilities by not signifying the specific balls to be drawn. The modified event space is given in Figure 6.3.

Note the changes in probabilities of drawing red or black on the branches of the second draw. Note also that the tree ends in only four events where the previous tree ended in six events and two of the six events were identical to two of the four mutually exclusive events.

The structuring of event trees is usually more laborious than the other techniques. Once it is completed correctly, the calculation of probabilities of the outcomes is vastly simplified. For example, to obtain the probability of event B_2R_1 we find the path through the tree to obtain that event as an outcome (shown by double lines in Figure 6.3) and multiply the probabilities assigned to the branches contained in the path. In this case,

$$\begin{aligned} p(B_2R_1/X) &= p(B_2/R_1X) \cdot p(R_1/X) \\ &= 2/3 \cdot 1/2 \\ &= 1/3 \end{aligned}$$

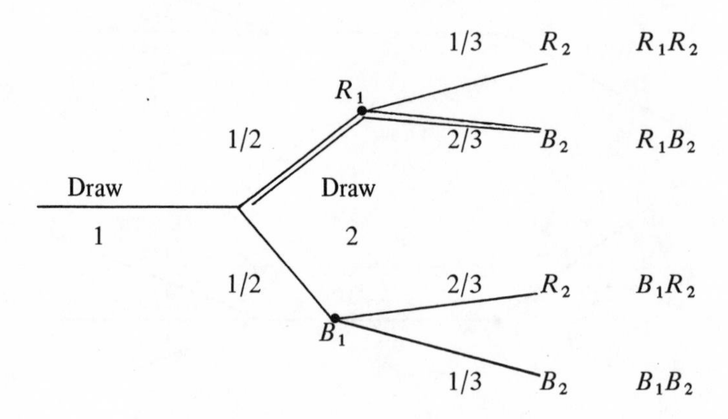

Figure 6.3 Modified event space.

It should be apparent that the structuring of an event tree is facilitated if a Venn diagram or listing of the events at each draw is available. The tree, with the probabilities of simple events assigned to each branch of the tree, is more specific than the listing or Venn diagram and contains more details on how the complex events occur. As such it is very useful in recording and calculating the probabilities of events at any stage (draw) of the experiment. The event tree is widely employed in fields of analysis other than probability. It is variously known as a decision tree, network, or probability tree.

The *third* step in the procedure is to assign probabilities to the simple events which make up the complex event. For example, we wish to find the probability $p(R_2R_1/X)$ of the complex event R_2R_1. If R_2 and R_1 were independent (balls were replaced after each draw $= \bar{X}$), then

$$p(R_2R_1/\bar{X}) = p(R_2/\bar{X}) \cdot p(R_1/\bar{X})$$

However, the events R_1 and R_2 are not independent, so we must use the formula given previously as

$$p(R_2R_1/X) = p(R_2/R_1X) \cdot p(R_1/X)$$

Thus we need to determine the probabilities for the simple events $p(R_1/X)$ and $p(R_2/R_1X)$. To do this we use the event spaces S_1 and S_2. On the first draw there are two red and two black balls in the box each having an equal probability (a priori) of being drawn. Thus, for S_1

$$p(R_1/X) = 1/2$$

The second event R_2/R_1X states that when R_1 is withdrawn on the first draw it is not replaced. Hence the event space for the second draw is S_2. In S_2 there are one red and two black balls. The probability of drawing a red ball from S_2 is

$$p(R_2/R_1X) = 1/3$$

These probabilities were previously computed for the event tree. They were repeated here to illustrate that the Venn diagram or listings could also be employed in finding these values.

The *fourth* and final step in the procedure is to substitute the probability values for the simple events into the correct formula for the complex event and complete the computation. This step normally is the simplest of the four required steps. In this case it results in the following.

$$\begin{aligned} p(R_2R_1/X) &= p(R_2/R_1X) \quad p(R_1/X) \\ &= 1/3 \cdot 1/2 \\ &= 1/6 \end{aligned}$$

A few words on the application of the above procedures in practice. For the reader who has little background in probability it is recommended that the steps be followed carefully for the first dozen or so probability calculations. After that, some of the steps in the procedures will be combined naturally as confidence builds with experience. The analyst should be aware that care must be exercised in accomplishing each of the steps. Special care should be taken to assure that all conditions (such as X above) are used in the calculations. Otherwise, errors will be made in assigning probabilities to the simple events, especially where the events are not independent or mutually exclusive.

Example 6.13 (Independent Events). In this example we will repeat the previous example with one modification. The draws will be independent of each other. To do this, we replace the ball drawn on the first draw prior to initiating the second draw. Thus the condition X will be replaced by the condition Y identified as

Y: Four balls in box, two red and two black.
Ball drawn is replaced.

Again we wish to determine the probability of drawing two red balls, i.e., we wish to evaluate the probability denoted by

$$p(R_2R_1/Y) = p(R_2/R_1Y) \cdot p(R_1/Y)$$

With the statement Y and the expression for the probability given above, we have completed the first step. In the second step we will again use the Venn diagram and listing methods and the event tree to describe the event spaces for the first and second draws. On the first draw the event space is denoted S_1 and its Venn diagram is depicted as

S_1

R	R
B	B

Since balls are replaced after each draw, the event space does not change from draw to draw. The Venn diagram for the event space for the second draw is depicted as

S_2

R	R
B	B

Now we will construct the event tree for the two draws noting again that the balls are replaced after each draw and each ball in the box has an equal probability of being selected on each draw. On the first draw either a red or a black ball, but not both, is drawn from the box. It is noted and replaced in box. The second draw is essentially identical to the first. The event tree, with probabilities on each event denoted, is given in Figure 6.4.

This tree has the same structure as the event tree for dependent events. The important difference is the assignment of probabilities to each of the events (draws). Thus the event spaces are identical as sets but the probabilities assigned to the occurrence of events in the space differ.

By completing the event tree with probabilities assigned, we have completed Step 3, i.e., obtained the probabilities for the simple events which make up the complex events. Having determined the probabilities, we are now ready to complete Step 4, the substitution of the probability values in the formula for the complex event given as

$$p(R_2R_1/Y) = p(R_2/R_1Y) \cdot p(R_1/Y)$$
$$= 1/2 \cdot 1/2 = 1/4$$

■■

Exercise 6.10 If R_1 and R_2 are independent events (Y), complete the formula

$$p(R_2R_1/Y) = \_\_\_\_\_\_\_\_\_\_\_\_$$

■■

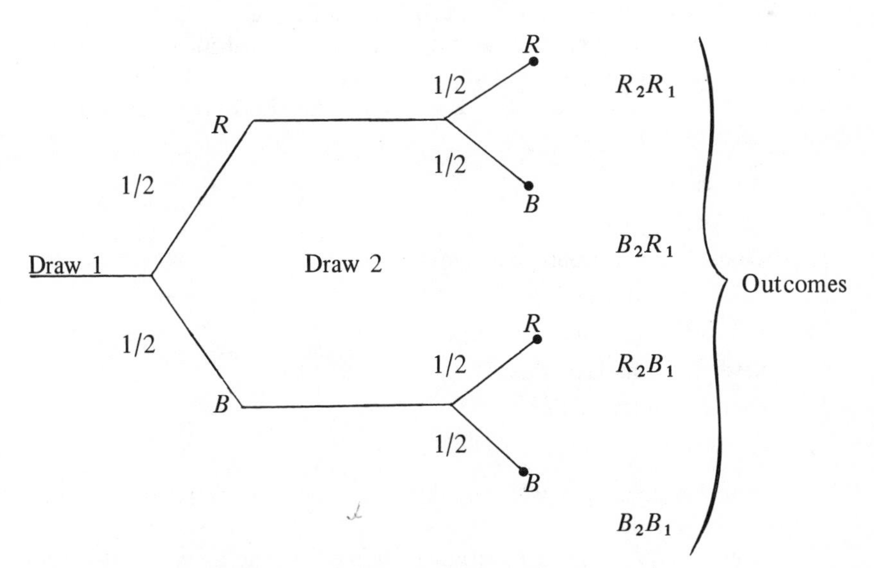

Figure 6.4 Event tree (independent events)

Exercise 6.11 If R_1 and R_2 are dependent events (X), complete the formula

$$p(R_2R_1/X) = \_\_\_\_\_\_\_\_\_\_\_\_\_\_\_\_\_\_\_\_$$ ■■

Exercise 6.12 For the following conditions:

X: Four red balls and four green balls are in box.
Y: Balls not replaced after draw.
Z: Balls are replaced after draw.
R_n: Red ball drawn on nth draw.
G_n: Green ball drawn on nth draw.

evaluate the following probabilities:

(a) $p(R_1/R_2XY)$
(b) $p(R_1/R_2XZ)$
(c) $p(R_1R_2/XY)$
(d) $p(R_1R_2/XZ)$
(e) $p(R_1R_2R_3/XY)$
(f) $p(R_1R_2R_3/XZ)$ ■■

Example 6.14 (Subjective Probabilities). Examples 6.12 and 6.13 dealt with event spaces wherein a priori probabilities are assigned to events and there is one and only one such assignment for the event space. This example discusses the situation wherein subjective probabilities are assigned to events but there is more than one interpretation of the composition of the event space and of the assignment of probabilities to the events in the event space.

Let us return to Example 6.10 where the event was identified as E: Country X's reaction to country Y's action. The event space listed eight possible outcomes or reactions to country Y's action. The list is repeated here for ready reference.

E_1: All out war.
E_2: Large-scale war.
E_3: Limited conventional war.
E_4: Guerrilla warfare.
E_5: Boycott, blockade, or seizure.
E_6: Sever diplomatic relations.
E_7: United Nations action.
E_8: Diplomatic protest.

It is assumed that country X will initiate one and only one action in retaliation for country Y's action.

The event space S for this situation is defined by the listing of the eight possible outcomes. Thus Steps 1 and 2 are completed when the outcomes are listed. Note that this is not a unique listing. Other listings and orders are possible. The same can be said about the assignment of probabilities to the events which still remains to be done. In determining the outcomes listed, two important considerations should guide the selection. The outcomes should be mutually exclusive, if at all possible. If two outcomes are not mutually exclusive, then the two should be combined with each other or with other outcomes and the list redefined accordingly. Also, the list of outcomes should be an exhaustive list. One may never be sure (subjectively) that the list is complete, but one will assume it to be until information to the contrary is available.

When the outcomes in the event space are mutually exclusive and exhaustive, then it is possible to assign probabilities to the events so that

$$p(E_1) + p(E_2) + \cdots + p(E_8) = 1$$

In Chapter 3 we described how subjective probabilities can be assigned to a set of mutually exclusive and exhaustive set of events. The reader should review the appropriate sections of Chapter 3 if necessary. This will complete Step 3 which is the determination of probabilities for the simple events. To complete Step 4, one will substitute the probabilities of the simple events into the formula defining the probability for the complex event and perform the indicated calculations. ■■

6.11 UNION (SUM) OF EVENTS

In this section we will consider a set of events $A, B, C, \ldots,$ which may or may not be mutually exclusive. We are interested in determining the probability of two or more of the events occurring simultaneously. Specifically, for two events A and B we would like to know with what probability A or B will occur when we know the probabilities for A occurring and B occurring. If A and B are simple events then the complex event: A or B or both occurring is denoted $A + B$ or $A \cup B$. The complex event A or B but not both occurring is denoted $A\bar{B} + \bar{A}B$. Thus the probabilities are denoted as

$p[(A + B)/X]$ = probability that event A or event B will occur

$p[(A\bar{B} + \bar{A}B)/X]$ = probability that events A or B, but not both, will occur

The Venn diagrams for the two complex events are given in Figure 6.5.

In Figure 6.5 (a) the cross-hatched area represents the complex event $(A + B)/X$. In Figure 6.5 (b) the cross-hatched area represents $(A\bar{B} + \bar{A}B)/X$. Note that the intersection $A \cdot B$ is not included in $A\bar{B} + \bar{A}B$. The part not included represents the events in A and B which occur simultaneously in S.

Hence $A \cdot B$ is contained in diagram (*a*) but not in (*b*). Note also that the condition X restricts the original event space to the X section of the space designated by horizontal lines. If the condition X was not included in the probability statement, then the complete event space S would be considered and the complex events $A + B$ and $AB + AB$ enlarged to the complete circles accordingly.

When the events A and B are mutually exclusive, then the probability of the union of the events occurring is calculated by

$$p[(A+B)/X] \quad = \quad p(A/X) + p(B/X)$$

When the events A and B are not mutually exclusive, then the probability of the union of the events occurring is calculated by

$$p[(A+B)/X] \quad = \quad p(A/X) + p(B/X) - p(AB/X)$$

The term $-p(AB/X)$ enters into this calculation to account for the instances when events A and B both occur and hence contribute to the sum through both $p(A/X)$ and $p(B/X)$. Recall that the complex event $A + B$ is read "at least one event will occur."

To understand the formula for $p(A + B/X)$ more fully, let us return to the counting formula used in sets and logic. The formula stated that for two sets A and B, the number of elements in the union of the two sets is given by

$$N(A+B) \quad = \quad N(A) + N(B) - N(AB)$$

If S is the universal set and $N(S)$ designates the number of elements in S, then the probabilities that an element randomly selected from S belongs to A, B, AB, or $A + B$ are given by

$$\frac{N(A)}{N(S)}, \frac{N(B)}{N(S)}, \frac{N(AB)}{N(S)} \quad \text{and} \quad \frac{N(A+B)}{N(S)}$$

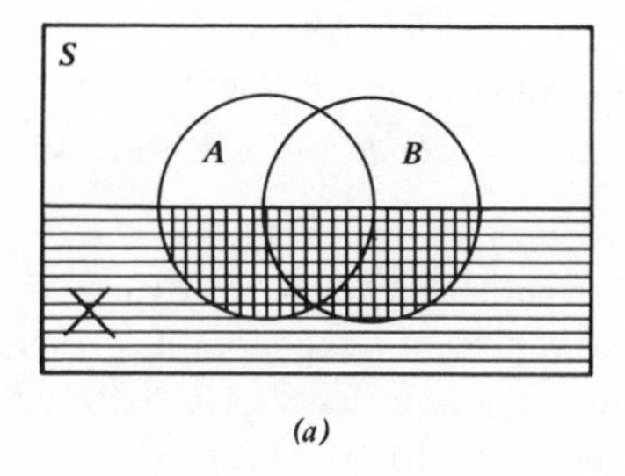

(*a*)

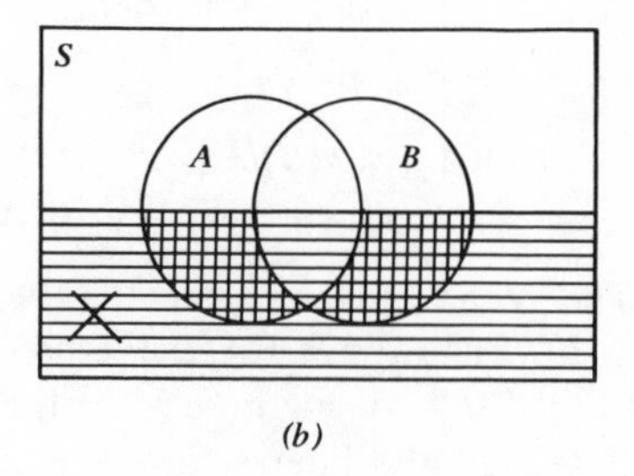

(*b*)

Figure 6.5 Venn diagrams for (*a*) $(A + B)/X$ and (*b*) $(A\bar{B} + \bar{A}B)/X$.

respectively. The above ratios are relative frequencies for the occurrence of the indicated events A, B, AB, and $A + B$. If we divide both sides of the equation for $N(A + B)$ by $N(S)$, we obtain

$$\frac{N(A + B)}{N(S)} = \frac{N(A)}{N(S)} + \frac{N(B)}{N(S)} - \frac{N(AB)}{NS}$$

which can also be expressed as

$$p[(A + B)/X] = p(A/X) + p(B/X) - p(AB/X)$$

The formulas tell us that for finite event spaces, wherein each event has an equal probability of occurring, probabilities for simple or complex events can be assigned by counting the number of elements in the appropriate event spaces and determining the relative frequencies of occurrence. When the events do not all have equal probability of occurring, then a listing or event tree usually is better suited for the assignment of probabilities to the simple events. We now illustrate how to use the formula in several simple situations.

Example 6.15 (Counting Formula). Let the space S consist of 16 points arranged in a square as follows:

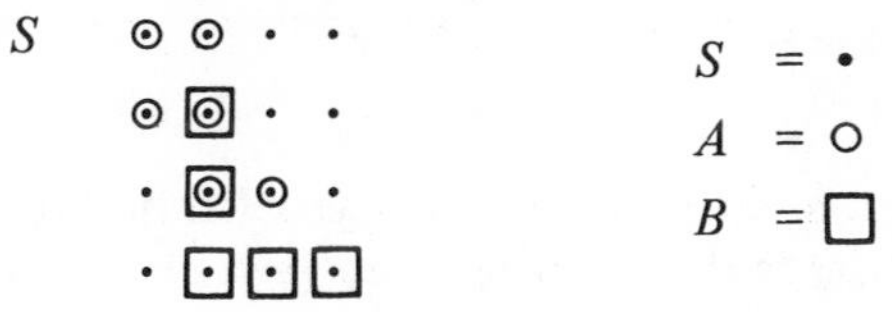

S ⊙ ⊙ • •
⊙ ⊡̥ • •
• ⊡̥ ⊙ •
• ⊡ ⊡ ⊡

S = •
A = ○
B = □

Let the sets A and B be the points of S enclosed by circles and squares, respectively. Then by actual count we have

$N(A) = 6 =$ number of points of S circled

$N(B) = 5 =$ number of points of S squared

$N(AB) = 2 =$ number of points of S both circled and squared

Hence the number of points in either A or B (at least one of the sets A and B) is given by

$$N(A + B) = 6 + 5 - 2 = 9$$

This number can be checked by actual count.

The probability that a point randomly selected from S will be contained in either A or B is

$$\frac{N(A + B)}{N(S)} = \frac{N(A)}{N(S)} + \frac{N(B)}{N(S)} - \frac{N(AB)}{N(S)}$$

$$\begin{aligned} p(A + B/X) &= p(A/X) + p(B/X) - p(AB/X) \\ &= 6/16 + 5/16 - 2/16 = 9/16 \end{aligned}$$

■■

Example 6.16 (Missile Hit Probabilities). The probability that a missile will hit its target is given as 0.9. If two missiles, from different launchers, are fired at a target, what is the probability that:

(a) Both will hit the target.
(b) At least one will hit the target.
(c) Neither will hit the target.

Let A, B be the events "missile will hit its target." Then $p(A)$ and $p(B)$ are the probabilities that the missiles (designated A and B) will hit their targets. Thus we have given

$$p(A) = 0.9 \qquad p(B) = 0.9$$

The event "both A and B will hit their targets" is denoted AB. The event "at least one will hit its target" is denoted $A + B$. The event "neither will hit its target" is denoted $\bar{A} \cdot \bar{B}$. Thus we seek the probabilities

(a) $p(AB)$
(b) $p(A + B)$
(c) $p(\bar{A} \cdot \bar{B})$

Before one should attempt to solve any problem in probability, one should determine the answer to the following three questions:

Are the events mutually exclusive?
Are the events exhaustive?
Are the events independent?

In this example the events are *not mutually exclusive* because the firing of one missile does not preclude the firing of the other; the events are *independent* because the firing of one missile has no affect on the firing or nonfiring of the other, and it cannot be determined from the problem statement whether the events are exhaustive. The answers are as follows:

(a) Since the events are independent, we write

$$\begin{aligned} p(AB) &= p(A)p(B) \\ &= 0.9 \cdot 0.9 \\ &= 0.81 \end{aligned}$$

(b) Since A and B are not mutually exclusive, we use the formula

$$\begin{aligned} p(A + B) &= p(A) + p(B) - p(A)p(B) \\ &= 0.9 + 0.9 - 0.81 \\ &= 0.99 \end{aligned}$$

(c) If A and B are independent, so also are $\bar{A}$ and $\bar{B}$. Thus we use the formula

$$\begin{aligned} p(\bar{A} \cdot \bar{B}) &= p(\bar{A}) \cdot p(\bar{B}) \\ &= [1 - p(A)] \cdot [1 - p(B)] \\ &= 0.1 \cdot 0.1 \\ &= 0.01 \end{aligned}$$

■■

Example 6.17 (Torpedo Hit Probabilities). The probability that a torpedo salvo will hit a ship is 3/4 on the first firing. Due to time lags, evasive action, etc., the probability of a torpedo salvo hit drops to 1/2 on the second firing. If a submarine fires two salvos of torpedoes at a ship, what are the probabilities that:

1. Neither salvo will hit.
2. Both salvos will hit.
3. At least one salvo will hit.
4. Only one salvo will hit.
5. At most one salvo will hit.

Let H_1, H_2 be the events "torpedo salvo will hit on the first and second firings, respectively. Let X represent the condition X: Fire two salvos of torpedoes in quick succession, the firings being independent of each other.

$p(H_1/X)$ = probability that the first salvo will hit target given that two salvos are fired

$p(H_2/X)$ = probability that the second salvo will hit target given that two salvos are fired

Hence, we are given that

$$p(H_1/X) = 3/4 \qquad \text{and} \qquad p(H_2/X) = 1/2$$

As stated before, the event "neither salvo will hit target" is interpreted as not H_1 and not H_2 or $\bar{H}_1 \cdot \bar{H}_2$. The event "both salvos will hit target" is interpreted as H_1 and H_2 or $H_1 \cdot H_2$. The event "at least one salvo will hit target" is interpreted as H_1, or H_2, or H_1 and H_2, i.e., $H_1 + H_2$. Saying "at least one" will occur is equivalent to saying "it is false that neither will occur," i.e., $\overline{\bar{H}_1 \cdot \bar{H}_2}$.

The event "only one salvo will hit" is interpreted as H_1, or H_2, but not H_1 and H_2, and not $\bar{H}_1$ and $\bar{H}_2$, i.e., $H_1\bar{H}_2 + \bar{H}_1 H_2$. Saying "only one" is equivalent to saying "exactly one." The event "at most one salvo will hit the target" is interpreted as H_1, or H_2, or $\bar{H}_1$ and $\bar{H}_2$, and not H_1 and H_2, i.e., $\bar{H}_1 + \bar{H}_2$. Saying "at most one," is equivalent to saying "none or no more than one."

The key conditions on the events are as follows: The events are not mutually exclusive since the occurrence of the first firing does not preclude the occurrence of the second firing. The events are considered independent of each other by agreement, and the exhaustivity of the events cannot be determined from the problem statement. The answers to the five questions are

1. Since the salvos are independent,

$$\begin{aligned} p(\bar{H}_1\bar{H}_2/X) &= p(\bar{H}_1/X)\ \ p(\bar{H}_2/X) \\ &= 1/4 \cdot 1/2 \\ &= 1/8 \end{aligned}$$

2. By the independence of the salvos,

$$\begin{aligned} p(H_1H_2/X) &= p(H_1/X)\ \ p(H_2/X) \\ &= 3/4 \cdot 1/2 \\ &= 3/8 \end{aligned}$$

3. Since the salvos are *not* mutually exclusive,

$$\begin{aligned} p(H_1 + H_2/X) &= p(H_1/X) + p(H_2/X) - p(H_1H_2/X) \\ &= 3/4 + 1/2 - 3/8 \\ &= 7/8 \end{aligned}$$

4. Since $H_1\bar{H}_2$ and $\bar{H}_1 H_2$ are mutually exclusive,

$$\begin{aligned} p(H_1\bar{H}_2 + \bar{H}_1H_2/X) &= p(H_1\bar{H}_2/X) + p(\bar{H}_1H_2/X) \\ &= 3/4 \cdot 1/2 + 1/4 \cdot 1/2 \\ &= 1/2 \end{aligned}$$

5. Since the salvos are not mutually exclusive,

$$\begin{aligned} p(\bar{H}_1 + \bar{H}_2/X) &= p(\bar{H}_1/X) + p(\bar{H}_2/X) - p(\bar{H}_1\bar{H}_2/X) \\ &= 1/4 + 1/2 - 1/8 \\ &= 5/8 \end{aligned}$$

■■

Example 6.18 (Simplifying Probability Calculations). There are many equivalences among complex statements which can be employed advantageously to simplify some of the calculations required in determining probabilities of complex statements. For example, if A and B are statements, then

1. $\overline{A + B} = \bar{A}\bar{B}$ (none will occur)
2. $\bar{A} + \bar{B} = \overline{AB}$ (at most one will occur)
3. $\overline{AB + \bar{A}\bar{B}} = A\bar{B} + \bar{A}B$ (exactly one will occur)
4. $p(\bar{A}) = 1 - p(A)$ (complementary probability)

Equivalence (4) is especially useful where the probability of one or more events occurring out of many events is to be calculated. To illustrate a typical calculation, we compute the probability that if you select two cats at random at least one of them will be a female. We assume that the probability of selecting a female cat is 0.52. Hence the probability of selecting a male cat is 0.48.

We note first that the events are independent and not mutually exclusive. We seek the probability of the statement "at least one will be female." This is denoted $p(A_1 + A_2/X)$. We will find the probability of selecting no females, i.e., $p(\overline{A_1 + A_2}/X)$ or $p(\bar{A}_1\bar{A}_2/X)$. The formula for the latter probability is

$$\begin{aligned} p(\bar{A}_1\bar{A}_2/X) &= p(\bar{A}_1/X) \cdot p(\bar{A}_2/X) \\ &= 0.48 \cdot 0.48 \\ &= 0.23 \end{aligned}$$

The probability of selecting at least one female is complementary to selecting no females, i.e.,

$$\begin{aligned} p(A_1 + A_2/X) &= 1 - p(\bar{A}_1\bar{A}_2/X) \\ &= 1 - 0.23 \\ &= 0.77 \end{aligned}$$

We could have calculated the probability directly from the formula

$$\begin{aligned} p(A_1 + A_2/X) &= p(A_1/X) + p(A_2/X) - p(A_1A_2/X) \\ &= 0.52 + 0.52 - 0.52 \cdot 0.52 \\ &= 1.04 - 0.27 \\ &= 0.77 \end{aligned}$$

As we shall see later on, where many events are involved, the latter calculation will be considerably more complex than the former. Hence, where lengthy probability calculations are involved, it will pay to determine whether simplifying statement equivalences are available.■■

Exercise 6.13. Write out the probability statements for the following formulas.

$$p(R_1R_2/X) \qquad p(R_1\bar{R}_2 + \bar{R}_1R_2/X)$$
$$p(\overline{R_1R_2}/X) \qquad p(R_1 + R_2/X)$$

Let the condition X and events R_n be defined as follows:

X: There are two red and two green balls in an urn. Balls are drawn and not replaced.

R_n: A red ball is drawn on the nth draw.

Find the probability that,

(a) At least one red ball will be drawn in two draws.
(b) Exactly one red ball will be drawn in two draws.

6.12 THREE OR MORE EVENTS

Up to now we have been considering the probability of complex events containing only two simple events. We considered the union and product of two events A and B and determined the probabilities $p(A + B/X)$ and $p(AB/X)$ from the probabilities $p(A)$ and $p(B)$. We now consider the situation where more than two events make up the complex event. For example, the following are typical probability statements involving more than two events.

1. Drawing four aces in five card poker.
2. Making three hits out of five firings.
3. Having at least two girls in a family of five children.
4. Having a stock market advance, the prime interest rate increase, and federal spending decrease simultaneously.

To compute the probabilities of three or more events occurring in any of their possible combinations, we will require an extension (generalization) of the formulas employed for two events to the case where three or more events are involved. These generalizations are as follows:

1. If A, B, and C are *independent events,* then the probability of the three events occurring simultaneously is given by

$$p(ABC/X) = p(A/X)\ p(B/X)\ p(C/X)$$

2. If any pair of the events are *mutually exclusive,* then

$$p(ABC/X) = 0$$

3. If the three events are *not mutually exclusive,* then the probability of at least one of the events occurring is given by

$$p(A + B + C) = p(A/X) + p(B/X) + p(C/X) - p(AB/X) - p(AC/X) - p(BC/X) + p(ABC/X)$$

4. If four or more events are *not mutually exclusive,* then the probability of at least one of the events occurring is given by

$$p(A + B + C + \cdots + N/X) = \text{(sum of probabilities of events taken an odd number at a time)} - \text{(sum of probabilities of events taken an even number at a time)}$$

For four events A, B, C, and D, the probability of at least one occurring is

$$p(A+B+C+D/X) = [p(A/X)+p(B/X)+p(C/X)+p(D/X)+p(ABC/X) + p(ABD/X)+p(ACD/X)+p(BCD/X)] - [p(AB/X) + p(AC/X)+p(AD/X)+p(BC/X)+p(BD/X)+ + p(CD/X)+p(ABCD/X)]$$

5. When the events A, B, C, and D are *mutually exclusive,* then the probability that at least one event will occur is the sum of the probabilities that the simple events A, B, C, and D will occur.

$$p(A+B+C+D/X) = p(A/X) + p(B/X) + p(C/X) + p(D/X)$$

■■

Example 6.19 (Complex Statement Probabilities). Let us compute the probability that at least one card in a five card poker hand will be a heart. The sample or event space consists of four suits of 13 cards each as follows:

Hearts	Diamonds	Clubs	Spades
13	13	13	13

In poker, draws are made without replacement. Hence the events will not be independent. Note also that the event "draw at least one heart" is complementary to the event "draw no hearts." Thus we will find the probability of drawing no hearts and then obtain the probability of its complementary event. We now wish to compute the probability given by

$$p(\bar{H}_5\bar{H}_4\bar{H}_3\bar{H}_2\bar{H}_1/X)$$

where H_i is the event "draw a heart on the ith draw" and X is the condition that the draws are made without replacement. The event $\bar{H}_i$ is "a heart is *not* drawn on the ith draw." Since the events are not independent, we must use the conditional probability formula. This is

$$p(\bar{H}_5\bar{H}_4\bar{H}_3\bar{H}_2\bar{H}_1/X) = p(\bar{H}_5/\bar{H}_4\bar{H}_3\bar{H}_2\bar{H}_1X) \cdot p(\bar{H}_4/\bar{H}_3\bar{H}_2\bar{H}_1X) \cdot p(\bar{H}_3/\bar{H}_2\bar{H}_1X) \cdot p(\bar{H}_2/\bar{H}_1X) \cdot p(\bar{H}_1/X)$$

On the first draw, the probability of not drawing a heart ($\bar{H}_1$) is

$$p(\bar{H}_1/X) = 39/52$$

On the second draw, the probability of not drawing a heart ($\bar{H}_2$), given that a heart was not drawn on the first draw and the card was not replaced ($\bar{H}_1/X$), is

$$p(\bar{H}_2/\bar{H}_1X) = 38/51$$

For the third draw we obtain

$$p(\bar{H}_3/\bar{H}_2\bar{H}_1X) = 37/50$$

For the fourth draw we obtain

$$p(\bar{H}_4/\bar{H}_3\bar{H}_2\bar{H}_1X) = 36/49$$

For the fifth draw we obtain

$$p(\bar{H}_5/\bar{H}_4\bar{H}_3\bar{H}_2\bar{H}_1X) = 35/48$$

Hence

$$\begin{aligned} p(\bar{H}_5\bar{H}_4\bar{H}_3\bar{H}_2\bar{H}_1/X) &= 39/52 \cdot 38/51 \cdot 37/50 \cdot 36/49 \cdot 35/48 \\ &= 0.22 \end{aligned}$$

Thus

$$\begin{aligned} p(H_5 + H_4 + H_3 + H_2 + H_1/X) &= 1 - p(\bar{H}_5\bar{H}_4\bar{H}_3\bar{H}_2\bar{H}_1/X) \\ &= 1 - 0.22 \\ &= 0.78 \end{aligned}$$

■■

Exercise 6.15. Find the probability of drawing at least one heart in a five card poker hand without using complementary events as in Example 6.19. Compare the amount of work required.

Hint:

$$p(H_5 + H_4 + H_3 + H_2 + H_1/X) = \sum_{i=1}^{5} p(H_i/X) - \sum_{\substack{i>j \\ 1 \leqslant i,\, j \leqslant 5}} p(H_iH_j/X)$$

$$+ \sum_{\substack{i>j>k \\ 1 \leqslant i,\, j,\, k \leqslant 5}} p(H_i H_j H_k/X) - \sum_{\substack{i>j>k>l \\ 1 \leqslant i,\, j,\, k,\, l \leqslant 5}} p(H_i H_j H_k H_l/X)$$

$$+ p(H_5 H_4 H_3 H_2 H_1/X).$$

Note that sums contain 5, 10, 10, and 5 terms, respectively. Also the terms entering each of the sums are equal. Hence the formula can be rewritten as

$$5 \cdot p(H_1/X) - 10p(H_2 H_1/X) + 10p(H_3\, H_2\, H_1/X) - 5p(H_4\, H_3\, H_2\, H_1/X)$$
$$+\, p(H_5\, H_4\, H_3\, H_2\, H_1/X)$$

■■

6.13 SUMMARY

At this point we list and associate probability statements of complex events with their Venn diagrams and probability notations. Table 6.1 contains all the important probability statements containing two events and also the statements, diagrams, and notation for unconditional probabilities. It is included here for ready reference to all the formulas included in the preceding sections.

6.14 BAYES' FORMULA

There is one more probability formula which is of special interest to the substantive analyst. The formula is known as Bayes' rule. It specifies how probability estimates should be modified on the basis of new information. Bayes' rule is written

$$p(A/BX) \quad = \quad p(A/X)\,\frac{p(B/AX)}{p(B/X)}$$

An interpretation of this formula is as follows: We consider A as an event whose probability of occurrence is $p(A/X)$. We estimate or compute the probability $p(A/X)$ that the event A occurs from the available information signified by the condition X. The event B is relevant to the event A. When event B occurs, we know that the probability that A occurs changes. The question is: What is the change in probability that A occurs when the information that B has occurred is known or hypothesized? Bayes' rule states that the original probability estimate $p(A/X)$ should be modified by the multiplicative factor

$$\frac{p(B/AX)}{p(B/X)}$$

The numerator of this factor is the probability that B occurs conditioned on the occurrence of the event A and previous history X. The denominator of this factor is the estimated or computed probability that B occurs under the event conditions given by X.

Some examples of the use of Bayes' formula should clarify the role of the four probabilities in the formula and illustrate how to apply the formula in different situations.

Table 6.1

Notation and statement	Venn diagram	Probability formulas
$A + B$ (A or B or both)	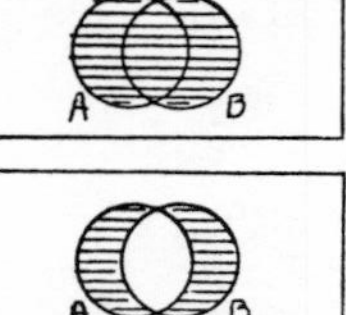	$p(A + B) = p(A) + p(B) - p(AB)$
$A\bar{B} + \bar{A}B$ (A or B but not both)		$p(A\bar{B} + \bar{A}B) = p(A\bar{B}) + p(\bar{A}B)$
$A \cdot B$ (A and B)	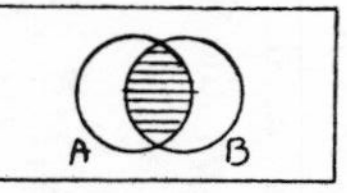	$p(AB) = p(A/B)\, p(B)$
$A \cdot B = \phi$ (A and B mutually exclusive)	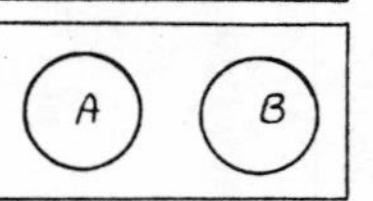	$p(AB) = \phi$
A and B independent		$p(AB) = p(A)\, p(B)$
$\overline{AB} = \bar{A} + \bar{B}$ not (A and B)	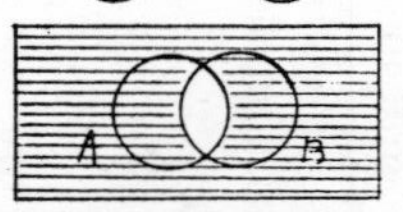	$p(\overline{AB}) = p(\bar{A}) + p(\bar{B}) - p(\bar{A}\,\bar{B})$
$\bar{A}\,\bar{B} = \overline{A + B}$ (not A and not B)	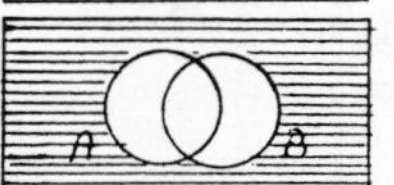	$p(\bar{A}\,\bar{B}) = p(\bar{A}/\bar{B})\, p(\bar{B})$
$AB + A\bar{B} + \bar{A}B + \bar{A}\,\bar{B}$ = universe	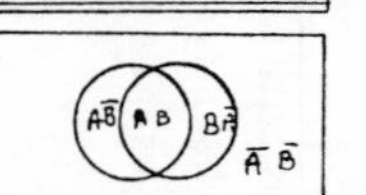	$p(AB + A\bar{B} + \bar{A}B + \bar{A}B) = \perp$

Example 6.20 (Simple Bayes' Rule). Suppose that information collected indicates that country Z will take one, but only one, of three actions against a second country. These actions are:

A: Diplomatic protest.

B: United Nations action.

C: Sever diplomatic relations.

After careful study of country Z's international relations, it is estimated that country Z will initiate the actions with the following probabilities:

$$p(A/X) = 0.4 \qquad p(B/X) = 0.2 \qquad \text{and} \qquad p(C/X) = 0.4$$

Then, through new sources, additional information is obtained. The information concludes that country Z will not issue a diplomatic protest ($\bar{A}$). The problem is: Based on the new information, what adjustments should be made in the current probabilities on country Z's actions. Specifically, if $\bar{A}$ is accepted as true, then what is the probability that the country Z will initiate action B. This probability is expressed as

$$p(B/\bar{A}X)$$

Bayes' rule states that the original probability estimate $p(B/X)$ is to be changed according to the following formula:

$$p(B/\bar{A}X) = p(B/X) \frac{p(\bar{A}/BX)}{p(\bar{A}/X)}$$

Evaluating the factors in the formula, we obtain from the given probabilities that $p(B/X) = 0.2$ and $p(\bar{A}/X) = 0.6$. The probability for $p(\bar{A}/BX)$ remains to be computed. But this probability is 1 since the condition BX states that action B has been taken, and since only one action will be taken, we are sure that actions A and C will *not* be taken. Thus, with $p(\bar{A}/BX) = 1$ we have that

$$p(B/\bar{A}X) = 0.2 \cdot \frac{1}{0.6} = 1/3$$

We see that the information given by A leads us to change $p(B/X)$ from 0.2 to $p(B/\bar{A}X) = 0.33$.

If we check the change the information $\bar{A}$ occasions in the probability for action C, we obtain

$$\begin{aligned} p(C/\bar{A}X) &= p(C/X) \frac{p(\bar{A}/CX)}{p(\bar{A}/X)} \\ &= 0.4 \ \frac{1}{0.6} \\ &= 2/3 \end{aligned}$$

Thus on the basis of the information $\bar{A}$ the original probabilities for A, B, and C are changed to

$$p(A/\bar{A}X) = 0$$
$$p(B/\bar{A}X) = 1/3$$
$$p(C/\bar{A}X) = 2/3$$

■■

Example 6.21 (Bayes' Rule Application). In this example the probabilities that enter into the Bayes' formula are obtained by analyzing the event space and using the ratio of successes to total number of events as the event's probability.

You receive ten transfer students from a school whose records were destroyed in a fire. You are told that the group consists of three A, six B, and one C students but you are not told which are the A, B, or C students. To determine the grade levels you test the ten students. The first student finishing the test achieves a grade higher than C. What is the probability that he is an A student? What is the probability that the second student completing the test will be an A student? ■■

Solution: We assign the notation A_1, A_2, and C_1 to the following events:

A_1: First student finished achieves an A grade.
A_2: Second student finished achieves an A grade.
C_1: First student finished achieves a C grade.

We also know that the event space consists of three A students, six B students, and one C student. We list or enumerate the event space as

Event Space	
$3A$, $6B$, $1C$	Before test
$3A$, $5B$, $1C$ $2A$, $6B$, $1C$	After first student

Note that since the first student finished is not a C student he can be either an A or B student. Hence the two possible listings for the event space.

We first seek the probability that the first student finished was an A student given that he scored higher than the C level. Thus, we use Bayes' formula to determine the modifying factor for the a priori estimate that any student finishing first is an A student. Specifically,

$$p(A_1/\bar{C}_1) = p(A_1) \frac{p(\bar{C}_1/A_1)}{p(\bar{C}_1)}$$

The a priori probabilities $p(A_1)$ and $p(\overline{C}_1)$ are determined by referencing the event space before the test. We note that

$$p(A_1) = 3/10 \qquad \text{and} \qquad p(\overline{C}_1) = 9/10$$

We want to determine the probability that the first student finished is not a C student $(\overline{C}_1)$ given that the first student finished is an A student. This probability is obviouslv equal to 1. Thus

$$\begin{aligned} p(A_1/\overline{C}_1) &= p(A_1)\ \frac{p(\overline{C}_1/A_1)}{p(\overline{C}_1)} \\ &= 3/10 \cdot \frac{1}{9/10} \\ &= 1/3 \end{aligned}$$

To determine the probability that the second student completing the test will be an A student (given that the first student is not a C student) we apply Bayes' formula again.

$$p(A_2/\overline{C}_1) = p(A_2)\ \frac{p(\overline{C}_1/A_2)}{p(\overline{C}_1)}$$

Again, by the event space listing

$$p(A_2) = 3/10 \qquad p(\overline{C}_1) = 9/10$$

However, the probability $p(\overline{C}_1/A_2)$ is a little trickier to calculate. It asks for the probability that the first student will not be a C student given that the second student finished is an A student. To obtain this probability we must modify the event space listing before the test in order to reserve one A student as the second student completing the test. The event space listing becomes

$$2A \qquad 6B \qquad 1C \qquad \text{before test}$$

These are eight non-C students and nine to choose from. Thus

$$p(\overline{C}_1/A_2) = 8/9$$

Substituting in Bayes' formula

$$\begin{aligned} p(A_2/\overline{C}_1) &= p(A_2)\ \frac{p(\overline{C}_1/A_2)}{p(\overline{C}_1)} \\ &= 3/10 \cdot \frac{8/9}{9/10} \\ &= 8/27 \end{aligned}$$

■■

Example 6.22 (Target Analysis–Bayes' Rule). Military air reconnaisance is used to pinpoint military targets for artillery fire or air strikes. In one military situation four out of ten targets hit, without being previously reconnoitered, turned out to be good targets. The military commander was not satisfied with these results. He initiated target reconnaisance which resulted in six out of ten targets reconnoitered being recommended as targets to strike. When the targets actually hit were inspected more closely, it was determined that eight out of ten of the good targets were identified through the new target reconnaisance procedures. Being briefed on these results, the commander asked the question, "What is the probability that targets identified as potentially good targets by reconnaisance will actually be good targets to strike?"

This question is answered through a simple application of Bayes' formula.

$$p(G/RX) = p(G/X) \frac{p(R/GX)}{p(R/X)}$$

where

$p(G/X)$ = probability of a target being a good target (before reconnaisance) = 4/10

$p(R/X)$ = probability that a reconnoitered target will be identified as a good target = 6/10

$p(R/GX)$ = probability that a good target will have been identified in the reconnaisance process = 8/10

$p(G/RX)$ = probability that a target selected by reconnaisance turns out to be a good target

Substituting the values for each of the first three probabilities into Bayes' formula, we obtain that

$$p(G/RX) = 4/10 \cdot \frac{8/10}{6/10} = 8/15$$

Thus the new evaluation criteria improved target selection from 4 successes in 10 to 8 successes in 15. ■■

Exercise 6.16

(a) An urn contains nine red balls and seven green balls. Balls drawn are not replaced. What is the chance of drawing a red ball on the second draw after a green ball was drawn on the first draw?

(b) Repeat Exercise (a) with the change that balls are replaced after a draw.

(c) An urn contains three red, four blue, and five white balls. Balls drawn are not replaced. We are told that in two draws the second draw was a white ball. What is the probability that the first ball drawn was not a red ball?

(d) Repeat Exercise (c) with the change that balls are replaced after a draw.

6.15 BAYES' RULE (RECURSIVE FORM)

Bayes' formula can be used to assess the changes in probability of an event due to a sequence of relevant events occurring. For if we know the probability that event A will occur is $p(A/X)$, then when event B occurs and is relevant to the occurrence of A, we use Bayes' rule to determine the new probability

$$p(A/BX) = p(A/X) \frac{p(B/AX)}{p(B/X)}$$

If another event C relevant to A occurs, we can consider the conditions B and X jointly as the more current condition X_1. The effect of C on the probability of A is then given as

$$p(A/CX_1) = p(A/X_1) \frac{p(C/AX_1)}{p(C/X_1)}$$

where

$$p(A/X_1) = \frac{p(A/X) \; p(B/AX)}{p(B/X)}$$

As more events $D, E, F, \ldots,$ occur, we can assess the effect of each in turn on the probability of event A in the manner described above. Thus we can apply Bayes' rule recursively as new evidence or events relevant to A are identified.

There is another version of Bayes' rule which is more useful in substantive analysis than the one discussed. Before we describe this second version, which we will call *recursive Bayes,* we introduce the concept of odds and discuss some relationships which exist between odds and probabilities.

6.16 ODDS AND PROBABILITY

The concept of odds is closely related to that of probability. This relationship is illustrated by means of the following simple example. Suppose four aces from a deck of cards are randomly mixed and placed face down on a table. All should agree that the probability of picking the ace of spades is equal to the probability of picking any one of the other three aces, all having the probability of 1/4 of being picked. Now suppose that someone wants to bet you that he can pick the spade over the other three aces. He figures that he has one chance of picking the ace of spades and three chances of picking one of the other three aces. He rationalizes (or should rationalize) that if he is to bet against you, he would expect you to put up $3 to every dollar he is willing to bet. What he is saying is that the *odds* in his favor are one-to-three. Intuitively, he defines his betting odds to be,

$$\text{odds} = \frac{\text{number of favorable outcomes}}{\text{number of unfavorable outcomes}} = 1/3$$

The above definition of odds is applicable in situations where the number of events in the event space is finite. It is the one used in the remainder of this chapter. The definition for the a priori or statistical probability of events such as drawing a specific ace is,

$$\text{probability} = \frac{\text{number of favorable outcomes}}{\text{total number of outcomes}}$$

Notice the difference in the denominators of the ratios for odds and probability.

Example 6.23 (Probability and Odds)

1. What are the odds of obtaining a 7 in one roll of two die?

To answer this, we need determine the number of favorable and unfavorable events in the event space. Using the matrix representation of the event space (Example 6.8) we note that there are six ways of obtaining a total of 7. These are, 1 + 6, 2 + 5, 3 + 4, 4 + 3, 5 + 2 and 6 + 1. There are 30 ways of not obtaining a total of 7. Hence

$$\begin{aligned} \text{odds} &= \frac{\text{number of favorable outcomes}}{\text{number of unfavorable outcomes}} \\ &= 6/30 \\ &= 1/5 \end{aligned}$$

and the bet is one dollar to five dollars that a roll will result in a 7.

2. What are the odds that a coin will come up heads?

In this situation the event space is simple. It consists of two events, heads which is favorable, and tails which is unfavorable. Thus the odds are 1/1, or even, for obtaining a head.■■

Many analysts prefer to work with probabilities rather than odds. In some instances the probability of an outcome is provided but the odds for the outcome occurring are required. For example, weather reports give the probability of rain on a day-to-day basis. However, in some instances the odds in favor of rain is the required datum. Hence, on the basis of preference or need, we require a means of converting from probabilities to odds and vice versa. This is a simple conversion and it is accomplished by means of the following formulas:

$$\text{probability} = \frac{\text{odds}}{\text{odds} + 1}$$

$$\text{odds} = \frac{\text{probability}}{1 - \text{probability}}$$

Example 6.24 (Probability and Odds–Conversion)

1. The odds for obtaining a 7 in a roll of two die is 6/30. What is the probability of obtaining a 7?

$$\text{probability} = \frac{\text{odds}}{\text{odds} + 1} = \frac{6/30}{6/30 + 1} = 6/36 = 1/6$$

2. The probability of rain today is 0.7. What are the odds that it will rain today?

$$\text{odds} = \frac{\text{probability}}{1 - \text{probability}} = \frac{0.7}{1 - 0.7} = 0.7/0.3 = 7/3$$

3. An urn contains three red, four green, and five black balls. What are the odds that a green ball is withdrawn on the first draw of a ball? What are the odds that on a draw of two balls, both balls will be red? What is the probability of drawing a red ball and then a green ball if the first ball drawn is replaced before the second draw?

To answer the first question in item 3, we use the relative frequency definition for odds,

$$\begin{aligned}
\text{odds} &= \frac{\text{number of favorable outcomes}}{\text{number of unfavorable outcomes}} \\
&= \frac{4 \text{ green}}{3 \text{ red} + 5 \text{ black}} \\
&= 4/8
\end{aligned}$$

To answer the second question in item 3, we first find the probability of drawing two red balls. Then we convert the probability to the odds for the event.

$$\begin{aligned}
p(R_2R_1/X) &= p(R_2/R_1X) \ \ p(R_1/X) \\
&= 2/11 \cdot 3/12 \\
&= 6/132 \\
&= 1/22
\end{aligned}$$

$$\begin{aligned}
\text{odds} &= \frac{\text{probability}}{1 - \text{probability}} \\
&= \frac{1/22}{1 - (1/22)} \\
&= 1/21
\end{aligned}$$

To answer the third question in item 3, we again first find the event's probability and then convert to odds.

$$\begin{aligned} p(G_2R_1/Y) &= p(R_1/Y)\ p(G_2/Y) \\ &= 3/12 \cdot 4/12 \\ &= 1/12 \\ \text{odds} &= 1/11 \end{aligned}$$

Figure 6.6 gives a direct comparison between odds and the equivalent values of probability. The scale is a logarithmic scale which is used to compress the length of line needed to display large ranges of odds. Probabilities range from 0 to 1 but odds range from 0 to ∞. Notice the small changes in probability relative to the large changes in odds for large values of odds. This ability to associate large changes in odds, say from 100/1 to 1000/1 against the smaller probability change from 0.990 to 0.999 when events indicate a large change is in order, is one reason why most analysts should prefer to work with odds rather than probabilities.

Up to this point, we have concerned ourselves with determining odds that a single event will occur. We defined odds as the ratio of favorable outcomes to unfavorable outcomes, i.e.,

$$\text{odds} = \frac{\text{number of favorable outcomes}}{\text{number of unfavorable outcomes}}$$

We had no restrictions on the number of types of unfavorable outcomes. For example, we found the odds of throwing a 7 in a single throw of two die to be 6/30. There were six ways of obtaining a 7 and 30 ways of not obtaining a 7. We did not specify the other events by types (2, 3, 4, 5, 6, 8, 9, 10, 11, 12). Rather, we placed all unfavorable events into one class and accounted for them accordingly.

6.17 ODDS FAVORING ONE OF TWO EVENTS

There is great utility in being able to express odds in terms of one event rather than another event occurring. For example, we are interested in evaluating the

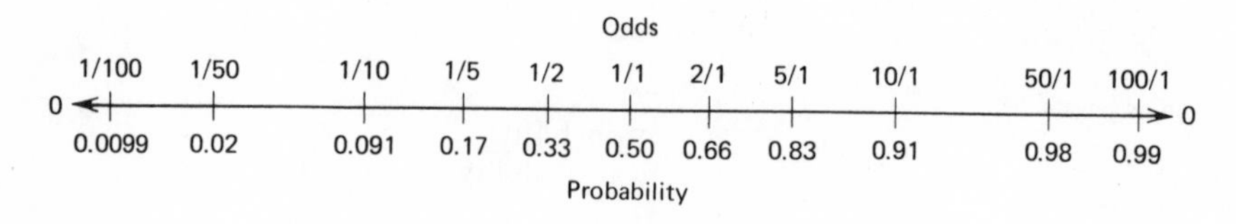

Figure 6.6 Odds and probability—log scale.

odds that the Republicans rather than the Democrats will win the next presidential election. Or we are interested in the odds that a 7 will be thrown before a 6 in the dice game. Where event spaces are easy to list or diagram, odds of this type are easy to evaluate. When the occurrence or nonoccurrence of events in an event space is subjective and the event space is not well-defined, there are several problems to be considered in the estimation of odds. First, let us consider odds relative to the occurrence of two or more events where the event space is well-defined.

We define the odds that event A will occur rather than event B as the ratio of the number of favorable outcomes for A to the number of favorable outcomes for B in the event universe E.

$$\text{odds}\,(A/B) \quad = \quad \frac{\text{number of favorable } A \text{ occurrences}}{\text{number of favorable } B \text{ occurrences}}$$

Example 6.25 (Two-Event Odds–Balls in Urn). Consider the event space consisting of four red, two yellow, and six black balls. We want the odds favoring the draw of a red ball before drawing a yellow one. We draw balls until we obtain a red or yellow ball. A draw of a black ball is ignored in that it is not relevant to our odds. We note that there are four draws which will result in a red ball being drawn and there are two draws for yellow. Hence the odds are

$$\text{odds}\,(R/Y) \quad = \quad 4/2 \quad = \quad 2/1$$

■■

Example 6.26 (Two-Event Odds–Voting). Consider the event space as a town consisting of 52 Republican, 48 Democrat, and 16 Socialist voters. We want the odds that the Republicans will beat the Democrats in the next election. If the Socialists realign their votes, ten going Democratic and six Republican, what are the odds that the Republicans will beat the Democrats?

The answer to the first question is obtained by a head count of the voters (just as polls do with their samples prior to election). This results in

$$\text{odds}\,(R/D) \quad = \quad 52/48 \quad = \quad 13/12$$

This says that Republicans should put up \$13 for every \$12 a Democrat is willing to bet.

The odds change after the Socialists realign themselves ten to six in favor of the Democrats. This results in

$$\text{odds}\,(R + S_1/D + S_2) \quad = \quad (52 + 6)/(48 + 10) \quad = \quad 58/58 \quad = \quad 1/1$$

This states that odds are even and if some voter does not change his mind, the election will result in a dead heat.

There is still another way of looking at odds favoring one event occurring over another. This viewpoint expresses odds as the ratio of the probabilities that the two events will occur. We write this as

$$\text{odds}\ (A/B) = \frac{\text{probability that } A \text{ occurs}}{\text{probability that } B \text{ occurs}}$$

When we express the probabilities of A or B occurring as the ratio of frequencies, we obtain

$$\text{odds}\ (A/B) = \frac{\text{frequency of } A/\text{total frequency}}{\text{frequency of } B/\text{total frequency}} = \frac{\text{frequency of } A}{\text{frequency of } B}$$

From this we see that expressing odds as the ratio of frequencies of success reduces the need to count the number of events contained in the event space. Of course, the two events whose odds are being determined must belong to the same event space.

Expressing odds as the ratio of two probabilities is quite useful in practice. In many instances we can estimate probabilities (subjectively or intuitively) where we cannot count the frequency with which events will occur. The next example illustrates this point.

Example 6.27 (Two-Event Odds–Subjective). What are the odds that country X will recall its ambassador rather than initiate a censure motion in the United Nations? To answer this question, we can attempt to use either of the two definitions for odds. If we want to use the ratio of frequencies, then we will have to look to previous situations wherein country X was faced with a similar decision and selected one of the two actions. If there were some such previous situations, then we could use the frequency of occurrences obtained therein to calculate the odds. However, normally there will be few, if any, historical situations analogous to the current situation. When the latter is the case, it is easier to estimate probabilities for the occurrence of each event. Naturally, these probabilities will be subjective probabilities and will involve a considerable amount of judgment on the part of the analyst. Yet they will be the only estimates available. In this case, the ratio of the probabilities will result in the odds that country X will initiate one action rather than another. ■■

There is an interesting point to note here. Analysts differ in the sense that one may assign uniformly higher probabilities for events to occur than a second analyst. For example, analyst 1 may assign probabilities of 0.6 and 0.3 to the two events (leaving a probability of 0.1 for other actions). Analyst 2 may assign probabilities of 0.4 and 0.2 to the two events (leaving a probability of 0.4 for other actions). Note that each will have assigned an odds of 2/1 in favor of event 1 over event 2. The assignment of probabilities by the two analysts may differ considerably. But, as long as each remains consistently high or low in assigning probabilities, their assignment of odds will be in closer agreement.

6.18 RECURSIVE BAYES

We now return to the recursive form of Bayes' rule and illustrate its use with problems where subjectivity enters into the substantive analysis process. We admit that subjectivity cannot be entirely removed from the process of evaluating information or evidence, but we equally admit that through recursive Bayes' rule, much of the subjectivity can be identified, managed, organized, and quantified to produce better analyses. While the results achieved by different analysts with Bayes' may vary from analyst to analyst, the methods of analysis will be uniform so that differences in results can be equated to subjective differences which can be more readily identified and clarified.

The recursive form of Bayes' rule guides the analyst in his evaluation of new information or evidence. A substantive analyst normally has accumulated a wealth of knowledge on his subject. On the basis of his knowledge, he normally forms many reasonably well-founded opinions on whether event A or event B will occur. For example, on the basis of previous voting history, current events, etc. a political analyst may state at any time that the odds are 9/8 that the Republicans will win the next presidential election (event A) rather than the Democrats (event B). Now suppose that the Gallop and Harris polls indicate a switch in voters preferences. How does the political analyst take this new information (or evidence) into account? Need he revise his odds because of it? If he feels that he must, how will he do it? Will he be able to explain explicitly the whys and hows for the change? Recursive Bayes' will help the analyst do all of these things, and more. It will help him organize his information in such a way that he can make his changes on the basis of his evaluation of specific items of information rather than from his evaluation of all his information simultaneously. Before we illustrate the application of the recursive form of Bayes' formula, we need to introduce a notation for and an explanation of the parts of the formula.

We shall assume that the analyst is concerned with the problem of predicting which one of two events or statements will occur or is true, respectively. For example, the political analyst may be concerned with which of two parties will win the election, the doctor with which of two diseases the symptoms might indicate, the judge or jurist trying to decide whether, on the basis of the evidence, the defendent is guilty or not guilty, and the military commander deciding whether to attack or maintain his ground. We shall call the two alternatives *hypotheses* and label them H_1 and H_2. We shall label the processed data base accumulated and analyzed by the analyst X. The analyst's current opinion on whether H_1 or H_2 is true based on the information X is denoted as

$$p(H_1/X) \quad \text{and} \quad p(H_2/X)$$

Note that these are stated as probabilities. In many situations they will be subjective probabilities or beliefs and hence will be evaluated as indicated in Chapter 3.

On the basis of these probabilities, the analyst can state odds on whether alternative hypothesis H_1 or H_2 is true. The odds are written as

$$\text{odds}\,(H_1{:}H_2/X) = \frac{p(H_1/X)}{p(H_2/X)}$$

These odds are called the prior (or "a priori") odds. They are based on the information X which includes all that is known to the analyst to date. The analyst receives some new information which is relevant to the analysis at hand. For example, the political analyst receives the results of voters' poll or is told that the incumbent party's platform will call for an increase in personal income taxes. The doctor receives the results from a blood test. And a witness states that he saw the defendent near the scene of the crime at the time of the crime. The military commander receives word of enemy reinforcements taking positions against him. The question is, how do these analysts change their current opinions (odds) to include and account for the new information at hand. In most instances there will be a reassessment or an "estimate of the situation" on the basis of the new information. Also, the reassessment will be done on the basis of human judgement, experience, set procedures, prejudices, individual value systems, etc. Let us agree that the analyst does change his original odds based on information X alone to the new odds denoted by

$$\text{odds}\,(H_1{:}H_2/XE) = \frac{p(H_1/XE)}{p(H_2/XE)}$$

where E represents the new information or evidence presented to the analyst. The quantities

$$p(H_1/XE) \quad \text{and} \quad p(H_2/XE)$$

represent the new probabilities for H_1 and H_2 as determined by the analyst after he has evaluated the new evidence E. Analysts work differently at evaluating evidence. Some will start with odds $(H_1{:}H_2/X)$ and estimate new odds $(H_1{:}H_2/XE)$ directly from the evidence E. Other analysts will estimate the original probabilities $p(H_1/X)$ and $p(H_2/X)$. Then when new evidence E is present, they will modify the original probabilities to obtain the new probabilities $p(H_1/XE)$ and $p(H_2/XE)$. Either approach is a matter of personal preference. In either case, the findings of these analyses can be expressed in the form of odds. Thus, we shall say that the analyst changes his original estimate of odds from

$$\frac{p(H_1/X)}{p(H_2/X)} \quad \text{to} \quad \frac{p(H_1/XE)}{p(H_2/XE)}$$

The question is: "What evaluative process did the analyst use in changing from his original odds to his new odds?" Ask any set of analysts who have been assigned to evaluate the same information. The answers obtained are quite diverse and in some cases the analysts cannot provide you with an answer. However, the simple form of Bayes' rule, recursive Bayes, provides the analyst with a modification factor which he can apply to his old odds to obtain his new odds. The modification factor is

$$\frac{p(E/XH_1)}{p(E/XH_2)}$$

This factor is called a *likelihood ratio* and states how the analyst should modify his original odds to account for the gain in information provided by the evidence E.

6.19 LIKELIHOOD RATIO

The likelihood ratio is the ratio of the probability that the information or evidence E appeared as a result of hypothesis H_1 being true, to the probability that the information or evidence E appeared as a result of hypothesis H_2 being true. In the latter probability, the analyst assumes that hypothesis H_2 is true and estimates the probability that the evidence E would appear. In the former probability, the analyst assumes that hypothesis H_1 is true and estimates the probability that the evidence E would appear. Bayes' rule thus provides the analyst with a diagnostic process to use in assessing the worth of new information. The ratio of the two probabilities is the modification factor used in changing the original odds $(H_1{:}H_2/X)$ to the new odds $(H_1{:}H_2/XE)$. The recursive form of Bayes is written as

$$\text{new odds} = [\text{modification factor}] \cdot [\text{old odds}]$$

$$\frac{p(H_1/XE)}{p(H_2/XE)} = \left[\frac{p(E/XH_1)}{p(E/XH_2)}\right] \cdot \left[\frac{p(H_1/X)}{p(H_2/X)}\right]$$

The application of this formula in specific problems requires that all hypotheses and information be clearly identified and stated. Where possible, event spaces should be listed or diagrammed. When new information or evidence is received, the analyst must estimate the credibilities of the two hypotheses based on the evidence. In most problems, the evidence and hypotheses are such that one item of information or evidence will seldom guarantee the truth or falsity of either hypothesis. Hence the analyst must state his degree of belief that the evidence supports the truthhood of each hypothesis. His degree of belief is expressed as a probability. One estimate is $p(H/X)$, the a priori probability, and the others are $p(H/XE)$ and $p(E/XH)$.

Looking a little closer at the latter probability, $p(E/XH)$, this is another way of asking: "What is your belief that the hypothesis H implies or caused the evidence E to appear, i.e., $b(H \rightarrow E)$?" The former probability $p(H/XE)$ asks the converse: "What is your belief that the evidence implies or caused the hypothesis H to appear, i.e., $b(E \rightarrow H)$?" Both of these beliefs are premised on the conditions and information provided by X.

6.20 APPLICATION OF RECURSIVE BAYES

Now let us consider what happens during the application of Bayes' formula to a specific problem. First the analyst estimates his prior odds based on his historical information X and the two stated hypotheses H_1 and H_2. He labels these odds

$$\text{odd}\ (H_1{:}H_2/X) = \frac{p(H_1/X)}{p(H_2/X)} = \theta_1$$

These odds remain fixed until some new information E_1 is uncovered. If E_1 is relevant, the analyst estimates two probabilities given by

$$p(E_1/XH_1) \qquad \text{and} \qquad p(E_1/XH_2)$$

and forms the likelihood ratio L_1

$$L_1 = \frac{p(E_1/XH_1)}{p(E_1/XH_2)}$$

Some analysts prefer to estimate the likelihood ratio without determining the probabilities $p(E_1/XH_1)$ and $p(E_1/XH_2)$. This is an acceptable procedure but the analysts must *not* consider their initial estimate as final. Rather, they must be ready to make modifications in their estimates if they feel that the new odds does not fully represent the impact the new information possesses. To obtain the new odds, the analyst multiplies the likelihood ratio L_1 by the old odds θ_1 to obtain a new set of odds, labeled θ_2

$$\theta_2 = L_1\theta_1$$

At this point the analyst compares his old odds with the new odds and determines whether the change carries the full impact of the information E_1. If he is not satisfied, he reexamines the likelihood ratio to determine where adjustments are necessary. Working between the new odds and the likelihood ratio, he eventually gets to the point where he considers the information E_1 accurately accounted for.

Not all analysts like to work with the likelihood ratio. Some analysts prefer to work directly with the new odds; they estimate the probabilities in the ratio

$$L_1 = \frac{p(E_1/XH_1)}{p(E_1/XH_2)}$$

or state the odds θ_2 directly. These analysts then compare the old odds with the new odds through the ratio $\theta_2/\theta_1 = L_1$ and determine whether the likelihood ratio represents the level of diagnosticity for the information E_1. If not, the analyst adjusts his odds θ_2 until he feels comfortable that both θ_2 and L_1 reflect the information in E_1.

If additional new information E_2 is received and is relevant, the analyst again estimates a likelihood ratio L_2 which accounts for the new information E_2.

$$L_2 = \frac{p(E_2/XH_1)}{p(E_2/XH_2)}$$

At this point in the process the analyst considers θ_2 as his a priori odds, i.e., the odds covering all his information to date and can now modify his old odds by means of a direct estimate θ_3 or the likelihood ratio L_2 computed for the new information E_2. In either case he computes

$$\theta_3 = L_2\theta_2 = L_2L_1\theta_1$$

In proceeding the analyst can again adjust θ_3 and L_2 until he is satisfied with his results.

As more and more information or evidence is received, the analyst can repeat the process until all evidence has been consumed in the analysis process. If there were k units of evidence, then the final odds will be θ_{k+1} which is expressed as

$$\theta_{k+1} = L_k \cdot L_{k-1} \cdots L_2 \cdot L_1 \cdot \theta_1$$

Note that a likelihood ratio L_i is computed for each unit of evidence E_i. These likelihood ratios are multiplied together with the initial odds to yield the final odds for the totality of evidence favoring one hypothesis over the other.

Example 6.28 (Recursive Bayes'–Political Analysis). This example considers an analysis situation where the analyst has assessed the a priori odds for one hypothesis over another and is then provided with three additional units of information. The steps taken in the analysis are as follows:

Step 1. The two (or more) hypotheses of current interest are stated. In this example they are

H_1: The President will recommend an increase in taxes in his next message to Congress.

H_2: The President will not recommend an increase in taxes in his next message to Congress. ■■

Step 2. The main facts leading to an initial odds estimate are stated and the odds given. The facts are

X: President has repeatedly stated that he will do everything possible to avoid a tax increase.

: The President, Congress, and the public have expressed concern over the increased rate of inflation.

: Government expenditures currently continue at the same deficit spending level of the previous three years. Debt ceiling will be exceeded this year.

: President has asked Congress for an increase in the debt ceiling and has been refused.

The initial odds are

$$\frac{p(H_1/X)}{p(H_2/X)} = 1.1/1$$

The analyst feels that the President must do something to force Congress to remove the debt ceiling and control inflation. His odds are dated September 5. ■■

Step 3. New information relevant to the President's decision to ask for a tax increase is recorded and ordered as it is received. Out of many items of information, the following three are deemed relevant and recorded:

September 12: Congress votes a three billion increase in federal expenditure for health, welfare, and environmental programs. (E_1)

September 19: Secretary of Defense states that new offensive and defensive weapon systems costing an additional \$3.5 billion over current defense spending levels must be purchased over the next 2 yr. (E_2)

September 21: Federal Reserve Board recommends an increase in the prime interest rate to help control inflation. (E_3) ■■

Step 4. A likelihood ratio or a final odds is estimated for each item of information E_1, E_2, E_3. If likelihood ratios are computed, the analyst determines the following ratios:

$$\frac{p(E_1/XH_1)}{p(E_1/XH_2)} = 1.2/1 = L_1$$

In this case the analyst feels that Congress is more likely to vote for increased federal expenditures if tax increases are favored by the President.

$$\frac{p(E_2/XH_1)}{p(E_2/XH_2)} = 1.3/1 = L_2$$

In this case the analyst feels that the Secretary of Defense also will attempt to strengthen the defense posture of U.S. if he felt that additional federal revenue was being raised through taxes.

$$\frac{p(E_3/XH_1)}{p(E_3/XH_2)} = 1/1.2 = L_3$$

In this case the analyst feels that the Federal Reserve Board has asked for an increase in prime interest rate because they feel that the President does not favor the tax route for the control of inflation.■■

Step 5. The three likelihood ratios L_1, L_2, and L_3 are multiplied together with the original odds to obtain the new odds.

$$\begin{aligned} \theta_4 &= L_3 \cdot L_2 \cdot L_1 \cdot \theta_1 \\ &= 1/1.2 \cdot 1.3/1 \cdot 1.2/1 \cdot 1.1/1 \\ &= 1.43/1 \end{aligned}$$

The interpretation is that the evidence E_1, E_2, and E_3 has caused the analyst to change his odds in favor of H_1 over H_2 from 1.1 to 1.43.■■

6.21 RECURSIVE BAYES–SUMMARY OF WHAT IT DOES

Now let us evaluate what the recursive form of Bayes' offers the analyst. *First,* it forces him to state explicitly what it is he is trying to determine. The statements or events are stated as hypotheses. *Second,* it forces him to itemize this information. *Third,* it forces the analyst to evaluate the credibility of his information, i.e., weight his information according to its worth. *Fourth,* it eliminates biases in that he applies the evidence equally to all hypotheses rather than using it to defend or defeat a particular hypothesis. *Fifth,* it provides sets of analysts with a common method of problem analysis wherein they can compare differences in results and know that the differences are substantive rather than methodological. *Sixth,* it allows the analyst to see the results of his analysis process on a step-by-step basis and provides him with feedback which guides him in making adjustments in his analysis at each step in the process. And *seventh,* it allows him to explain to himself and others his reasons for evaluations and decisions at each step in the analysis process.

REFERENCES

Commission on Mathematics. *Introductory Probability and Statistical Inference.* New York: College Entrance Exam Board, 1959.

Dixon, John R. *A Programmed Introduction to Probability.* New York: Wiley, 1964.

Edwards, Ward. *Dynamic Decision Theory and Probabilistic Information Processing. Human Factors* **4** (2), Apr. 1962.

Edwards, Ward. *Nonconservative Probabilistic Information Processing Systems.* Ann Arbor: Univ. of Michigan, Institute of Science & Technology, Dec. 1966 (AD 647 092).

Feller, W. *An Introduction to Probability Theory and Its Application.* New York: Wiley, 1950.

Good, I. J. *Probability and the Weighing of Evidence.* New York: Hafner, 1950.

Good, I. J. *The Estimation of Probabilities.* Cambridge: MIT Press, 1965.

Kelly, C. W. III, and Peterson, C. R. *Probability Estimates and Probabilistic Procedures in Current Intelligence.* Gaithersburg, Md.: IBM, Federal Systems Div., Jan. 1971.

Luckie, Peter T., Smith, Dennis E., and Wright, Grace H. *Investigation of a Bayesian Approach Applied to a Specific Intelligence Problem.* State College, Pa.: HRB-Singer Inc., June 1968.

Miller, Charles D., and Heeren, Vern E. *Mathematical Ideas on Introduction.* Glennew, Ill.: Scott, Foresman, 1969.

Richardson, W. H. *Finite Mathematics.* New York: Harper & Row, 1968.

Chapter 7 / Inferential Statistics

7.1 INTRODUCTION

In Chapter 4 we described the construction of frequency distributions and showed how to calculate distribution parameters such as the mean and variance.

All the universes considered in Chapter 4 contained only a few elements and hence were easy to tabulate. In those instances we dealt with the whole universe. However, in practice it is more frequent that only a sample of the universe is utilized for two reasons. First, due to time and access considerations, it is impractical to sample the whole universe. For example, in medicine the sampling of the whole universe of people suffering from terminal cancer would entail a worldwide effort of tremendous magnitude. In manufacturing, the sampling of every unit of production wherein millions of units (such as light bulbs, tires, etc.) are produced per month would require a large staff and entail considerable expense. Second, the nature of the operation to be performed on the sample may dictate sampling only a small percentage of the universe. For example, if the sample selected must be destroyed to obtain the information required, as in destructive testing, then to reduce costs and time, the smallest sample yielding significant results is selected. Drugs, clothing, tires, light bulbs, electronic parts, mechanical parts, and metals are but a few of the many items which are only partially sampled and tested in industry and government.

From the discussion above it is obvious that using only a small portion of the universe, usually a random sample, to make estimates on properties possessed by the whole universe (population) is necessary and desirable. Using only a sample of the universe to draw conclusions about properties possessed by the whole universe raises some interesting questions. The most important question is: "What methods, procedures, and techniques of sampling will allow us to reach orderly, objective, and repeatable conclusions?" As we shall see, the drawing of the samples must satisfy certain conditions. The selection of specific statistical tests to be employed must satisfy prescribed criteria and the confidence we place in the conclusions drawn are determined by the statistical test and sample size selected.

Some questions relate to determining the probability that observed differences between two randomly drawn samples are due to chance or signify that two different populations are involved. Other questions relate to whether it is likely that parameters such as the mean and variance estimated from samples are representative of the same parameter for the universe. Yet another question relates to whether it is likely that a sample was drawn from a specified population.

These questions posed above are all answered in the study of inferential statistics. There are two main areas of application for inferential statistics. These are called *estimation* and *tests of hypotheses. Estimation* is concerned with the use of (sets of) samples to estimate parameters or characteristics of the universe from which the sample was drawn. For example, we can estimate the average life of a population of light bulbs by selecting a random sample and determining the average life of the light bulbs in the sample. *Tests of hypotheses* are concerned with the use of (sets of) samples to determine whether it is likely that assumptions or hypotheses made on some characteristics of a population are true. For example, we can hypothesize that Bufferin is more effective than Aspirin in relieving headaches. We test this hypothesis by taking samples of users of both (under controlled conditions) and from the information contained in the samples determine the level of confidence we can place in the assertion.

In dealing with estimation and the tests of hypotheses, knowledge of the properties of the population is hypothesized, or sought, and samples are drawn with the object of determining, within probability limits, whether the assertions (hypotheses) about the properties are acceptable. Thus in inferential statistics one usually assumes or knows something about the population, its defining characteristics (or parameters), and the accepted methods of sampling. In the sections to follow we will restrict our discussion to two probability distributions, the normal and the binomial, and their three parameters, the mean, median, and variance. In later sections we shall consider other distributions which are useful when only the ranking or ordering of samples is known, or the proportions in which elements of the population fall into prespecified categories or classes. For example, our measurements may be limited to the ranks (not scores) achieved by ninth grade students in the fifty states. Or our measurements may determine two or more classes such as honors, pass, fail, and did not complete, and the percentage of students which fall in each class.

There are two classes of statistical techniques used with populations and samples. The first class is called *parametric* because its objective is to establish the credibility of assertions made about properties of the parameters (mean, median, variance) of the sample, the population, or both. The second class of statistical techniques used with populations and samples is called *nonparametric* or *distribution-free* because its objective is to establish the credibility of assertions made about properties of the sample, the population, or both which are not associated with the parameters (mean, median, variance) which characterize the population or sample. In this sense, fewer properties such as the shape, mean, and variance of the population need be known in order to apply these techniques. In the sequel it will be apparent that substantive analysts will be more concerned with nonparametric than with parametric statistical techniques.

In the sections to follow we shall cover both parametric and nonparametric statistical techniques. Under parametric techniques we will cover a few statistical inference techniques which make use of the normal, binomial, and t distributions leaving other techniques and distributions for the reader to cover elsewhere if the need for them arises. The statistical techniques covered in this chapter apply as well to other distributions. The mean and variance will be the two parameters for which estimates will be made and hypotheses stated. Under nonparametric statistical techniques, we will cover techniques which make use of the binomial and the χ^2 (chi-squared) distributions. Both one and two related samples tests will be covered. In the binomial and χ^2 tests the population is assumed to be partitioned into various categories. In the χ^2 test for two related samples, we test whether two samples differ with respect to the relative frequency with which sample members fall in various categories.

First, we will review some important properties of the normal and binomial distribution and show how to obtain probabilities associated with standard variables Z i.e., to find $p(Z \geqslant Z_1)$, $p(Z \leqslant Z_2)$, or $p(Z_1 \leqslant Z \leqslant Z_2)$. Second, we will discuss some techniques of sampling and measurement. We will cover briefly the types of samples and measurements and illustrate their use in parametric and nonparametric tests. Third, we will cover some basic concepts and results of the theory of estimation. Biased and unbiased estimators of the mean and variance for normal and binomial distributions will be described. Finally, we will cover the tests of hypotheses using both parametric and nonparametric techniques.

7.2 BINOMIAL DISTRIBUTIONS

Some of the estimates and tests of hypotheses we will discuss require some knowledge of the properties of the binomial and normal distributions. This knowledge will lead to a better understanding of methods used and results obtained. First, we will cover the properties and uses of the binomial distribution.

A set of n measurements or a sample of size n will have a binomial distribution whenever each measurement or trial in the sample can assume either of two values such as yes–no, success–failure, either–or, heads–tails, good–bad with fixed probabilities–i.e., a two-level measurement or a dichotomous choice is to be made where one choice or measurement has a probability of p of occurring on every trial and the other choice or measurement has a probability of $q = 1 - p$ of occurring. The n measurement or n choices must be independent. The result sought is the probability that a particular choice or measurement occurs (number of successes) out of all the possible choices or measurements. The binomial distribution yields the set of probabilities of obtaining any number of successes, from 0 to n, in n trials, with p as the probability of obtaining a success in any one trial. A few examples should help bring into focus how a binomial distribution is generated in theoretical and practical situations.

Example 7.1 (Binomial Distribution–Coin Tosses). Let us consider tossing a fair coin 6 times. We would like to know the probability of obtaining x heads in the 6 tosses. The probability of obtaining a head on any toss is $P(H) = 1/2$ since the coin is fair. The probability of obtaining $x = 0, 1, 2, 3, 4, 5$, or 6 heads in six tosses is dependent on the number of ways one can obtain 0, 1, 2, 3, 4, 5, or 6 heads in six tosses, and the total number of outcomes possible in six tosses. Let us calculate the latter number first. What we seek is the total number of ways we can write a string of six Hs or Ts. For example HTTHTH and HHHTHT are outcomes, where H stands for heads and T for tails. In each of six positions we can write either H or T, i.e., we have two choices. We have the same two choices in six positions. Thus we have $2 \cdot 2 \cdot 2 \cdot 2 \cdot 2 \cdot 2 = 2^6 = 64$ possible outcomes. All these outcomes are equally probable. To obtain the former numbers, we again appeal to writing Hs and Ts in each of six independent positions. To write zero Hs, we have only 1 choice, TTTTTT. To write six Hs, we have only one choice, HHHHHH. To write one H, we have six choices, namely HTTTTT, THTTTT, ..., TTTTTH. The same is true on writing five Hs rather than one H. To write two Hs, we have a more difficult time enumerating all the possibilities. One can see what is involved by listing the first four in the enumeration, i.e., HHTTTT, HTHTTT, HTTHTT, and HTTTHT, One can see that the problem reduces to determining the number of different ways two Hs can be written in six different positions. We assert that the answer for this problem is given by the formula,

$$\left\{ \begin{matrix} n \\ r \end{matrix} \right\} = \frac{n!}{r!\,(n-r)!}$$

This formula yields the number of combinations of n objects taken r at a time. The order of the arrangement is not considered in the combinations. The term $\left\{ \begin{matrix} n \\ r \end{matrix} \right\}$ designates the statement "number of combinations of n objects taken r at a time." The term $n!$ designates the quantity "factorial n," where factorial n is equal to the product of all positive integers up to and including the integer n.

$$n! = n(n-1)(n-2) \cdots 1 \qquad 0! = 1! = 1$$

For example,

$$5! = 5 \cdot 4 \cdot 3 \cdot 2 \cdot 1 = 120$$
$$2! = 2 \cdot 1 = 2$$
$$7! = 7 \cdot 6 \cdot 5 \cdot 4 \cdot 3 \cdot 2 \cdot 1 = 5040$$
$$\begin{Bmatrix} 7 \\ 5 \end{Bmatrix} = \frac{7!}{5!\,2!} = \frac{5040}{120 \cdot 2} = 21$$

Returning to our problem of writing two Hs in six positions, we have

$$\begin{Bmatrix} 6 \\ 2 \end{Bmatrix} = \frac{6!}{2!\,4!} = \frac{720}{2 \cdot 24} = 15$$

Thus there are 15 ways of obtaining 2 heads out of 6 tosses. In a similar fashion we can determine that there are 15 ways of obtaining 4 heads (and hence 2 tails) out of 6 tosses. To obtain the number of ways we can obtain 3 heads out of 6 tosses we evaluate

$$\begin{Bmatrix} 6 \\ 3 \end{Bmatrix} = \frac{6!}{3!\,3!} = 20$$

Each of the 64 combinations of heads and tails is equally likely. For example, the combination of 2 heads and 4 tails (HTHTTT) has a probability $p(\text{HTHTTT}) = p \cdot q \cdot p \cdot q \cdot q \cdot q$ of occurring. Since $p = q = 1/2$, we rearrange the sequence and obtain,

$$\underbrace{p \cdot p}_{\text{two heads}} \cdot \underbrace{q \cdot q \cdot q \cdot q}_{\text{four tails}} = (1/2)^2 \cdot (1/2)^4 = 1/64$$

There are 64 outcomes and each outcome has a probability of 1/64 of occurring. The formula for $\begin{Bmatrix} 6 \\ 2 \end{Bmatrix}$ gives the number of outcomes containing exactly 2 heads. The outcomes are mutually exclusive so that the probability of obtaining 2 heads and 4 tails on six tosses is the sum of the probabilities of obtaining 2 heads and 4 tails independent of the order of heads and tails. There are $\begin{Bmatrix} 6 \\ 2 \end{Bmatrix}$ such events. Thus the probability of obtaining exactly 2 heads is

$$\underbrace{\begin{Bmatrix}6\\2\end{Bmatrix}}_{\substack{\text{number of}\\\text{orders for}\\\text{2 heads in}\\\text{6 tosses}}} \quad \underbrace{(p^2q^4)}_{\substack{\text{probability}\\\text{of 2 heads in}\\\text{6 tosses}}} = \begin{Bmatrix}6\\2\end{Bmatrix}(1/2)^6 = 15/64$$

The results for the six other probabilities are given in Table 7.1.

Table 7.1 Binomial probabilities.

Number of heads, r	0	1	2	3	4	5	6
Number of combinations of r heads, $\begin{Bmatrix}n\\r\end{Bmatrix}$	1	6	15	20	15	6	1
Probability of an outcome having r heads, p^rq^{n-r}	1/64	1/64	1/64	1/64	1/64	1/64	1/64
Probability of r heads out of n tosses, $\begin{Bmatrix}n\\r\end{Bmatrix} p^rq^{n-r}$	1/64	6/64	15/64	20/64	15/64	6/64	1/64

Note that in this example the probability of obtaining one sequence of heads and tails is equal to the probability of obtaining any other sequence of heads and tails. This result is obtained because $p = q = 1/2$ and we can write the probability of any sequence as $(1/2)^6 = 1/64$. However, when the probability of success (heads) does not equal the probability of failure (tails), the probability for any sequence of events is dependent on the number of successes and failures it contains. The probability remains independent of the order in which successes and failures occur. For example, the probability of obtaining the sequence HTHTHT is $pqpqpq = p^3q^3$. For the sequence HHTHTT the probability is $ppqpqq = p^3q^3$ which equals the probability of obtaining the first sequence. There are three heads and three tails in each sequence. One sequence is a reordering of the other. As previously indicated, there are $\begin{Bmatrix}6\\3\end{Bmatrix} = 20$ ways of ordering 3 heads and 3 tails in a sequence of six. Each has a probability of p^3q^3 of occurring and are mutually exclusive. Thus the probability of obtaining a sequence containing three heads and three tails is

$$\begin{Bmatrix} 6 \\ 3 \end{Bmatrix} p^3 q^3$$

We can generalize the specialized results obtained above into the following statement. The probability of obtaining a sequence of n events containing exactly r successes and $n - r$ failures, is given by the binomial probability function

$$f(r/n,p) = \begin{Bmatrix} n \\ r \end{Bmatrix} p^r (1-p)^{n-r} = \frac{n!}{r!\,(n-r)!} \cdot p^r q^{n-r}$$

where p and $q = 1-p$ are the probabilities for success and failure, respectively. The notation $f(r/n,p)$ represents the probability of having r successes in n trials, where the probability of success is p. The number of successes r ranges from 0 to n. The probability of success p ranges from 0 to 1. For must purposes, n will be a positive integer. From the binomial probability function $f(r/n,p)$ we can graph binomial probability distributions for selected values of n, r, and p. The binomial probability distributions are shown in Figure 7.1. Note the different shapes of the binomial distribution for different sets of values for n, r, and p. We illustrate the difference for $n = 6$ and $p = 1/10$, $p = 1/4$, $p = 1/2$, $p = 3/4$, and $p = 9/10$. Note that the distribution for $p = 1/2$ is symmetrical and resembles the shape of a normal distribution. The two distributions for $p = 1/4$ and $p = 3/4$ are slightly skewed right and left, respectively. For $p = 1/10$ and $p = 9/10$, the skew becomes more exaggerated.

From the formula for the binomial probability function or the graphs of the resulting probability distributions, we can make some general observations.

First, the binomial probability distribution does not assume a single "standard" or "normal" shape as does the normal distribution.

Second, there are basically three shapes to the binomial probability distribution. If the probability of success p is less than half, then the distribution is skewed to the right. If p is approximately 1/2, the distribution is approximately symmetrical. If p is greater than 1/2, then the distribution is skewed to the left. The skew increases to the right or left as the probability of success p tends toward 0 and 1, respectively.

Third, if the number of trials n is large and p is close to 1/2, then the binomial probability distribution approximates the normal distribution.

The distributions in Figure 7.1 are drawn as if the variable r, the number of successes, is an interval variable. Such is not the case. The variable r takes on integral values and can be considered as an ordinal variable. Properly, the probability distribution should be a discrete line graph with lines drawn at the values 0, 1, 2, 3, 4, 5, and 6 for r.■■

Example 7.2 (Binomial Distribution–Baseball). Suppose that you are the manager of a baseball team. The middle third of your batting order (batters 4, 5, and 6) are .300 hitters. You, as manager, would like to know the probability of none, one, two, or three of these batters obtaining a hit during any inning the three batters all come to bat. Being a .300 hitter means that the probability that the hitter will get a hit each time at bat is 0.3. Thus there are eight possible outcomes

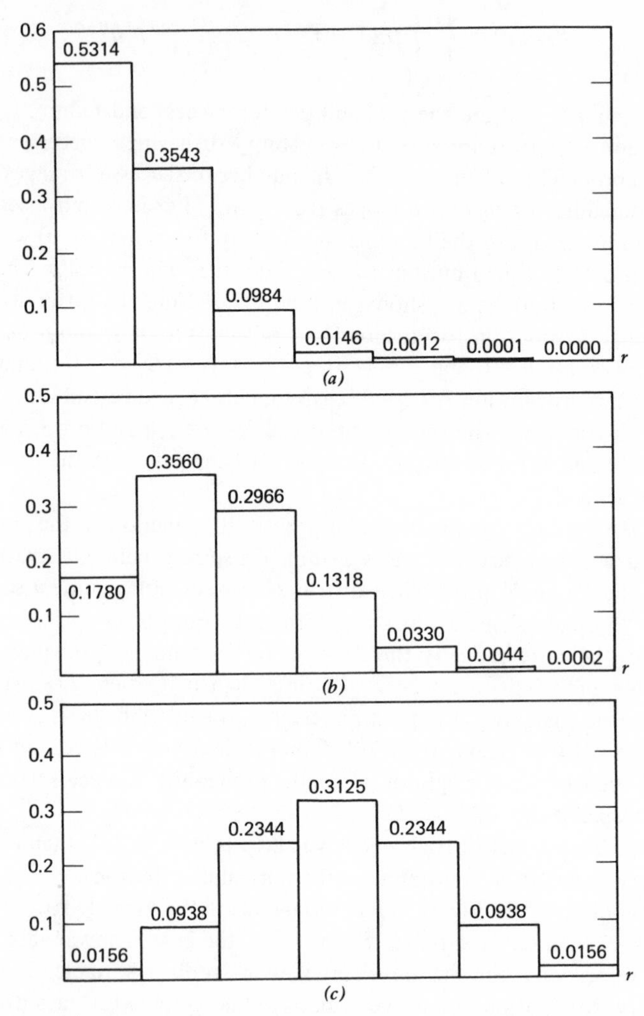

Figure 7.1 Binomial probability distributions for (*a*) $f(r/6, 1/10)$; (*b*) $f(r/6, 1/4)$; and (*c*) $f(r/6, 1/2)$.

HHH HHN HNH HNN NHH NHN NNH NNN

where H signifies success or a hit an N signifies failure to get a hit.

We know that the probabilities that the batters do or do not obtain a hit are, respectively

$$P(\text{H}) = p = 0.3$$
$$P(\text{N}) = q = 0.7$$

For this example, we shall assume that the batters are independent of each other. In practice, this is not always the case, as demonstrated by the occasional action of intentionally walking a batter when the previous batter has placed base runners in good scoring positions. We need the independence among batters to

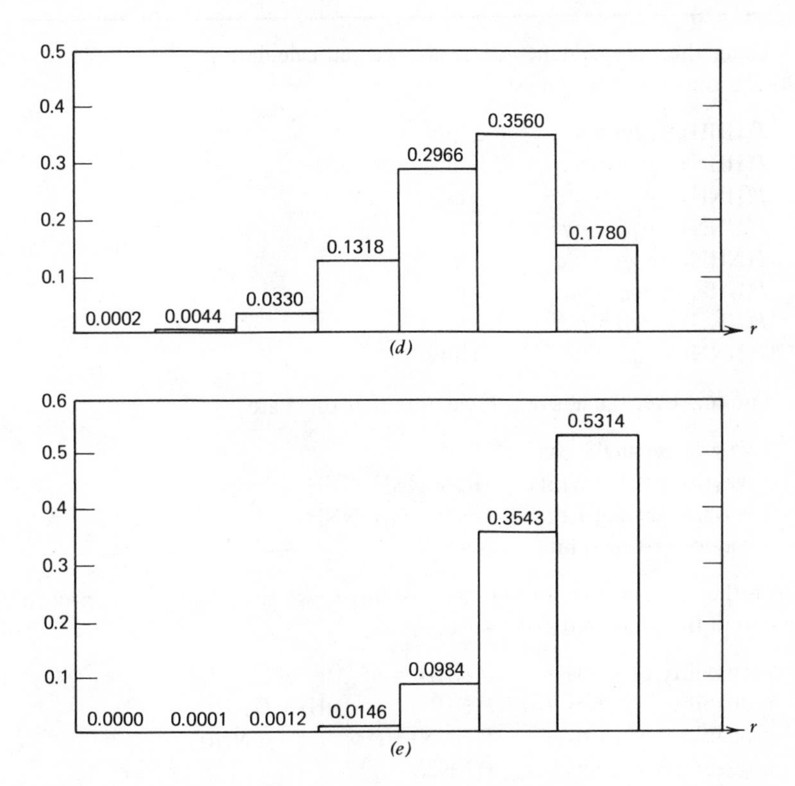

Figure 7.1 (continued) Binomial probability distributions for (*d*) $f(r/6, 3/4)$; and (*e*) $f(r/6, 9/10)$.

be able to multiply the probabilities of each getting a hit. For example, the probability that the second better is the only one to get a hit is

$$P(\text{NHN}) = P(\text{N})\,P(\text{H})\,P(\text{N}) = qpq = pq^2$$

Table 7.2 Binomial probabilities.

Number of hits, r	0	1	2	3
Combinations of hits $\left\{ {3 \atop r} \right\}$	1	3	3	1
Probability of r hits, $p^r q^{n-r}$	p^0q^3	p^1q^2	p^2q^1	p^3q^0
Probability of r hits in 3 tries $\left\{ {3 \atop r} \right\} p^r q^{n-r}$	$q^3 =$ 0.343	$3pq^2 =$ 0.441	$3p^2q =$ 0.189	$p^3 =$ 0.027

Using the independence of events we can calculate probabilities for each of the eight possible outcomes:

$P(\text{HHH}) = ppp = p^3$ 3 hits
$P(\text{HHN}) = ppq = p^2q$ 2 hits
$P(\text{HNH}) = pqp = p^2q$ 2 hits
$P(\text{HNN}) = pqq = pq^2$ 1 hit
$P(\text{NHH}) = qpp = p^2q$ 2 hits
$P(\text{NHN}) = qpq = pq^2$ 1 hit
$P(\text{NNH}) = qqp = pq^2$ 1 hit
$P(\text{NNN}) = qqq = q^3$ 0 hits

Looking over the above list, we note that there are

1 way of getting 3 hits HHH
3 ways of getting 2 hits HHN, HNH, NHH
3 ways of getting 1 hit HNN, NHN, NNH
1 way of getting 0 hits NNN

Since the eight events are mutually exclusive, we sum the separate probabilities to obtain the probability of

probability of 3 hits = $P(\text{HHH}) = p^3$
probability of 2 hits = $P(\text{HHN}) + P(\text{HNH}) + P(\text{NHH}) = 3p^2q$
probability of 1 hit = $P(\text{HNN}) + P(\text{NHN}) + P(\text{NNH}) = 3pq^2$
probability of 0 hits = $P(\text{NNN}) = q^3$

The four probabilities are listed in Table 7.2.

We should note that the sum of the probabilities of obtaining exactly 0, 1, 2, and 3 hits is

$$p^3 + 3p^2q + 3pq^2 + q^3 = (p+q)^3$$

This is the expansion for the binomial $(p+q)^3$. The coefficients 1, 3, 3, 1 for the four terms are called the binomial coefficients. Hence, the reason for calling the probability distribution the binomial probability distribution. One other point should be noted. We know that $p + q = 1$, so that $(p+q)^3 = 1$ and the sum of the binomial probabilities is 1, i.e.,

$$p^3 + 3p^2q = 3pq^2 + q^3 = 1$$

Referring back to the last row of Table 7.2, we can answer some specific questions for the baseball manager. For example, what is the probability that the three batters get at least one hit in any inning the three batters come to bat? We denote this probability by

$$f(r \geqslant 1/3, 0.3)$$

where

$$\begin{aligned} f(r \geqslant 1/3, 0.3) &= f(1/3, 0.3) + f(2/3, 0.3) + f(3/3, 0.3) \\ &= 0.441 + 0.189 + 0.027 \\ &= 0.657 \end{aligned}$$

This probability is complementary to the probability that the three batters obtain no hits, i.e.,

$$\begin{aligned} f(r \geqslant 1/3, 0.3) &= 1 - f(r=0/3, 0.3) \\ &= 1 - \left\{ \begin{matrix} 3 \\ 0 \end{matrix} \right\} p^0 q^3 \\ &= 1 - q^3 \\ &= 1 - 0.343 \\ &= 0.657 \end{aligned}$$

The manager could also find the probability of obtaining exactly 0, 1, 2, or 3 hits. These probabilities are listed in the last row of Table 7.2.■■

7.3 BINOMIAL PROBABILITY TABLES

In computing the probabilities for the outcomes of a set of binomial trials, the use of the binomial formula should be reserved as a last resort. There are extensive tables available which contain the binomial probability for most combinations of parameters r, n, and p that one might require. There are two types of binomial tables and one should be able to use either. The first type is given in Table 7.3 and is the Individual Term Binomial Distribution. The second type is

Table 7.3 Binomial distribution of individual terms.

Table entries obtained by evaluating $f(r/n,p) = \left\{ \begin{matrix} n \\ r \end{matrix} \right\} p^r (1-p)^{n-r}$ for the values of p, r and n indicated in the table.

n	r	$p = 0.1$	$p = 0.25$	$p = 0.5$
5	0	.5905	.2373	.0312
	1	.3280	.3955	.1562
	2	.0729	.2637	.3125
	3	.0081	.0879	.3125
	4	.0004	.0146	.1562
	5	.0000	.0010	.0312
10	0	.3487	.0563	.0010
	1	.3784	.1877	.0098
	2	.1937	.2816	.0439
	3	.0574	.2503	.1172
	4	.0112	.1460	.2051
	5	.0015	.0584	.2461
	6	.0001	.0162	.2051
	7	.0000	.0031	.1172
	8	.0000	.0004	.0439
	9	.0000	.0000	.0098
	10	.0000	.0000	.0010
25	0	.0718	.0008	.0000
	1	.1994	.0063	.0000
	2	.2659	.0251	.0000
	3	.2265	.0641	.0001
	4	.1384	.1175	.0004
	5	.0646	.1645	.0016
	6	.0239	.1828	.0053
	7	.0072	.1654	.0143
	8	.0018	.1241	.0322
	9	.0004	.0781	.0609
	10	.0001	.0417	.0974
	11	.0000	.0189	.1328
	12	.0000	.0074	.1550
	13	.0000	.0025	.1550
	14	.0000	.0007	.1328
	15	.0000	.0002	.0974
	16	.0000	.0000	.0609
	17	.0000	.0000	.0322
	18	.0000	.0000	.0143
	19	.0000	.0000	.0053
	20	.0000	.0000	.0016
	21	.0000	.0000	.0004
	22	.0000	.0000	.0001
	23	.0000	.0000	.0000
	24	.0000	.0000	.0000
	25	.0000	.0000	.0000

given in Table 7.4 and is the Cumulative Binomial Distribution. Table 7.3 gives the probability of obtaining exactly r successes in n trials when the probability of success in a single trial is p. Table 7.4 gives the probability of obtaining r or more (at least r) successes in n trials when the probability of success in a single trial is p. Either table can be used to obtain the probabilities sought. However, one is easier to use than the other, depending on the types of probabilities sought. The differences in usage will be illustrated by means of a series of examples in which both tables will be used to find the required probabilities.

Example 7.3 (Expected Probabilities). Find the probability that a batter with a .500 batting average (an average of 5 hits in 10 times at bat) will obtain at least two hits on his next five times at bat.■■

Solution: The probability we seek is given by the formula,

$$\begin{aligned} f(r \geqslant 2/5, 0.5) &= \Sigma_{i=2}^{5} f(i/5, 0.5) \\ &= f(2/5, 0.5) + f(3/5, 0.5) + f(4/5, 0.5) + f(5/5, 0.5) \end{aligned}$$

Using Table 7.3, we enter the table at $n = 5$ and $p = 0.5$. There we find that the probability for exactly 2 hits is 0.3125, for 3 hits is 0.3125, for 4 hits is 0.1562, and for 5 hits is 0.0312. We total the column from $r = 2$ to $r = 5$ to obtain the total probability 0.8124. Using Table 7.4 we find $n = 5$ and enter at $r = 2$ and $p = 0.5$. There we find that

$$f(r \geqslant 2/5, 0.5) = 0.812$$

■■

Example 7.4 (Hit Probabilities). Using the same batter as in the above example, find the probability of his obtaining exactly 3 hits in his next 5 times at bat.■■

Solution: We use Table 7.3 and enter at $n = 5, r = 3$, and $p = 0.5$. We read the probability entry as 0.3125. Using Table 7.4, we seek two table entries for the two cumulative probabilities given by

$$f(r \geqslant 3/5, 0.5) \qquad \text{and} \qquad f(r \geqslant 4/5, 0.5)$$

The first entry is the sum of probabilities for 3, 4, and 5 hits. The second entry is the sum of probabilities for 4 and 5 hits. Hence their difference will be the probability of obtaining exactly 3 hits, i.e.,

$$f(r \geqslant 3/5, 0.5) - f(r \geqslant 4/5, 0.5) = f(r{=}3/5, 0.5)$$

From Table 7.4 we obtain 0.500 – 0.188 = 0.312. This agrees with the result obtained from Table 7.3.■■

Example 7.5 (Election Probabilities). Prior to a recent election, 100 samples of 25 voters each were taken. It was determined that 85 out of the 100 samples resulted in voters favoring candidate A over candidate B. Assuming that the poll was conducted properly, what percentage of voters can be expected to vote for candidate A?■■

Table 7.4 Brief binomial probability of binomial distribution.

Table entries obtained by evaluating $f(i \geqslant r/n,p) = \Sigma_{i=r}^{n} f(i/n,p) = \Sigma_{i=r}^{n} \left\{ {n \atop i} \right\} p^i (1-p)^{n-i}$ for the values of p, r, and n indicated in the tables.

Three digit entries should be read with a decimal preceding: 1– is a number greater than 0.9995, 0+ a number less than 0.0005.

$n = 5$

r \ p	.01	.05	.10	.20	.30	.40	.50	.60	.70	.80	.90	.95	.99	p / r
0	1	1	1	1	1	1	1	1	1	1	1	1	1	0
1	049	226	410	672	832	922	969	990	998	1–	1–	1–	1–	1
2	001	023	081	263	472	663	812	913	969	993	1–	1–	1–	2
3	0+	001	009	058	163	317	500	683	837	942	991	999	1–	3
4	0+	0+	0+	007	031	087	188	337	528	737	919	977	999	4
5	0+	0+	0+	0+	002	010	031	078	168	328	590	774	951	5

$n = 10$

r \ p	.01	.05	.10	.20	.30	.40	.50	.60	.70	.80	.90	.95	.99	p / r
0	1	1	1	1	1	1	1	1	1	1	1	1	1	0
1	096	401	651	893	972	994	999	1–	1–	1–	1–	1–	1–	1
2	004	086	264	624	851	954	989	998	1–	1–	1–	1–	1–	2
3	0+	012	070	322	617	833	945	988	998	1–	1–	1–	1–	3
4	0+	001	013	121	350	618	828	945	989	999	1–	1–	1–	4
5	0+	0+	002	033	150	367	623	834	953	994	1–	1–	1–	5
6	0+	0+	0+	006	047	166	377	633	850	967	998	1–	1–	6
7	0+	0+	0+	001	011	055	172	382	650	879	987	999	1–	7
8	0+	0+	0+	0+	002	012	055	167	383	678	930	988	1–	8
9	0+	0+	0+	0+	0+	002	011	046	149	376	736	914	996	9
10	0+	0+	0+	0+	0+	0+	001	006	028	107	349	599	904	10

Table 7.4 (continued).

$n = 15$ r \ p	.01	.05	.10	.20	.30	.40	.50	.60	.70	.80	.90	.95	.99	$n = 15$ p / r
0	1	1	1	1	1	1	1	1	1	1	1	1	1	0
1	140	537	794	965	995	1–	1–	1–	1–	1–	1–	1–	1–	1
2	010	171	451	833	965	995	1–	1–	1–	1–	1–	1–	1–	2
3	0+	036	184	602	873	973	996	1–	1–	1–	1–	1–	1–	3
4	0+	005	056	352	703	909	982	998	1–	1–	1–	1–	1–	4
5	0+	001	013	164	485	783	941	991	999	1–	1–	1–	1–	5
6	0+	0+	002	061	278	597	849	966	996	1–	1–	1–	1–	6
7	0+	0+	0+	018	131	390	696	905	985	999	1–	1–	1–	7
8	0+	0+	0+	004	050	213	500	787	950	996	1–	1–	1–	8
9	0+	0+	0+	001	015	095	304	610	869	982	1–	1–	1–	9
10	0+	0+	0+	0+	004	034	151	403	722	939	998	1–	1–	10
11	0+	0+	0+	0+	001	009	059	217	515	836	987	999	1–	11
12	0+	0+	0+	0+	0+	002	018	091	297	648	944	995	1–	12
13	0+	0+	0+	0+	0+	0+	004	027	127	398	816	964	1–	13
14	0+	0+	0+	0+	0+	0+	0+	005	035	167	549	829	990	14
15	0+	0+	0+	0+	0+	0+	0+	0+	005	035	206	463	860	15

Table 7.4 (continued).

$n = 20$													$n = 20$
r \ p	.01	.05	.10	.20	.30	.40	.50	.60	.70	.80	.90	.95	.99	p / r
0	1	1	1	1	1	1	1	1	1	1	1	1	1	0
1	182	642	878	988	999	1–	1–	1–	1–	1–	1–	1–	1–	1
2	017	264	608	931	992	999	1–	1–	1–	1–	1–	1–	1–	2
3	001	075	323	794	965	996	1–	1–	1–	1–	1–	1–	1–	3
4	0+	016	133	589	893	984	999	1–	1–	1–	1–	1–	1–	4
5	0+	003	043	370	762	949	994	1–	1–	1–	1–	1–	1–	5
6	0+	0+	011	196	584	874	979	998	1–	1–	1–	1–	1–	6
7	0+	0+	002	087	392	750	942	994	1–	1–	1–	1–	1–	7
8	0+	0+	0+	032	228	584	868	979	999	1–	1–	1–	1–	8
9	0+	0+	0+	010	113	404	748	943	995	1–	1–	1–	1–	9
10	0+	0+	0+	003	048	245	588	872	983	999	1–	1–	1–	10
11	0+	0+	0+	001	017	128	412	755	952	997	1–	1–	1–	11
12	0+	0+	0+	0+	005	057	252	596	887	990	1–	1–	1–	12
13	0+	0+	0+	0+	001	021	132	416	772	968	1–	1–	1–	13
14	0+	0+	0+	0+	0+	006	058	250	608	913	998	1–	1–	14
15	0+	0+	0+	0+	0+	002	021	126	416	804	989	1–	1–	15
16	0+	0+	0+	0+	0+	0+	006	051	238	630	957	997	1–	16
17	0+	0+	0+	0+	0+	0+	001	016	107	411	867	984	1–	17
18	0+	0+	0+	0+	0+	0+	0+	004	035	206	677	925	999	18
19	0+	0+	0+	0+	0+	0+	0+	001	008	069	392	736	983	19
20	0+	0+	0+	0+	0+	0+	0+	0+	001	012	122	358	818	20

Table 7.4 (continued).

$n = 25$ $r \backslash p$	.01	.05	.10	.20	.30	.40	.50	.60	.70	.80	.90	.95	.99	$n = 25$ p / r
0	1	1	1	1	1	1	1	1	1	1	1	1	1	0
1	222	723	928	996	1–	1–	1–	1–	1–	1–	1–	1–	1–	1
2	026	358	729	973	998	1–	1–	1–	1–	1–	1–	1–	1–	2
3	002	127	463	902	991	1–	1–	1–	1–	1–	1–	1–	1–	3
4	0+	034	236	766	967	998	1–	1–	1–	1–	1–	1–	1–	4
5	0+	007	098	579	910	991	1–	1–	1–	1–	1–	1–	1–	5
6	0+	001	033	383	807	971	998	1–	1–	1–	1–	1–	1–	6
7	0+	0+	009	220	659	926	993	1–	1–	1–	1–	1–	1–	7
8	0+	0+	002	109	488	846	978	999	1–	1–	1–	1–	1–	8
9	0+	0+	0+	047	323	726	946	996	1–	1–	1–	1–	1–	9
10	0+	0+	0+	017	189	575	885	987	1–	1–	1–	1–	1–	10
11	0+	0+	0+	006	098	414	788	966	998	1–	1–	1–	1–	11
12	0+	0+	0+	002	044	268	655	922	994	1–	1–	1–	1–	12
13	0+	0+	0+	0+	017	154	500	846	983	1–	1–	1–	1–	13
14	0+	0+	0+	0+	006	078	345	732	956	998	1–	1–	1–	14
15	0+	0+	0+	0+	002	034	212	586	902	994	1–	1–	1–	15
16	0+	0+	0+	0+	0+	013	115	425	811	983	1–	1–	1–	16
17	0+	0+	0+	0+	0+	004	054	274	677	953	1–	1–	1–	17
18	0+	0+	0+	0+	0+	001	022	154	512	891	998	1–	1–	18
19	0+	0+	0+	0+	0+	0+	007	074	341	780	991	1–	1–	19
20	0+	0+	0+	0+	0+	0+	002	029	193	617	967	999	1–	20
21	0+	0+	0+	0+	0+	0+	0+	009	090	421	902	993	1–	21
22	0+	0+	0+	0+	0+	0+	0+	002	033	234	764	966	1–	22
23	0+	0+	0+	0+	0+	0+	0+	0+	009	098	537	873	998	23
24	0+	0+	0+	0+	0+	0+	0+	0+	002	027	271	642	974	24
25	0+	0+	0+	0+	0+	0+	0+	0+	0+	004	072	277	778	25

Solution: The poll determined the probability that for any sample of the 100 taken a majority of the 25 voters sampled (13 or more) favored candidate A. Thus the poll determined that

$$f(r \geqslant 13/25, p) = 0.85$$

That is, if samples of 25 are considered as successes when 13 or more voters favor candidate A, and failures otherwise, then the binomial probability of success is 0.85. The problem now is to determine the probability p that the individual voters will favor candidate A. In this example, we have little choice but to use the cumulative probabilities in Table 7.4. We enter the table headed by $n = 25$, and the row headed by $r = 13$ (which means 13 or more), and proceed across the row until we reach the entry 0.846. The column heading for the column containing 0.846 is the individual probability $p = 0.6$. For most purposes, 0.85 is close enough to 0.846 to accept 0.6 as a good approximation. If the numbers are not close, then interpolation between columns can be used to obtain good estimates for the probability p. Thus we have determined that

$$f(r \geqslant 13/25, 0.6) = 0.846$$

which we accept as close enough for our purposes. Hence approximately 6 out of 10 voters will favor candidate A in the next election.■■

Example 7.6 (Effectiveness Probabilities). Prior to the introduction of a new drug, 100% of the people contracting a certain disease did not recover. With the new drug, there is a 50% chance of recovery. What is the probability that among 25 persons with the disease, that treatment with the drug will result in 8 to 17 recoveries, inclusive?■■

Solution: In this problem the use of Table 7.4, i.e., the cumulative probabilities, is more tractable than Table 7.3. We need to find the difference of two cumulative binomial probabilities, namely

$$f(r \geqslant 8/25, 0.5) \qquad \text{and} \qquad f(r \geqslant 18/25, 0.5)$$

The second probability uses $r \geqslant 18$ to retain the probability that exactly 17 will recover in the difference. Table 7.4, for $n = 25$, $r = 8$, and $p = 0.5$ yields 0.978. For $r = 18$, the table yields 0.022. Thus the resulting probability that between 8 and 17 will recover is

$$f(8 \leqslant r \leqslant 17/25, 0.5) = 0.978 - 0.022 = 0.956$$

Note that in this example, the sample size was reasonably large (15) and the probability of success was 1/2 so that the binomial distribution is symmetric and approximates the normal distribution. The range of successes 8–17, inclusive, is centered at the mean of the distribution ($r = 12.5$) so that one would expect a large percentage of the samples to fall in this range. The result 0.956 agrees with this expectation. ▪▪

Exercise 7.1. Use Table 7.3 to determine the binomial probabilities for

(a) $r = 8, \quad n = 10, \quad p = 0.1$
(b) $r = 5, \quad n = 10, \quad p = 0.3$
(c) $r = 3, \quad n = 5, \quad p = 0.5$
(d) $r = 4, \quad n = 8, \quad p = 0.8$ ▪▪

Exercise 7.2. Use Table 7.4 to determine the binomial probabilities for

(a) $r \geqslant 8, \quad n = 10, \quad p = 0.1$
(b) $r \leqslant 5, \quad n = 10, \quad p = 0.3$
(c) $12 \leqslant r \leqslant 16, \quad n = 25, \quad p = 0.4$
(d) $r \leqslant 8, \quad n = 25, \quad p = 0.5$ ▪▪

Exercise 7.3. The probability that a patient will recover when provided with treatment X is p. If five patients are given treatment X, what is the probability that

(a) Three will recover when $p = 0.2$?
(b) Three will recover when $p = 0.5$?
(c) Three will recover when $p = 0.8$?
(d) At least three will recover when $p = 0.2$?
(e) At least three will recover when $p = 0.5$?
(f) At least three will recover when $p = 0.8$? ▪▪

Exercise 7.4. A baseball world series is decided when one team has won four games. Assume that the national league team N has a probability $p = 0.6$ of winning any game, i.e., the odds are 6/4 in favor of N. What are the probabilities that the series ends in 4, 5, 6, and 7 games? ▪▪

Exercise 7.5. A theatre experiences a 10% "no show" on unpaid reservations. If the theatre "reserves" 22 seats for unpaid reservations but accepts 25, what is the probability that the 22 seats will accommodate all unpaid reservations which show? (Hint: Find the probability that at least 23 will show and calculate its complementary probability.) ▪▪

7.4 MEAN AND VARIANCE FOR BINOMIAL DISTRIBUTION

In Chapter 4 we introduced the concepts of the mean and variance as they apply to frequency distributions. We illustrated how one could calculate the mean and variance of distributions associated with selected sets of grouped or ungrouped samples. These calculations were specifically applied to data sets from normally distributed populations. We illustrated how the mean and variance characterize the normal distribution and allows one to make probability estimates on variables assuming specific values or ranges of values.

We now illustrate how to determine the mean and variance of a binomial distribution, recalling that the shape of "the" binomial is not "standard" or "normal" and depends on the value of p in the probability formula

$$f(r|n,p) = \left\{ \begin{matrix} n \\ r \end{matrix} \right\} p^r (1-p)^{n-r}$$

where r ranges from 0 to n to generate the frequency bars for the binomial distribution.

We can determine the mean and variance following the procedures given in Chapter 4. However, the calculations are long and tedious and will be omitted here. Instead we will state the results and show how to apply them in specific applications. In the following formulas we shall again assume that the probability for success in each binomial trial is p. We also assume that there are n independent trials. The resulting probability distribution will be symmetric, skewed right, or skewed left depending on whether p is close to 1/2, 0, or 1, respectively. The mean variance and standard deviation of the binomial distribution are

Mean	$u = np$
Variance	$\sigma^2 = np(1-p) = npq$
Standard Deviation	$\sigma = \sqrt{np(1-p)} = \sqrt{npq}$

Example 7.7 (Binomial Mean and Variance). Calculate and graph the means and standard deviations for the five binomial distributions given in Figure 7.1.

The five distributions are for $n = 6$ and $p = 1/10$, 1/4, 1/2, 3/4, and 9/10. Call these distributions (*a*), (*b*), (*c*), (*d*), and (*e*), respectively. The calculations are as follows.

1. $u = np = 6 \cdot 1/10 = 0.6$
   $\sigma^2 = np(1-p) = 6 \cdot 1/10 \cdot 9/10 = 0.54$
   $\sigma = \sqrt{0.54} = 0.73$

2. $u = np = 6 \cdot 1/4 = 1.5$
   $\sigma^2 = np(1-p) = 6 \cdot 1/4 \cdot 3/4 = 9/8$
   $\sigma = \sqrt{9/8} = 1.06$

3. $u = np = 6 \cdot 1/2 = 3$
   $\sigma^2 = np(1-p) = 6 \cdot 1/2 \cdot 1/2 = 1.5$
   $\sigma = 1.224$

4. $u = np = 6 \cdot 3/4 = 4.5$
   $\sigma^2 = np(1-p) = 6 \cdot 3/4 \cdot 1/4 = 9/8$
   $\sigma = \sqrt{9/8} = 1.06$

5. $u = np = 6 \cdot 9/10 = 5.4$
   $\sigma^2 = np(1-p) = 6 \cdot 9/10 \cdot 1/10 = 0.54$
   $\sigma = \sqrt{0.54} = 0.73$

7.5 NORMAL APPROXIMATION TO BINOMIAL DISTRIBUTION

It has been stated that when n is large and p is close to $1/2$, the shape of the binomial is very close to the shape of the normal distribution. This approximation allows one to use the more extensive and easier-to-use tables for the normal distribution in place of the tables for the binomial distribution. The procedure to follow is quite simple. First, we calculate the mean and standard deviation for the binomial distribution using the formulas listed previously, i.e.,

$$u = np \qquad \text{and} \qquad \sigma = \sqrt{np(1-p)}$$

Next, consider the mean and standard deviation, as calculated above, as the mean and standard deviation of the normal distribution which approximates the binomial distribution. Convert the variable x to standard normal units, i.e.,

$$Z_x = \frac{x - u}{\sigma}$$

The value of Z_x is used to enter cumulative normal distribution tables such as the one given in Table 7.6. The value *of* Z_x is read to one decimal place in the first column. The second decimal place in Z_x is read across the top row. Thus the table value for $Z_x = 2.62$ is located in the row headed by 2.6 and the column headed by 0.02. At the intersection of the selected row and column we read the number 0.9956 which is the probability that normal variables for the distribution will assume values less than or equal to Z_x. The shaded area under the curve for the normal distribution given in Figure 7.2 represents this probability.

The normal curve is a reasonably good approximation to the binomial distribution whenever the mean of the binomial distribution is at least 5 units away from either end of its range. The conditions for a good approximation are satisfied when

(a) $p \leqslant 1/2$ and $np > 5$
(b) $p \geqslant 1/2$ and $n(1-p) > 5$

Example 7.8 (Normal Approximation to Binomial). From experience it is known that 2 out of 10 shells fired are duds. What is the probability that 50 or fewer shells out of a firing of 200 will be duds? ▪▪

Solution: If binomial tables for $n = 200$, $r = 50$, and $p = 0.2$ were available, we could obtain the value for

$$f(r \leqslant 50/200, 0.2)$$

from the tables and the problem would be solved. Since such tables are not readily available we check to see if the normal distribution can be used to approximate the binomial distribution. If so, we will use the tables for the standard normal distribution to find the required probabilities. To check the "goodness" of the approximation we note that $p < 1/2$ and $np = 200 \cdot 0.2 = 40$. Thus the approximation will be good. The mean and standard deviation of the binomial are

$$u = np = 0.2 \cdot 200 = 40$$
$$\sigma = \sqrt{npq} = \sqrt{200 \cdot 0.2 \cdot 0.8} = 5.65$$

Next we convert to standard units

$$Z_x = \frac{x - u}{\sigma} = \frac{x - 40}{5.65}$$

with $x = 50$, we have that $Z_x = 10/5.65$ or $Z_x = 1.77$. Entering the cumulative normal distribution table (Table 7.6) at $Z_x = 1.77$, we read the value $p = 0.9616$. Thus the probability that 50 or fewer duds will result in 200 rounds is 0.9616. ▪▪

Example 7.9 (Normal Approximation). Assume that inclement weather causes a postponement of 5 out of every 50 baseball games. In a season of 162 games, what is the probability that no more than 20 doubleheaders will be required to make up the postponed games? ▪▪

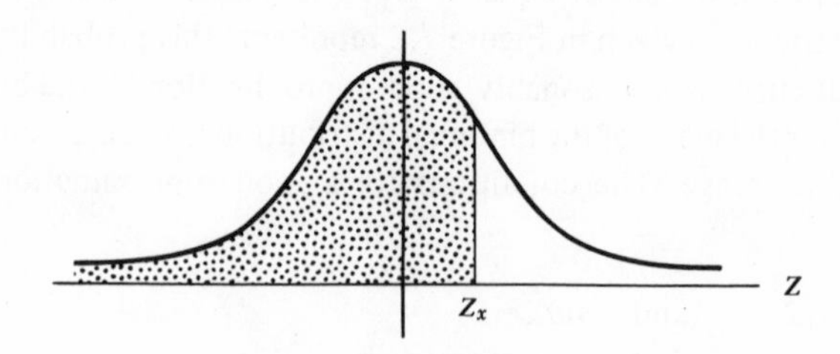

Figure 7.2

Solution: We assume that $p = 5/50 = 0.1$ and $n = 162$. The result is the probability that 20 or fewer doubleheaders will be sufficient to make up the rained out games. The quantity is given by

$$f(r \leqslant 20/162, 0.1)$$

We check to see if the normal distribution can be used. Since $p < 1/2$, and $np > 5$, we can use a normal approximation to the binomial distribution. Computing the mean and standard deviation for the binomial distribution, we obtain

$$u = np = 16.2 \qquad \sigma_u = \sqrt{npq} = 3.82$$

Converting to a standard normal variable, we obtain

$$Z_x = \frac{x - u}{\sigma_u} = \frac{20 - 16.2}{3.82} \cong 1$$

Entering the cumulative normal distribution tables at $Z_x = 1$, we find that the probability that 20 doubleheaders will suffice is 0.84.■■

We shall illustrate further applications of the binomial distribution when we discuss the subject areas of estimation and tests of hypotheses.

Exercise 7.6. In the following exercises, use either the normal or binomial tables (or both) to determine the probabilities sought:

(a) What is the mean and variance of the binomial distribution for $n = 10$ and $p = 0.25$? For this distribution, determine the probability of obtaining no more than 3 successes out of 10 tries.

(b) What is the mean and variance of the binomial distribution for $n = 25$ and $p = 0.2$? For this distribution, determine the probability of having at least 21 successes in 25 tries.

(c) During certain months bad weather causes 3 out of every 10 airstrikes to be canceled. If 30 airstrikes are planned in each month, what is the probability that between 7 and 11 airstrikes, inclusive, will be canceled in a bad weather month?

(d) Voter registrations indicate that 60% of the voters prefer the R party. Assuming that the population of voters is large and that voters vote as they register, what is the probability that out of 100 voters, 51 or more (a majority), will vote for the R party? Determine the probability for 101 or more R votes out of 200 and 501 or more out of 1000. Compare the three probabilities and state your reasons for the increase in probability with increasing sample size n. ■■

7.6 NORMAL DISTRIBUTION

In this section, we shall review some facts covered in Chapter 4 on distributions and especially on the properties of the normal distribution. We shall introduce some additional material and illustrate how probability distributions or their associated tables can be used to shorten or simplify probability calculations in estimation and the tests of hypotheses.

First, we recall that a frequency distribution or table can be constructed for any sample. The variable sampled may take on nominal, ordinal, interval, or ratio values. The size and shape of the distribution depends on the number of measurements in the sample, how the measurements were taken, the properties of the universe from which the sample was taken, etc. As the sample size increases, the frequency bars of the distribution become closer to approximating a smooth curve.

Frequency distributions are changed to relative frequency distributions by changing the interval frequencies to relative frequencies through the use of the following formula:

$$\text{relative frequency} = \frac{\text{interval frequency}}{\text{total frequency}}$$

This conversion to relative frequencies, or proportions, allows us to deal directly with the probability that a sample measurement will fall in a particular interval. For example, in Table 7.5, the relative frequency of interval 6 is 30/216. The (statistical) probability that a sample measurement will have the value 6 is given by the relative frequency 30/216.

We can plot the frequency distribution from the frequency table and can use either the frequency or the relative frequency as the ordinate of the graph. When plotting distributions in relative frequencies, the area under the frequency curve is equal to 1. This fact allows us to estimate the probability that a sample measurement will fall in a prespecified interval. For the smoothed relative frequency distribution constructed in Figure 7.3 from Table 7.5, we can estimate the probability that a randomly selected measurement will have a value between a and b. This value is equal to the area under the curve between the two values a and b. The area under consideration is shadowed in Figure 7.3.

The relationship between the probability and area is an important one and is used repeatedly in statistics. From Figure 7.3, we can state that

$p(a \leqslant x \leqslant b)$ = area under curve between a and b
$p(x \leqslant c)$ = area under curve to the left of c
$p(x \geqslant d)$ = area under curve to right of d

If the probability curves were constructed accurately, we could use a planimeter (a device which measures area) and determine the probabilities directly

Table 7.5 Relative frequency.

Interval	Frequency	Relative Frequency
2	6	6/216
3	12	12/216
4	18	18/216
5	24	24/216
6	30	30/216
7	36	36/216
8	30	30/216
9	24	24/216
10	18	18/216
11	12	12/216
12	6	6/216
	$\Sigma = 216$	$\Sigma = 1$

from the probability curves. But the area relationship is used so frequently in statistics that extensive tables have been constructed which directly relate probabilities to variable values for the more often used probability distributions such as the normal, binomial, and χ^2 distributions. In this section we shall illustrate the use of tables constructed for the normal distribution, reserving later sections to discuss and illustrate the use of tables constructed for the binomial and χ^2 distributions.

7.7 SOME PROPERTIES OF STANDARD NORMAL DISTRIBUTION

First we review some properties of the normal distribution. The normal distribution, also called the normal curve or gaussian distribution, is a smooth, bell-

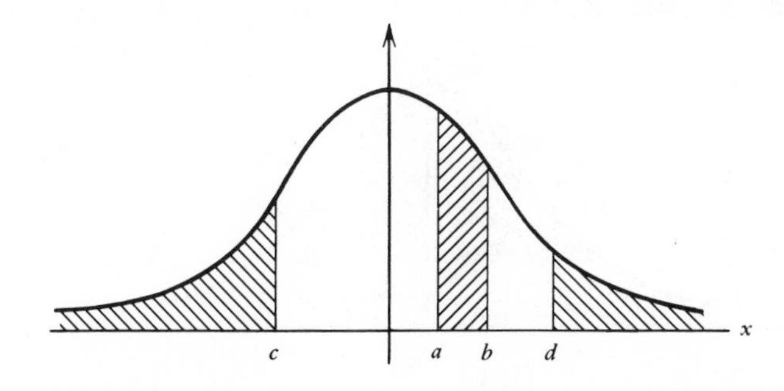

Figure 7.3 Areas and probabilities in probability distributions.

shaped curve, symmetrical about its mean. When the variable for a normal curve is expressed in standard units by the formula

$$Z_x = \frac{x - \bar{x}}{\sigma_x}$$

where $\bar{x}$ and σ_x are the mean and standard deviation of the normal curve, the resulting curve is called the standard normal curve. It has a mean of 0 and a standard deviation of 1. The height of the standard normal curve at the mean is 0.399, and the mean, median, and mode all coincide. As stated in Chapter 4, 68.27% of the total area under the standard normal curve lies within the range $-\sigma_x$ to σ_x, 95.45% within the range $-2\sigma_x$ to $2\sigma_x$, and 99.73% within the range $-3\sigma_x$ to $3\sigma_x$. The height of the standard normal curve at $Z_x = \pm\sigma_x$ and $\pm 2\sigma_x$ are 0.242 and 0.054, respectively.

7.8 AREAS OF STANDARD NORMAL CURVE

To assist in determining probabilities associated with normal distributions, Table 7.6 has been prepared to give the areas (probabilities) under the standard normal curve between the ordinate line drawn at variable value Z_x and $-\infty$, i.e., the left most extremity of the left tail of the curve which extends indefinitely. Table 7.6 contains a set of probabilities associated with variable values given to two decimal places. The leftmost column contains the variable value to one decimal place. The top row identifies the second decimal place of the variable value. For example, the value $Z_x = 2.67$ is located in the first column at the entry 2.6 and in the first row at the entry 0.07. At the intersection of the row and column is the entry 0.9962. This entry is the probability that random standard normal measurements assume values equal to or less than 2.67. Figure 7.4 illustrates the area under the standard normal curve associated with the probability. When it is understood that x is the original variable, we shall drop the subscript x on the standard normal variable Z_x.

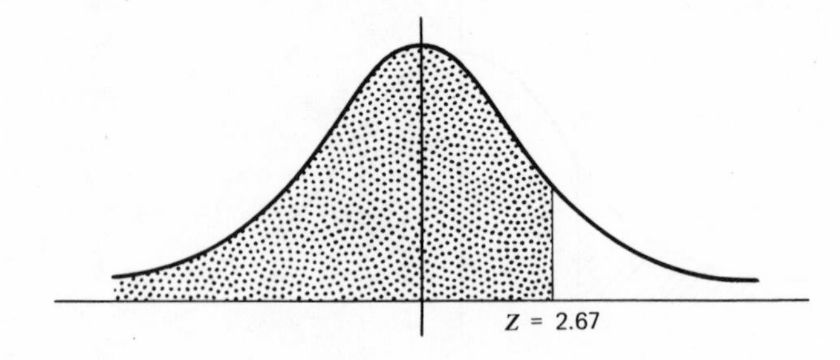

Figure 7.4 Area corresponding to Z.

Note that the table contains values for positive values of Z only. To find the probability associated with a negative value $-Z$, one uses the magnitude of $-Z$, namely $|-Z| = Z$ to enter the table and find the probability p associated with Z. The probability associated with $-Z$ is the complementary probability $1 - p$. Figure 7.5 illustrates the areas and the probabilities involved.

The area sought under the normal curve lies to the left of $-Z$, is cross-hatched and is denoted by A in Figure 7.5. To find the value of the area A, we use the table entry for the standard normal variable with value Z. This gives the area to the left of Z (indicated by lines with positive slopes). The area complementary to the latter area is designated B (single negative slope lines). By symmetry of the normal curve, area B is equal to area A.

In all of the above discussions with the normal curve, it was assumed that the variable Z is the value associated with the variable value x through the transform

$$Z = \frac{x - \bar{x}}{\sigma_x}$$

where $\bar{x}$ is the mean of the original normal distribution and σ_x its standard deviation. Thus, before the tables can be employed, all variable values must be converted to the standard normal values by the formula given above.

Example 7.10 (Standard Normal Variables). For a population having a normal distribution of mean $\bar{x} = 6$ and a standard deviation $\sigma_x = 3$, determine the probability that random samples from the population will not exceed the value $x = 12$.■■

Solution: First we convert to standard variables

$$Z = \frac{x - \bar{x}}{\sigma_x} = \frac{12 - 6}{3} = 2$$

Using Table 7.6, we obtain the value 0.9772 for the probability that randomly selected values for x will not exceed the value 12.■■

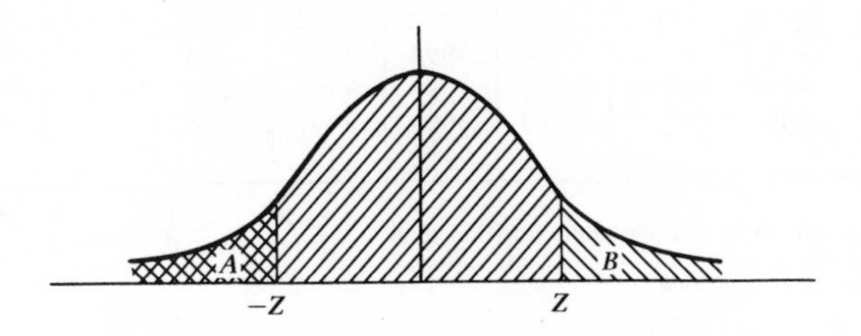

Figure 7 5 Negative standard variables.

Table 7.6* Cumulative normal distribution.

Values of p corresponding to Z_x for the normal curve; Z_x is the standard normal variable. The value of p for $-Z_x$ equals 1 minus the value of p for $+Z_x$, e.g., the p for -1.62 equals $1 - 0.9474 = 0.0526$.

Z_x	.00	.01	.02	.03	.04	.05	.06	.07	.08	.09
.0	.5000	.5040	.5080	.5120	.5160	.5199	.5239	.5279	.5319	.5359
.1	.5398	.5438	.5478	.5517	.5557	.5596	.5636	.5675	.5714	.5753
.2	.5793	.5832	.5871	.5910	.5948	.5987	.6026	.6064	.6103	.6141
.3	.6179	.6217	.6255	.6293	.6331	.6368	.6406	.6443	.6480	.6517
.4	.6554	.6591	.6628	.6664	.6700	.6736	.6772	.6808	.6844	.6879
.5	.6915	.6950	.6985	.7019	.7054	.7088	.7123	.7157	.7190	.7224
.6	.7257	.7291	.7324	.7357	.7389	.7422	.7454	.7486	.7517	.7549
.7	.7580	.7611	.7642	.7673	.7704	.7734	.7764	.7794	.7823	.7852
.8	.7881	.7910	.7939	.7967	.7995	.8023	.8051	.8078	.8106	.8133
.9	.8159	.8186	.8212	.8238	.8264	.8289	.8315	.8340	.8365	.8389
1.0	.8413	.8438	.8461	.8485	.8508	.8531	.8554	.8577	.8599	.8621
1.1	.8643	.8665	.8686	.8708	.8729	.8749	.8770	.8790	.8810	.8830
1.2	.8849	.8869	.8888	.8907	.8925	.8944	.8962	.8980	.8997	.9015
1.3	.9032	.9049	.9066	.9082	.9099	.9115	.9131	.9147	.9162	.9177
1.4	.9192	.9207	.9222	.9236	.9251	.9265	.9279	.9292	.9306	.9319
1.5	.9332	.9345	.9357	.9370	.9382	.9394	.9406	.9418	.9429	.9441
1.6	.9452	.9463	.9474	.9484	.9495	.9505	.9515	.9525	.9535	.9545
1.7	.9554	.9564	.9573	.9582	.9591	.9599	.9608	.9616	.9625	.9633
1.8	.9641	.9649	.9656	.9664	.9671	.9678	.9686	.9693	.9699	.9706
1.9	.9713	.9719	.9726	.9732	.9738	.9744	.9750	.9756	.9761	.9767
2.0	.9772	.9778	.9783	.9788	.9793	.9798	.9803	.9808	.9812	.9817
2.1	.9821	.9826	.9830	.9834	.9838	.9842	.9846	.9850	.9854	.9857
2.2	.9861	.9864	.9868	.9871	.9875	.9878	.9881	.9884	.9887	.9890
2.3	.9893	.9896	.9898	.9901	.9904	.9906	.9909	.9911	.9913	.9916
2.4	.9918	.9920	.9922	.9925	.9927	.9929	.9931	.9932	.9934	.9936
2.5	.9938	.9940	.9941	.9943	.9945	.9946	.9948	.9949	.9951	.9952
2.6	.9953	.9955	.9956	.9957	.9959	.9960	.9961	.9962	.9963	.9964
2.7	.9965	.9966	.9967	.9968	.9969	.9970	.9971	.9972	.9973	.9974
2.8	.9974	.9975	.9976	.9977	.9977	.9978	.9979	.9979	.9980	.9981
2.9	.9981	.9982	.9982	.9983	.9984	.9984	.9985	.9985	.9986	.9986
3.0	.9987	.9987	.9987	.9988	.9988	.9989	.9989	.9989	.9990	.9990
3.1	.9990	.9991	.9991	.9991	.9992	.9992	.9992	.9992	.9993	.9993
3.2	.9993	.9993	.9994	.9994	.9994	.9994	.9994	.9995	.9995	.9995
3.3	.9995	.9995	.9995	.9996	.9996	.9996	.9996	.9996	.9996	.9997
3.4	.9997	.9997	.9997	.9997	.9997	.9997	.9997	.9997	.9997	.9998

*Table values obtained by evaluating approximation to $f(Z_x) = \int_{-\infty}^{Z_x} \frac{1}{\sigma\sqrt{2\pi}} \exp\left\{-\frac{t^2}{2\sigma^2}\right\} dt$, for the values of Z_x indicated in the table.

Example 7.11 (Standard Variable Probabilities). For the last example, determine the probability that random samples from the population will lie between the values $x = 0$ and $x = 12$. ■■

Solution: Again we convert to standard variables

$$Z = \frac{x - \bar{x}}{\sigma_x} = \frac{0 - 6}{3} = -2$$

$$Z = \frac{x - \bar{x}}{\sigma_x} = \frac{12 - 6}{3} = 2$$

Thus we seek the probability that the standard variable Z lies between the values -2 and 2. Note that this is equivalent to finding the area under the normal curve from $-2\sigma_x$ to $2\sigma_x$. From Chapter 4 we know this to be 0.954. Using Table 7.6 to calculate the probability, we find

$$P(Z \leqslant 2) = 0.9772$$

This value includes the probability that Z is less than -2. Hence we need find the value of $P(Z \leqslant -2)$ and subtract it from $P(Z \leqslant 2)$, i.e.,

$$P(-2 \leqslant Z \leqslant 2) = P(Z \leqslant 2) - P(Z \leqslant -2)$$

The probability $P(Z \leqslant -2)$ is the complementary probability $1 - P(Z \leqslant 2) = 0.0228$. Thus the answer we seek is

$$P(-2 \leqslant Z \leqslant 2) = 0.9772 - 0.0228 = 0.9544$$

■■

Example 7.12 (Tables, Areas, and Probabilities). Tables for the cumulative normal distribution exist which give the areas A under the standard normal curve between the vertical line through the mean of the normal curve and a vertical line drawn through the value Z. Figure 7.6 illustrates the areas involved. In other words, the table gives the probability that the variable Z takes on a

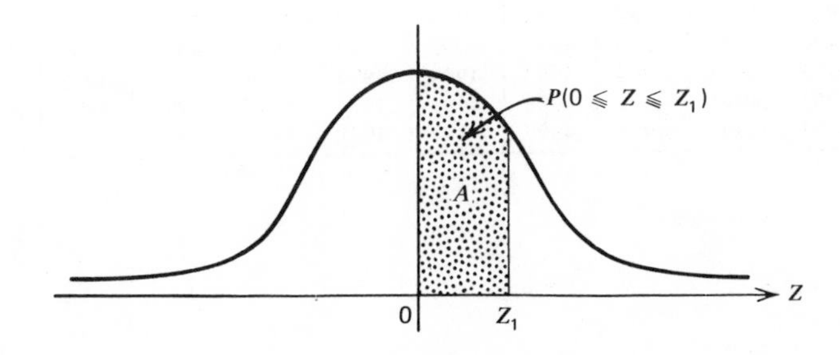

Figure 7.6 One-sided areas.

Table 7.7 Cumulative normal distribution.

Table values obtained from Table 7.6. Each entry in Table 7.6 is reduced by 0.5 to account for the areas under the normal curve from $-\infty$ to 0.

Z_x	.00	.01	.02	.03	.04	.05	.06	.07	.08	.09
0.0	.0000	.0040	.0080	.0120	.0159	.0199	.0239	.0279	.0319	.0359
0.1	.0398	.0438	.0478	.0517	.0557	.0596	.0636	.0675	.0714	.0753
0.2	.0793	.0832	.0871	.0910	.0948	.0987	.1026	.1064	.1103	.1141
0.3	.1179	.1217	.1255	.1293	.1331	.1368	.1406	.1443	.1480	.1517
0.4	.1554	.1591	.1628	.1664	.1700	.1736	.1772	.1808	.1844	.1879
0.5	.1915	.1950	.1985	.2019	.2054	.2088	.2123	.2157	.2190	.2224
0.6	.2257	.2291	.2324	.2357	.2389	.2422	.2454	.2486	.2518	.2549
0.7	.2580	.2612	.2642	.2673	.2704	.2734	.2764	.2794	.2823	.2852
0.8	.2881	.2910	.2939	.2967	.2995	.3023	.3051	.3078	.3106	.3133
0.9	.3159	.3186	.3212	.3238	.3264	.3289	.3315	.3340	.3365	.3389
1.0	.3413	.3438	.3461	.3485	.3508	.3531	.3554	.3577	.3599	.3621
1.1	.3643	.3665	.3686	.3708	.3729	.3749	.3770	.3790	.3810	.3830
1.2	.3849	.3869	.3888	.3907	.3925	.3944	.3962	.3980	.3997	.4015
1.3	.4032	.4049	.4066	.4082	.4099	.4115	.4131	.4147	.4162	.4177
1.4	.4192	.4207	.4222	.4236	.4251	.4265	.4279	.4292	.4306	.4319
1.5	.4332	.4345	.4357	.4370	.4382	.4394	.4406	.4418	.4430	.4441
1.6	.4452	.4463	.4474	.4485	.4495	.4505	.4515	.4525	.4535	.4545
1.7	.4554	.4564	.4573	.4582	.4591	.4599	.4608	.4616	.4625	.4633
1.8	.4641	.4649	.4656	.4664	.4671	.4678	.4686	.4693	.4699	.4706
1.9	.4713	.4719	.4726	.4732	.4738	.4744	.4750	.4756	.4762	.4767
2.0	.4773	.4778	.4783	.4788	.4793	.4798	.4803	.4808	.4812	.4817
2.1	.4821	.4826	.4830	.4834	.4838	.4842	.4846	.4850	.4854	.4857
2.2	.4861	.4865	.4868	.4871	.4875	.4878	.4881	.4884	.4887	.4890
2.3	.4893	.4896	.4898	.4901	.4904	.4906	.4909	.4911	.4913	.4916
2.4	.4918	.4920	.4922	.4925	.4927	.4929	.4931	.4932	.4934	.4936
2.5	.4938	.4940	.4941	.4943	.4945	.4946	.4948	.4949	.4951	.4952
2.6	.4953	.4955	.4956	.4957	.4959	.4960	.4961	.4962	.4963	.4964
2.7	.4965	.4966	.4967	.4968	.4969	.4970	.4971	.4972	.4973	.4974
2.8	.4974	.4975	.4976	.4977	.4977	.4978	.4979	.4980	.4980	.4981
2.9	.4981	.4982	.4983	.4983	.4984	.4984	.4985	.4985	.4986	.4986
3.0	.4987	.4987	.4987	.4988	.4988	.4989	.4989	.4989	.4990	.4990
3.1	.4990	.4991	.4991	.4991	.4992	.4992	.4992	.4993	.4993	.4993
3.2	.4993	.4993	.4994	.4994	.4994	.4994	.4994	.4995	.4995	.4995
3.3	.4995	.4995	.4995	.4996	.4996	.4996	.4996	.4996	.4996	.4997
3.4	.4997	.4997	.4997	.4997	.4997	.4997	.4997	.4997	.4997	.4998

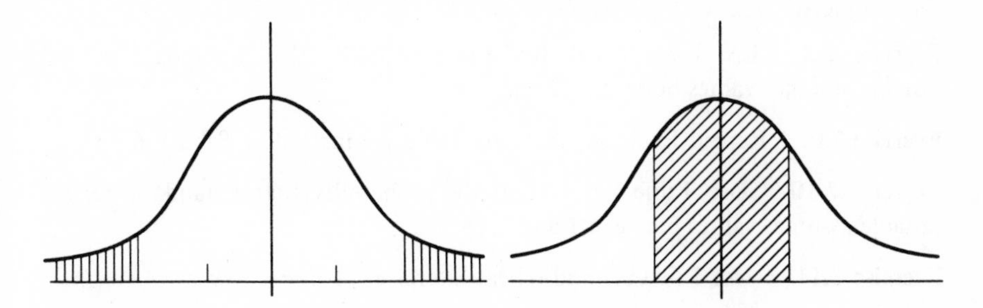

Figure 7.7 Symmetric areas.

value between 0 and Z_1. Such tables are easier to use when we require probabilities for regions which are symmetric about the mean of the normal curve. Figure 7.7 illustrates two such areas which are given in terms of the number of standard deviations the variables are removed from the mean.

We will now illustrate how both tables can be used in finding the probability associated with a region which is not symmetric about the mean. We shall use Tables 7.6 and 7.7, where Table 7.7 is identical to Table 7.6, except that 0.5000 is subtracted from every entry in Table 7.6. This accounts for the area from $-\infty$ to 0 under the normal curve. Or, stated in probabilities, it accounts for

$$p(-\infty < Z \leqslant 0)$$

Let us find the probability that the standard normal variable assumes values between -1 and 2. The area is illustrated in Figure 7.8. We know that the area from -1 to 0 is equal to the area from 0 to 1. We use Table 7.6 and enter at $Z = 1$ where we read the value, 0.8413. Table 7.7 contains the value 0.8413–0.5000 = 0.3413 which is the probability associated with Area A_1. For Area A_2, we use Table 7.7 which contains the value 0.9772 – 0.5000 = 0.4772. The sum of the two, 0.3413 + 0.4772 = 0.8185, is the value of the probability sought.■■

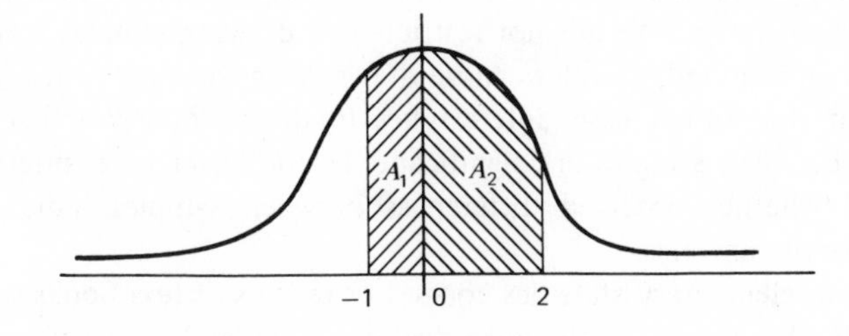

Figure 7.8 Asymmetric areas.

Exercise 7.7. Find the probability that a variable takes on values which yield the symmetric areas of Figure 7.7.

Exercise 7.8. Use Table 7.6 to find the probability that a standard normal variable assumes values between –2 and 1.■■

Exercise 7.9. Repeat Exercise 7.8, using Table 7.7 instead of Table 7.6.■■

Exercise 7.10. Use Table 7.6 to find the probability that a standard normal variable assumes values between 1 and 2.■■

Exercise 7.11. Use Table 7.7 and repeat Exercise 7.10.■■

7.9 SAMPLING

Because it is often impractical if not impossible to measure or observe all elements of a universe, one must rely on information gained from samples taken from the universe. In statistics, the term "sample" is used in any of three senses. In the first sense, when a ball is drawn from an urn, or we measure the characteristics of a person, or piece of equipment, the sample is considered as but one element of the universe. In the second sense, when we draw samples of 10 balls from an urn or measure some characteristic of 20 persons, the sample is considered as a single element of a new universe consisting of complex elements, namely, 10 balls or 20 persons. For example, the event can be recorded as "6 black and 4 red balls." For the 20 persons, we might record "14 males and 6 females." Other events or samples will consist of different mixtures of balls or persons. Our interests lie in learning something about the universe of samples consisting of 10 balls or 20 persons. Stated in other words, the sampling universe consists of sets of combinations of 10 balls or 20 persons taken from the original universes. Hence in this sense sampling is used to construct another universe whose elements are samples of 10 balls or 20 persons. In a third sense, we consider a sample of n elements as a selected sample or a representative portion of the universe. In this third sense we hope to use the sample to make estimates for some of the unknown parameters (mean, median, variance) of the universe from which the sample was drawn. We are not restricted to drawing samples from a single universe. When required, samples can be drawn from the same or from different universes. In the former case, interest lies in determining whether observed differences between samples are significant. In the latter case, interest lies in determining whether observed differences between samples indicate similar differences in the universes.

Students in elementary statistics courses or readers of text books such as this one, normally are spoon fed the universes or samples and are asked to make the calculations required to estimate the parameters required for the universe or samples provided. But, in practice, the analyst must identify the universe, collect or observe samples, make measurements, determine the parameters he wishes to estimate, and state the hypothesis he wishes to accept or reject. All these operations are important and necessary to good statistical analysis. At this point we will discuss these operations, starting with the different ways of sampling a population.

7.10 TYPES OF SAMPLES

The most important types of samples are *(a)* random, *(b)* stratified, and *(c)* judgment. A *random* sample is one in which each sample of some fixed size has an equal chance of being drawn from the universe. It is normally assumed that the individual samples are independent of each other. A *stratified* sample is one in which a heterogeneous universe is partitioned into homogeneous subsets. Samples are taken randomly from the subsets in proportion to the size of the subsets. For example, the universe of working families can be partitioned into subclasses defined as those earning below $5000 per annum, those earning between $5000 and $15,000, and those earning more than $15,000. If these subsets contain 55, 30, and 15%, respectively, of the universe, then random samples in proportions 55:30:15 of the total sample will be taken from the three subsets. A *judgment* sample is one in which samples are drawn to obtain a representative cross-section of the universe. For example, in forecasting the result of an election on the basis of early voting returns, samples of early returns are taken from voting precincts which are representative of the county or state containing them. The selection of these precincts is based on the "judgment" of the analysts who designed the voting forecast model.

In the remainder of this chapter, we shall employ random sampling except where it is specifically stated otherwise. We shall also assume that samples are independent of each other. We will be assuming that the draw of any sample will not materially affect the probability of drawing any other sample. We know that this assumption is not exactly true in sampling without replacement. However, if the universe is large and the number of samples drawn is small relative to the size of the universe, then the assumption of independence between samples results in a good approximation to true independence.

We now discuss some of the different types of measurements one employs in the process of sampling, keeping in mind that the terms "draw a sample" and "take a measurement" are often used synonomously in statistics.

7.11 OBSERVATIONS AND MEASUREMENTS

In Chapter 4 we discussed the following three types of variables: *(a)* those variables which when observed or measured are assigned numbers, names, or symbols to identify the groups, sets, classes, intervals, etc., to which various objects, persons, or characteristics belong; *(b)* those variables, which, when observed or measured, are assigned names or numbers to identify not only the groups, sets, classes, intervals, etc., to which various objects, persons, or characteristics belong, but also to identify an order relation among them; and *(c)* those variables which, when observed or measured, are assigned names or numbers to identify the various classes, the order relation between the classes, and the distance between any two classes in the order specified. The variables were called nominal, ordinal, and interval, respectively. We will now assert some important properties associated with measurements (or variables) of the three types and briefly discuss a fourth type called the ratio scale.

7.12 NOMINAL MEASUREMENTS

When a variable is observed or measured, one assigns it a name, a symbol, or a number. For example, let us sit by a roadside and count the number of automobiles which pass an observation point in a prespecified period of time. We would be observing one class of objects and we could assign it the classification—automobiles. We could also observe the automobiles and give them classifications corresponding to the manufacturer, model year, country of origin, size, etc. In doing this, we would be assigning either numbers (class 1, class 2, etc.), names (Ford, General Motors, Chrysler, Foreign, Sports, etc.), or symbols (Class A, Class B, etc.) to each of the values the variables could assume. These numbers, names, or symbols are assigned to the different groups of objects, one name per group, in accordance with values assumed by the objects in each group. Thus, in "measuring" automobiles, the measurement can be restricted to assuming any of six discrete values: Ford, American Motors, General Motors, Chrysler, Foreign, and all others. Each automobile, as it passes, will be observed (or measured) and will be placed in one of the six classes.

The automobiles could be stopped and the manufacturer's serial number could be observed. These numbers constitute an assignment of numerical values to a nominal variable and could be used to classify the universe of automobiles. Nominal classifications are used quite frequently in substantive analysis. In content analysis, a measure of content is the number of times certain words, semantically similar clauses, or themes occur in a series of communications. In military and diplomatic intelligence, troops, material, diplomats, and offensive and defensive weapons are assigned nominal classifications reflecting partitioning into many subclasses.

A nominal classificatory system provides an analyst little to work with. The analyst can only say whether an object belongs or does not belong to a specific subclass. Thus, the classification scheme partitions the universe into mutually exclusive subclasses in accordance with the properties assigned to the names, numbers, or symbols used in the classification. The assignment of names, numbers, or symbols to the subclasses (or values assumed by the variable) can be done in an arbitrary manner, except where the name or symbol carries with it a special meaning or measurement which identifies the subclass. For example, nominal classifications such as tanks, aircraft, guns, etc. should not be interchanged or assigned indiscriminately. There are accepted terminologies and these should be followed in the assignment of nominal values to the subclasses of a universe. However, we could use the letters A, B, C, etc., to identify classes of tanks, aircraft, guns, etc. In this case it does not matter which letter is assigned to which subclass. Further, once the symbols have been assigned, they could be interchanged among the subclasses, provided that the interchange is done completely and consistently. Assignment or interchange of names will not alter the information about the subclasses essential to statistical or other forms of analysis. The statistical descriptors associated with nominal variables will remain invariant under name changes; i.e., the mode, frequency count, or number of categories will not change with an interchange of names, provided the name change is done in a one-to-one fashion.

7.13 ORDINAL MEASUREMENTS

Nominal measurements subdivide the universe into mutually exclusive subclasses. There is no order among the subclasses, i.e., they can be arranged in any order and not diminish or contribute to the information contained in the categorization. There are instances when the subclasses stand in some relation to one another. For example, some classes may be "taller than," "heavier than," "more preferred than," "cheaper than," or "more informative than" other classes of the universe. Rank in the military or a graduating class is an example of such a relation. A private is lower than a corporal who in turn is lower than a sergeant. Graduates are scored or ranked by the decile or quintile assigned to them by "measuring" their grades over their academic careers.

When a measurement subdivides a universe into nominal subclasses and an order relation exists among the subclasses, the measurement is called *ordinal* or ordered. For example, people are subdivided in classes identified by social status, income, or education level. The subclasses are mutually exclusive and a distinct order is implied among the subclasses. The order is normally lowest-to-highest, e.g., for income groups we might have low, low-middle, middle, high-middle, and high categories. However, other orders are possible. But, whatever the order, the distance between adjacent classes is not implied by the measurement. For example, with income groups the median income for the income group may be

$2500 per annum for low, $5000 for low-middle, $9000 for middle, $15,000 for high-middle, and $25,000 for high. The order is obvious. The difference between low and low-middle income groups is $2500. It is $10,000 between the high-middle and high income group. The lack of a uniform measure between classes raises some interesting questions. The first is, can we average, i.e., find the mean, of a set of ordinal measurements? For example, five students will rank 1, 2, 3, 4, and 5 among themselves. The average rank is 3, but this imparts little or no information of value to an analyst. The students may have scored 70, 80, 85, 90, and 100 to achieve their ranks, or equally, they could have scored 0, 5, 10, 15, and 20, or perhaps 0, 25, 50, 75, and 100. The ranking remains the same, but the range and differences can vary widely. To answer the question we state the following: the averaging of some variables such as rank entails performing arithmetic operations which have no counterparts with the variables. We can average five numbers and the result is interpreted by all to mean the same thing regardless of the numbers involved. If we average five ranks, the so-called "average rank" is interpreted only in terms of the scores from which the ranks were derived. We can average the actual scores, but scores imply a higher level of measurement than do ranks. Thus, most ordinal measurements have but two distinguishing properties: First, the measurements subdivide the universe into mutually exclusive subclasses; second, the subclasses stand in an order relationship among themselves.

The second question is: What operations can we perform on the subclasses of an ordinal variable? Specifically, as with nominal variables, we are free to assign any numbers as identifiers of the subclasses, provided that the numbers reflect the order existing among the subclasses. For example, the upper class or top ranked will normally be assigned the highest number with the other classes being assigned numbers decreasing with the rank achieved. Often the numbers can be assigned in increasing order with rank. In this case, the top rank is called first or number 1, the next in rank is called second or number 2, and so on. It does not matter what numbers (or names) are assigned, as long as we give higher (or lower) to the higher ranked classes. Keep in mind that both names and numbers can be used to designate order among subclasses. Military officer grades are numbered 0-1 to 0-10. They are also given the names Second Lieutenant to General (Army) or Ensign to Admiral (Navy). Either assignment of value (measurement) to the variable (military grade) retains the class distinction and order of rank. Thus the names or numbers assigned to the subclasses or their members through an ordinal measurement process can be transformed to another set of names or numbers provided that the transforming is done completely, consistently, and does not change the order existing among the subclasses. Any transformation resulting in a change in order will result in a loss of information and hence is inadmissible.

As stated in Chapter 4, the median is the most appropriate descriptor of central tendency for distributions of ordinal variables. The median gives the central measurement in the order associated with the measurements and hence is not affected by changes in magnitude or scale which do not affect the order of the measurements.

Ordinal measurements differ from nominal measurements in one additional characteristic. Nominal measurements are always discrete; they assume unique distinct values. Ordinal measurements may be either discrete or from a continuous scale of measurements. Normally, ordinal measurements fall into discrete categories such as pass-fail, win-lose, quartiles, or deciles. The measurements can take on any value in a prespecified set of intervals. For example, a score of 75 will rate the measurement value "pass" as will any score 70 or above. It will place the score in the class labeled "passed." Although the name attached to this score or measurement is discrete, the score or measurement is drawn from a continuum of values, namely 70 to 100. This is also true of subjective measurements as well as physical or mathematical measures. The categories of bright and dull, agree and disagree, support and nonsupport are names attached to the two divisions or intervals of a continuum of values. Thus the category names can assume any of a discrete set of values each represents and can be derived from any of a continuum of values.

7.14 INTERVAL MEASUREMENTS

In the previous section we pointed out that the measurement of some ordinal variables ranges over a continuum of values yet assumes only the discrete set of values which are assigned to intervals or categories in the continuum. When an ordinal measurement ranges over a continuum of values and a unit of measure is associated with the measurement which assigns a "difference" or "distance", between any two measurements in the continuum, then the measurement is called an *interval measurement.* For example, a decile rank is an ordinal measurement because in decile ranking the rank difference between the grades 68 and 75, and the grades 67 and 76 can turn out to be the same. The measurement of height is an interval measurement because the difference between the heights 68 and 75 in. is 7 in., while the difference between the heights 67 and 76 in. is 9 in. Hence the order relation associated with the height measurement applies also to the difference of "distance" measurement.

Interval measurements have an additional property. The continuum of interval measurements can be translated (adding the same constant to all measurements in the continuum) or "expanded" (multiplying by the same constant) without disturbing the ratio which exists between any two intervals on the continuum.

For example, we can start with the interval 0–100, expand it by 3 to the interval 0–300, and then translate the result by 50 units to 50–350. The two scales subdivided into 10 intervals each, with the subdivisions corresponding on the two scales through the transformation

$$I_2 = 3I_1 + 50$$

are shown in Figure 7.9. On scale I_1 we find the ratio between the two intervals B_1 and A_1.

$$\frac{B_1}{A_1} = \frac{90 - 60}{30 - 20} = \frac{30}{10} = 3$$

On scale I_2, we find the ratio between the two intervals B_2 and A_2.

$$\frac{B_2}{A_2} = \frac{320 - 230}{140 - 110} = \frac{90}{30} = 3$$

Thus the ratio between intervals is preserved in interval measurement. This allows us to change units of measurements, say from centimeters to inches (2.54), from degrees Centigrade to degrees Fahrenheit ($F = \frac{9}{5}C + 32$) and other well-known unit changes. That these changes are permissible also indicates that with interval measurements, the unit of measurement and the origin (zero point) are arbitrary.

7.15 RATIO MEASUREMENTS

When an interval measurement has the additional property that the ratio of any two measurements is independent of the unit of measurement, then the measurement is a ratio measurement. The three basic physical measurements of time,

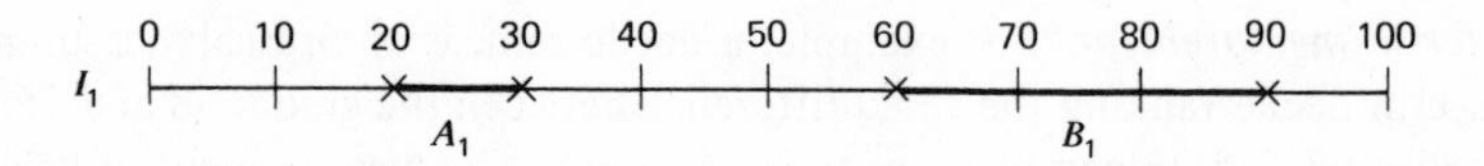

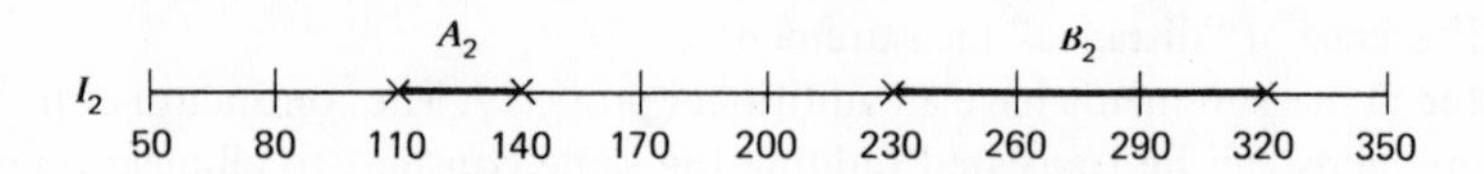

Figure 7.9 Interval ratios.

length, and mass are ratio measurements. The ratio between any two items, or weights or lengths is independent of the unit of measurement. The ratio of two times measured in hours, say 5–2 hrs., converts to 300 and 120 min. Likewise, the ratio between two length measurements remains the same whether measured in yards, meters, or inches.

A key to ratio measurements is the existence of a true zero point (or origin). For example, in Figure 7.9 the I_2 interval does not have a true zero point relative to the I_1 scale. Zero on the I_1 scale corresponds to 50 on the I_2 scale. Thus, the ratio between two measurements on the I_1 scale, say 60 and 30, is 2. The ratio between the corresponding points 230 and 140 is 23/14 on the I_2 scale.

7.16 SUMMARY

The process of measurement or observation is one of assigning names, symbols, or numbers to designate properties possessed by individuals or samples of a universe. Measurements can be subjective, such as measuring color, historical meaning, or political intention. Other measurements are objective, such as measuring physical quantities, some social, or some psychological parameters. Subjective measurements normally are not performed according to a prescribed set of rules. Objective measurements normally are performed in accordance with established procedures and rules. The procedures and rules usually include the conditions under which the measurements are taken, the instruments or standards used in making measurements, and the number of measurements to be taken.

Once a set of measurements is taken and recorded, the novice analyst is prone to compute averages and variances without regard to whether the arithmetic operations associated with averaging, etc., are permissible with the measurement set. For example, if the measurement is the power ranking of nations of the world, then the measurements are ordinal and averaging is not permissible. The "average rank," while it is a nice real number, has little or no meaning on the ranking scale. The median rank is more meaningful in this case. Only certain types of mathematical operations are permitted on each of the four types of measurements described above. The accompanying table lists the four types of measurements, the relationships which exist among measurements of the same type, and the arithmetic operations which are permissible with each type of measurement.

The table indicates that no arithmetic operations other than an equivalence mapping is permissible on sets of nominal measurements. If one wishes to preserve the ratio properties of ratio measurements, addition of nonzero constants is inadmissible. With sets of ordinal measurements, the median is meaningful but the mean is not. Many other statements can be made about the contents of the

Measurement	Relations	Permissible operations
Nominal	Equivalence within class, none between classes	Complete and consistent symbol interchange between classes
Ordinal	Equivalence within class; order between classes	Symbol interchange; order preserving transformations such as addition of constant and multiplication by a *positive* constant
Interval	Equivalence within class; order between classes and members; distance between classes and members	Symbol interchange; linear transformation such as addition of, and multiplication by, *any* constant; differences between classes and elements; ratios between intervals
Ratio	Equivalence within class; order between classes, members; distance between classes and members; true zero point	Symbol interchange; must preserve zero point; multiplication by any constant; differences between classes or elements; ratios between intervals; ratios between elements

table. The basic principle underlying the table is this: When numbers or symbols (names) are associated with measurements and arithmetic operations are performed on the measurements, then the arithmetic operations must bear a direct and meaningful relation on the quantity being measured. For example, if we measure the time to complete a task as t min., then adding 2 to the time numerically implies that the time to complete the task is increased by 2 min. Dividing by 2 would indicate that the time to complete the task was cut in half. Thus before performing any arithmetic operations on sets of measurements, one must be certain that the operations are admissible in the type of measurements under analysis.

7.17 ESTIMATION WITH SAMPLES

Up to this point we have been dealing with entire populations or universes of objects. We constructed frequency and probability distributions and learned how to determine their modes, means, medians, and variances. We learned that there are several ways of obtaining samples (or measurements) and that only certain arithmetic operations are permissible on the measurements, depending on the measurement type. We now employ this information to use samples to estimate characteristics associated with populations. Some of the characteristics will be well-known parameters such as the mean and variance and hence will be applicable only to interval or ratio measurements. Other characteristics will be nonparametric such as frequency counts, proportions of classes, and distribution shape and therefore will be especially applicable to nominal or ordinal measurements.

To better understand why and how population characteristics (parametric and nonparametric) can be estimated from samples, we now examine the anatomy of samples and sample spaces a little closer.

7.18 ANATOMY OF SAMPLING

A population or universe U will consist of M elements, $x_1, x_2, ..., x_M$. Each element x_i of the universe U can be considered as a sample of one taken from the universe U. Every set of N elements, denoted $S_i(N) = \left\{ x_{i_1}, x_{i_2}, ..., x_{i_N} \right\}$, can be considered as a sample of N taken from the universe U. The subscripts $i_1, i_2, ...,$ i_N denote a random selection of N elements from the M elements $x_1, ..., x_M$ of U. There are M^n samples of size N in the universe U of M elements. We are assuming that sampling is done with replacement and that the samples are ordered. Without replacement, the number of samples would be $M!/(M-N)!$. For example, if U contains 10 elements and we select all possible subsets of 3 elements, we find that there are $10^3 = 1000$ such subsets. If the selection is done without replacement, we find that there are $10!/(10-3)! = 10!/7! = 10 \cdot 9 \cdot 8 = 720$ such subsets.

Before we proceed, let us consider as an example a universe U consisting of four elements $x_1 = 2, x_2 = 4, x_3 = 6$, and $x_4 = 8$. Samples of one from U will result in the sets $\{2\}$, $\{4\}$, $\{6\}$, $\{8\}$, each set consisting of one element of U. Samples of two from U will result in 16 sets given in matrix form in Figure 7.10.

In each cell, defined by the intersection of a row and a column, three numbers are entered. The pair of numbers in the upper left half of the cell is one of the samples of two that can be drawn from U. The number in the lower right half of the cell is the mean value for the pair of numbers in the cell. We shall use these means in a continuation of this example later in this section. There are 16 ordered pairs ($M^n = 4^2 = 16$) of numbers (samples of two) listed in the matrix. This is an exhaustive listing of all samples of two with replacements. If we sampled without replacement, we would obtain 12 pairs, $M!/(M-N)! = 4!/2! = 12$. These are the 12 pairs of numbers remaining after the main diagonal of the matrix is deleted. From now on, we shall restrict our discussion to sampling with replacement. We can list and denote the samples as follows:

$$s_1 = \{x_1, x_1\} \qquad s_2 = \{x_1, x_2\} \qquad s_3 = \{x_1, x_3\}$$

$$s_4 = \{x_1, x_4\}, \ldots, s_{16} = \{x_4, x_4\}$$

The set of samples can be thought of as another universe, $S_u(2)$, derived from the original universe U. Likewise, the set of all samples consisting of 3 elements of U can be thought of as still another universe $S_u(3)$. Thus when we draw a sample of 2 or 3 elements, we can think of the sample as coming from either of two universes. The first universe is the original universe, U. The second is the universe $S_u(2)$ or $S_u(3)$ consisting of all samples of 2 or 3 elements drawn from the original universe. For the time being, we will use the second or latter interpretation, the sample consisting of an element from $S_u(2)$.

Let us return to the matrix representation of the universe (sample space) consisting of samples of 2 elements from U. Let us also label the elements of the new universe $S_u(2)$ as $S_1, S_2, \ldots, S_{16}$. Then each sample S_i of $S_u(2)$ has a mean

	$x = 2$	$x = 4$	$x = 6$	$x = 8$
$x = 2$	2,2 / 2	2,4 / 3	2,6 / 4	2,8 / 5
$x = 4$	4,2 / 3	4,4 / 4	4,6 / 5	4,8 / 6
$x = 6$	6,2 / 4	6,4 / 5	6,6 / 6	6,8 / 7
$x = 8$	8,2 / 5	8,4 / 6	8,6 / 7	8,8 / 8

Figure 7.10 Samples of two elements with sample means.

Table 7.8 Mean and variance of sample means.

i	u_i	f_i	$u_i f_i$	$u_i^2 f_i$	Mean and variance of U^*
1	2	1	6	4	$\bar{u} = \frac{1}{16} \Sigma_{i=1}^{7} u_i f_i$
2	3	2	6	18	
3	4	3	12	48	$= 5$
4	5	4	20	100	
5	6	3	18	108	$\sigma_{\bar{u}}^2 = \frac{1}{16} \Sigma_{i=1}^{7} f_i u_i - (\bar{u})^2$
6	7	2	14	98	
7	8	1	8	64	$= 27.5 - 25$
					$= 2.5$
		$\sum_{i=1}^{7} f_i = 16$	$\sum_{i=1}^{7} u_i f_i = 84$	$\sum_{i=1}^{7} u_i^2 f_i = 440$	

value. We shall label the means of the elements $u_1, u_2, \ldots, u_{16}$. Thus, associated with the universe $S_u(2)$ is still another universe, the universe U^* made up of the means of all the elements of $S_u(2)$. The values for these means are entered in the matrix of Figure 7.10. A frequency table for these sample means is given in Table 7.8.

The universe U^* is called the universe of sample means. Associated with U^* is the distribution of sample means. We shall use the term universe to mean the set of objects or measurements and its associated distribution. Thus, if the universe's distribution is not known, we shall be referring to the set of objects or measurements. Otherwise, the term universe will be used interchangeably with the term distribution.

The universe of sample means U^* has a mean value and a variance. The mean and variance of the sample means are computed in Table 7.8 following the procedures given in Chapter 4.

Now that we have the mean and variance of the distribution (or universe) of the sample means, what information have we gained about the original universe U? Before we answer that question, let us find the mean and variance of the four elements of the universe U. Again, we construct a frequency table and perform the calculations as previously indicated. The frequency table for U and the calculations for its mean m and the variance σ^2 are given in Table 7.9

At this juncture we can compare the parameter values obtained for the universe of samples U and the universe of sample means U^*. We note that the means of the two universes are equal and that the variance of the sample means is one-half that of the mean of the original universe U. These are interesting, important, and have significance beyond mere chance. We find that fixed relation-

Table 7.9 Mean and variance of universe *U*.

i	x_i	f_i	$x_i f_i$	$x_i^2 f_i$	Mean and variance
1	2	1	2	4	$m = 1/4 \cdot 20 = 5$
2	4	1	4	16	$\sigma^2 = 1/4 \cdot 120 - (5)^2$
3	6	1	6	36	$= 30 - 25$
4	8	1	8	64	$= 5$

ships hold between the means and variances of any original universe U and its universe of sample means U^*, provided the universe U contains a finite number of elements. We find that the two means are always equal and the variances differ by the multiplicative factor N, the number of elements in the samples. The distributions for the two universes, and the relationships which exist among the frequencies, the means and variances are illustrated in Figure 7.11.

Note in Figure 7.11 that the original samples 2, 4, 6, and 8 are distributed uniformly over the range of the distribution U. The sample means have the same

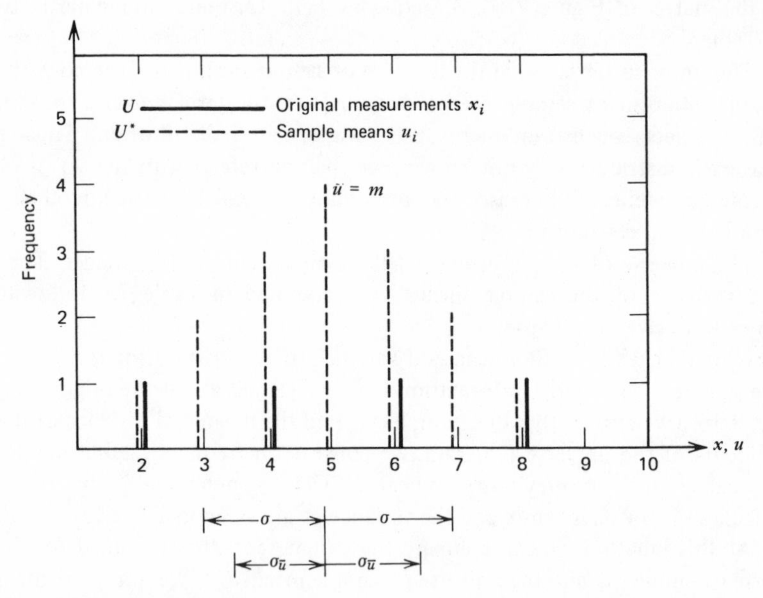

Figure 7.11 Distributions and parameters for original measurements and sample means where $\sigma = 2.23$, $\sigma_{\bar{u}} = 1.58$, and $\sigma_{\bar{u}}^2 = \sigma^2/N$.

range but tend to cluster more at the mean than do the original measurements. This observation is true for any finite distribution and its sample mean distribution. The reason for this stems from the fact that the mean of any set of numbers is always a number within the range of the sample. The movement of the mean of a sample toward the "center" of the sample range results in the distribution of the sample means having a smaller variance than the original measurements. The surprising, yet credible, relation is that the variance of the sample means is diminished by the factor $1/N$, where N is the number of samples from which the sample means are calculated.

Figure 7.12 is a schematic for the structuring of three distributions from the original set of M measurements. Distribution parameters (mean and variance) are associated with each distribution and the relationships existing among the parameters are indicated. If one keeps the schematic of Figure 7.12 in mind when dealing with samples and their means or other parameters associated with samples, then one should have little trouble in understanding the universes involved when estimates about the original universe's parameters are made from samples drawn from that universe. We shall now go through another example and then summarize the important findings.

Example 7.13 (Sample Means Distribution). In this example we again use the universe U of four elements $x_1 = 2$, $x_2 = 4$, $x_3 = 6$, and $x_4 = 8$. The sample space $S_u(N)$ will consist of all samples of 3 elements of U. Thus the sampling space is designated $S_u(3)$. For each element S_i in $S_u(3)$ we compute the mean u_i and form the universe U^* of sample means. There are $(M)^N = 4^3 = 64$ elements in

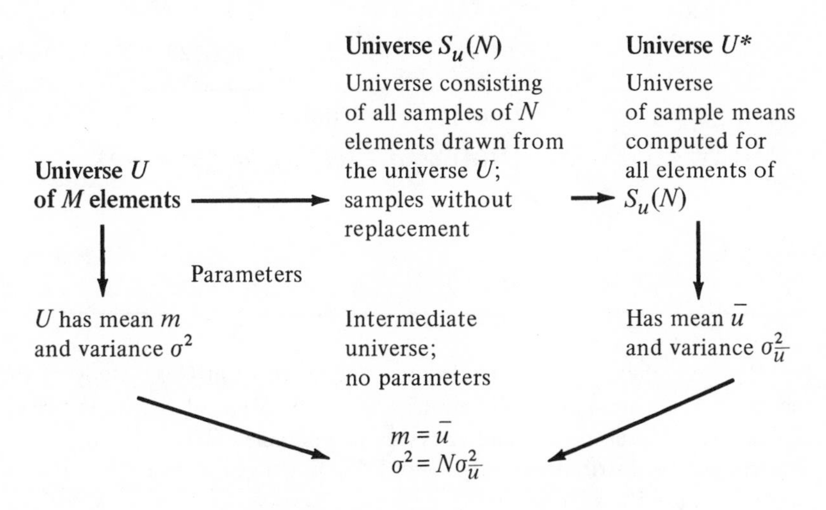

Figure 7.12 Schematic of universes.

Table 7.10 Mean and variance of universes U and U^* (sample means).

U	$S_u(3)$		U^*		
x_i	S_i	f_i	u_i	f_iu_i	$f_iu_i^2$
2	2,4,6	6	4	24	96
4	2,4,8	6	4.67	28	130.67
6	4,6,8	6	6	36	216
8	2,6,8	6	5.33	32	170.67
	2,2,2	1	2	2	4
	4,4,4	1	4	4	16
	6,6,6	1	6	6	36
	8,8,8	1	8	8	64
	2,2,4	3	2.67	8	21.33
	2,2,6	3	3.33	10	33.33
	2,2,8	3	4	12	48
	4,4,2	3	3.33	10	33.33
	4,4,6	3	4.67	14	65.33
	4,4,8	3	5.33	16	85.33
	6,6,2	3	4.67	14	65.33
	6,6,4	3	5.33	16	85.33
	6,6,8	3	6.67	20	133.33
	8,8,2	3	6	18	108
	8,8,4	3	6.67	20	133.33
	8,8,6	3	7.33	22	161.33
		$\Sigma = 64$		$\Sigma = 320$	$\Sigma = 1{,}706.64$

$m = (1/4)\ \Sigma x_i = 5$ $\qquad \bar{u} = (1/64)\ \Sigma f_i u_i = 320/64 = 5$

$\sigma^2 = (1/4)\ \Sigma x_i^2 - (5)^2$ $\qquad \sigma_{\bar{u}}^2 = (1/64)\ \Sigma f_i u_i^2 - (\bar{u})^2 = 1706.64/64 - (5)^2$

$= 5$ $\qquad \cong 1.64$

$= \sigma^2/N$

$= 5/3$

$S_u(3)$. However, there are only 10 distinct sample means in U^*. Many of the elements of $S_u(3)$ have the same means. The composition of the three spaces, U, $S_u(3)$, and U^*, their means, and variances are contained in Table 7.10.

The frequency distributions of U and U^*, constructed from Table 7.10, are shown in Figure 7.13.

Now we are in a position to summarize some of the key relationships between distributions and their samples which we have been illustrating by means of the previous examples and discussions.

The principal relationships are as follows: When all possible samples of size N are drawn with replacement from a universe U of M elements with mean m and variance σ^2, then the distribution of sample means U^* has a mean $\bar{u}$ and a variance $\sigma_{\bar{u}}^2$ which are related to the mean and variance of the original distribution as

$$\bar{u} = m$$
$$\sigma_{\bar{u}}^2 = \sigma^2/N$$

The relations tell us two things: The means of the two distributions are equal, and the variance of the sample means decreases with increasing sample size. Let us delve deeper into this last statement on the variance of the sample means.

First, let us consider drawing all samples of size 1 from the original universe. Each sample of one is its own mean. The resulting universe U^* is therefore identical with U. Also, there is no difference in the variances and we have $\sigma_{\bar{u}}^2 = \sigma^2$. Since $N = 1$, this is in accordance with the original relation. Next, let us consider drawing samples of size N. With increasing N, the variance of the sample means decreases, i.e., the sample means tend to cluster about the mean of the universe. Moreover, the sample means more frequently assume values near the universe mean. Thus, although the original measurements may be uniformly or randomly distributed over an interval, the distribution of the sample means more closely approximates a normal distribution over the original interval.

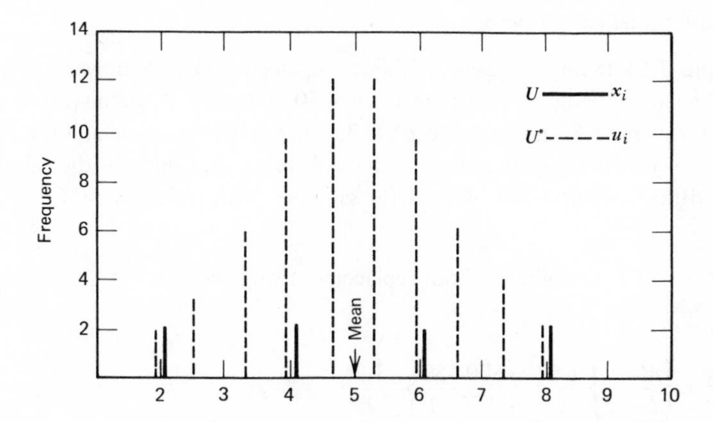

Figure 7.13 Distributions and parameters for original measurements and sample means where $m = \bar{u} = 5$ and $\sigma_{\bar{u}}^2 = \sigma^2/N = 5/3$.

This relationship between distribution shapes is illustrated in Figures 7.11 and 7.13. In each case, the universe of original measurements is uniformly distributed over the range 2–8, but the distribution of sample means U^* takes on the shape of a normal distribution over the same interval 2–8.

We have restricted our attention to the universe of sample means obtained from samples taken with replacement. There are situations where the sample selected cannot be replaced because the measurement may alter or destroy the sample. Hence, it is valuable to know what modifications in the variance formula must be made to account for the differences between sampling with and without replacement.

When samples of size N are drawn without replacement from a universe U of M measurements with mean m and variance σ^2, the distribution of sample means U^* has a mean $\bar{u}$ and a variance $\sigma_{\bar{u}}^2$, which are related to the mean and variance of the original universe U as follows:

$$\bar{u} = m$$

$$\sigma_{\bar{u}}^2 = \frac{(M-N)}{(M-1)} \cdot \frac{\sigma^2}{N}$$

In many applications, the universe will be large relative to the sample size. M will be very much larger than N. Then the factor $(M-N)/(M-1)$ will be close to 1 so that the original relationship

$$\sigma_{\bar{u}}^2 = \sigma^2$$

will serve as a good approximation. For example, if $M = 1000$ and $N = 30$, then $(M-N)/(M-1) = (1000-30)/(1000-1) = 970/999$, which is close enough to 1 for most statistical purposes.

Example 7.14 (Sample Means–Without Replacement). A universe U of $M = 500$ elements is sampled in groups of $N = 50$ elements. If the distribution of U has a mean $m = 5$ and variance $\sigma^2 = 3$, what is the mean $\bar{u}$ and the variance $\sigma_{\bar{u}}^2$ of the distribution of sample means? Assume that initially the sampling is done without replacement. Repeat for sampling with replacement. Compare the results.■

Solution: For sampling without replacement we have,

$$\bar{u} = m = 5$$

$$\sigma_{\bar{u}}^2 = \left\{\frac{M-N}{M-1}\right\} \frac{\sigma^2}{N} = \left\{\frac{500-50}{500-1}\right\} \frac{3}{50}$$

$$\sigma_{\bar{u}}^2 = \frac{450}{499} \cdot \frac{3}{50} = \frac{27}{499} = \frac{3}{55.4}$$

For sampling with replacement, we have

$$\bar{u} = m = 5$$

$$\sigma_{\bar{u}}^2 = \frac{\sigma^2}{N} = \frac{3}{50}$$

Comparing the two results, we note that the variance obtained for sampling without replacement is smaller than the variance obtained for sampling with replacement. Does this relationship hold for all samples? If so, can you give a reason for it?■■

Example 7.15 (Sample v. Universe Parameters). You know that country X plans to purchase 2000 personnel carriers of type Y from country Z. Through an intermediary you are able to purchase and test 25 of the personnel carriers. You measure the driving range of the carriers and you find that the mean driving range is 400 mi. with a variance of 100 (a standard deviation of 10 mi.). What can you say about the mean driving range and variance of the universe of 2000 personnel carriers?■■

Solution: Obviously the sampling is without replacement. To answer the questions, one is tempted to use the information on the variance of the sample as an "estimator" for the variance of the universe. One is also tempted to use the mean of the sample as an estimator for the mean of the universe. However, things are not quite that simple. In the sections on estimation and tests of hypotheses to follow, we shall determine how parametric and nonparametric measures associated with samples are related to their counterparts in the distribution from which the sample was taken. More important, we will show how to determine the confidence one can place in estimates which are based on samples. For the time being, we will leave the questions raised by Example 7.15 for the next section.■■

Exercise 7.12. Let U be the universe consisting of the three elements, 3, 5, and 7. Let $S_u(N)$ be $S_u(2)$ where we will consider all samples of size 2. Find m and σ^2 for U, and $\bar{u}$ and $\sigma_{\bar{u}}^2$ for U^*. Check the relationships of $\bar{u} = m$ and $\sigma^2 = N\sigma_{\bar{u}}^2$.■■

Exercise 7.13. For samples of size $N = 9$ it is determined that the variance $\sigma_{\bar{u}}^2$ of the sampling space U^* is 16. What is the variance of the original universe U?■■

Exercise 7.14. It is stated that the mean $\bar{u}$ of the sampling space U^* is 4. What is the mean of the universe?■■

Exercise 7.15. The variance of the universe U is found to be $\sigma^2 = 16$. What is the variance of the sampling universe U^* for $N = 9$? For $N = 25$?■■

7.19 CONFIDENCE INTERVALS

At this point we are in a position to use knowledge gained from samples to make assertions about the population from which the samples were taken. Naturally, we cannot expect the information gained from samples to precisely describe all or even any characteristics of the universe. We should suspect that large samples provide more information or at least information with greater precision, than do small samples. Further, we should suspect that the sample chosen and the method used to select the sample (random, stratified, judgment) has some effect on the ability of the sample to faithfully reproduce universe characteristics.

We accept that using a sample to estimate a characteristic or parameter of a universe will almost surely yield parameter values having some error. Hence it is very important that estimates made from samples should also state the accuracy and reliability one can ascribe to the estimate. For example, suppose that a sample of 30 is used to estimate the mean of a universe U. We find that the kth sample of 30 from U has a mean $u_k = 5$, and a variance of $\sigma^2_{u_k}$. We do not expect the mean m of U to be equal to 5. However, we should expect it to be somewhere near 5. We should be able to state the confidence we place in the mean lying in some interval centered at 5. For example, the interval from 3 to 7. Just how confident can we be? We can determine this if we are willing to work long at the tedious task of determining the means of many samples of size 30. A certain proportion P_1 of the sample means will have values between 3 and 7. The remaining proportion P_2 of sample means will have values outside of the range 3–7. Thus we can state that we are confident at a level of $P_1/(P_1 + P_2) \times 100\%$ that the mean m will take on a value in the range 3–7. This is equivalent to saying that the probability that the interval 3–7 will include the sample m is $P_1/(P_1 + P_2)$.

The selection of the interval length of 4 units (3–7) was arbitrary. However, in practice, the interval length is given in terms of the standard deviation of the sample or the universe. This is done to state the interval length in terms of the probability that the interval contains the mean of the universe. We now illustrate how this relation is achieved. We recall that for large samples, the universe U^* of sample means has a distribution which approximates a normal distribution with mean $\bar{u} = m$ and variance σ^2/N. The distribution of sample means is called the *sampling distribution* of the mean. Other sampling distributions exist for the median, variance, proportions, etc. Thus any sample of size N, taken from the universe U, has a mean which is an element of the distribution U^*. The probability that it falls within a certain interval about the mean of U^* is given in a table of central areas for normal distributions, a part of which is reproduced in Table 7.11 and is also in Table 7.6 which contains values for the cumulative normal distribution. The confidence level is equal to the percent area between the

Table 7.11 Central areas.

Confidence level, $(1-a)$ 100	Confidence factor	Confidence interval
10	0.126	$\bar{u}_k - 0.126\ \sigma/\sqrt{N} < m < \bar{u}_k + 0.126\ \sigma/\sqrt{N}$
20	0.253	$\bar{u}_k - 0.253\ \sigma/\sqrt{N} < m < \bar{u}_k + 0.253\ \sigma/\sqrt{N}$
30	0.385	$\bar{u}_k - 0.385\ \sigma/\sqrt{N} < m < \bar{u}_k + 0.385\ \sigma/\sqrt{N}$
40	0.524	$\bar{u}_k - 0.524\ \sigma/\sqrt{N} < m < \bar{u}_k + 0.524\ \sigma/\sqrt{N}$
50	0.674	$\bar{u}_k - 0.674\ \sigma/\sqrt{N} < m < \bar{u}_k + 0.674\ \sigma/\sqrt{N}$
60	0.842	$\bar{u}_k - 0.842\ \sigma/\sqrt{N} < m < \bar{u}_k + 0.842\ \sigma/\sqrt{N}$
70	1.036	$\bar{u}_k - 1.036\ \sigma/\sqrt{N} < m < \bar{u}_k + 1.036\ \sigma/\sqrt{N}$
80	1.282	$\bar{u}_k - 1.282\ \sigma/\sqrt{N} < m < \bar{u}_k + 1.282\ \sigma/\sqrt{N}$
90	1.645	$\bar{u}_k - 1.645\ \sigma/\sqrt{N} < m < \bar{u}_k + 1.645\ \sigma/\sqrt{N}$
95	1.960	$\bar{u}_k - 1.960\ \sigma/\sqrt{N} < m < \bar{u}_k + 1.960\ \sigma/\sqrt{N}$
99	2.576	$\bar{u}_k - 2.576\ \sigma/\sqrt{N} < m < \bar{u}_k + 2.576\ \sigma/\sqrt{N}$

two confidence factors $-Z = Z_{a/2}$ and $Z = Z_{1-(a/2)}$ and under the standard normal curve (normal curve with mean 0 and variance 1); a is called the confidence or *area index* of the interval.

Using values from Table 7.11, we note that for $\sigma/\sqrt{N} = 2$ and $\bar{u}_k = 5$, we must select an interval from $\bar{u}_k - 1.645\ \sigma/\sqrt{N}$ to $\bar{u}_k + 1.645\ \sigma/\sqrt{N}$ in order to have a confidence level of 90% that the mean m of the universe U will be in the interval selected. The resulting interval is $1.71 < m < 8.29$. Also, if we are willing to make the interval large enough, we are able to assign a very high level of confidence to having the interval contain the mean value m of the universe U.

Example 7.16 (Confidence Intervals). In a previously uncompleted example we obtained a sample of 25 personnel carriers from a universe of 2000 and found that the sample had a mean of 400 mi. for the cruising range. Suppose we learn through other sources that the manufacturer states the cruising range has a variability (standard deviation) of 50 mi., i.e., $\sigma = 50$. Determine the cruising range interval in which you will be 90% confident that the interval will contain the mean value m of the cruising range for the original universe of 2000 personnel carriers. ■■

Solution: The mean value of the sample of $N = 25$ is $\bar{u}_k = 400$. The standard deviation σ of the universe U is 50. The standard deviation $\sigma_{\bar{u}_k}$ of the universe of samples U^* for $N = 25$ is $\sigma_{\bar{u}_k} = \sigma/\sqrt{N} = 50/\sqrt{25} = 10$. The 90% confidence interval, according to Table 7.11, is determined by the confidence factor 1.645,

$$400 - 1.645\ \sigma/\sqrt{N} < m < 400 + 1.645\ \sigma/\sqrt{N}$$

or

$$400 - 1.645\ (10) < m < 400 + 1.645\ (10)$$
$$383.55 < m < 416.45$$

If a confidence level of 80% would suffice, then the confidence factor 1.282 is selected. For 95% confidence, the confidence factor 1.96 is selected. The confidence factor is the number of standard deviations $\sigma/\sqrt{N}$ from each side of the mean $\bar{u}_k$ of the sample we must extend the interval in order that the interval contains the universe mean m to the stated confidence level, where the confidence level is the area between the normal curve and the confidence interval. Figure 7.14 illustrates the sampling distributions of the sample means, confidence intervals, the associated areas, and dispersion of five confidence intervals for 5 different sample means $\bar{u}_1, ..., \bar{u}_5$.

Figure 7.14 specifically illustrates the distribution of sample means with a mean $\bar{u}_k = 400$. If the means of all other samples of 25 personnel carriers were available, 90% of them would be in the interval defined by the end points $\bar{u}_k - 1.645\ \sigma/\sqrt{N}$ and $\bar{u}_k + 1.645\ \sigma/\sqrt{N}$. The 90% intervals obtained from five groups of 25 samples are illustrated below the distribution of sample means. The mean $\bar{u}_4$ is one sample mean which by chance lies outside of the 90% confidence inter-

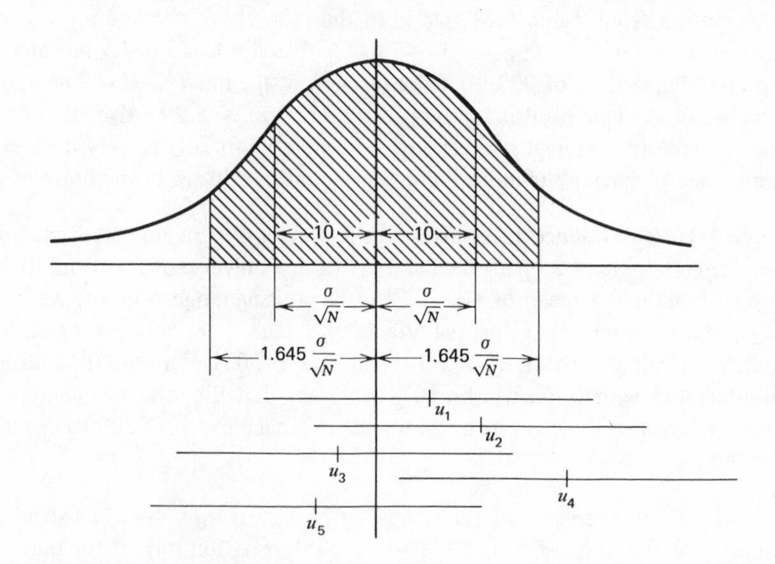

Figure 7.14 Distribution of sample means and 90% confidence interval.

val for m. From the analysis above, 10% of all sample means $\bar{u}_i$ will lie outside the interval illustrated on the sampling distribution. The end points of a confidence interval are called the *confidence limits* for the parameter under estimation. The percentage of samples which lie in the confidence interval is called the *confidence level.* ■■

In the above, we used the table of central areas for the standard normal distribution to determine confidence intervals and the associated confidence levels. We now show how to use the tables of one-sided areas for the cumulative normal distribution. We again use the notation a for the confidence index and define it as the value given by $a = 1 -$ (confidence level/100), or confidence level = $(1-a)100$. For a confidence level of 90, $a = 0.10$. For a confidence level of 95, $a = 0.05$. Since we are working with samples which are distributed normally, the mean of the distribution bisects the confidence interval. The confidence limits (end points of the confidence interval) are determined by the confidence factor. The confidence factor in turn is the number of standard deviations $(\sigma/\sqrt{N})$ the confidence interval extends on each side of the mean in order to achieve the confidence level specified. We identify a with the confidence factors through the notations $Z_{a/2}$ and $Z_{1-(a/2)}$. If a confidence level of 90 is specified, then $a = 10$ and the confidence factors are denoted $Z_{a/2} = Z_{0.05}$ and $Z_{1-(a/2)} = Z_{0.95}$. The confidence interval is now given in terms of the confidence factors by

$$\bar{u}_k + Z_{a/2}\ \sigma/\sqrt{N} < m < \bar{u}_k + Z_{1-(a/2)}\ \sigma/\sqrt{N}$$

or for a confidence level of 90, we have

$$\bar{u}_k + Z_{0.05}\ \sigma/\sqrt{N} < m < \bar{u}_k + Z_{0.95}\ \sigma/\sqrt{N}$$

where $Z_{0.05}$ and $Z_{0.95}$ are located in the cumulative standard normal distribution at approximately $Z_{0.05} = -1.64$ and $Z_{0.95} = 1.64$. The confidence factors and intervals for some widely used confidence levels are given in Table 7.12.

Table 7.12.

Confidence level, %	Confidence factors	Confidence interval
99	$Z_{0.005} = -2.58$	$\bar{u}_k - 2.58\ \sigma/\sqrt{N}$
	$Z_{0.995} = 2.58$	$\bar{u}_k + 2.58\ \sigma/\sqrt{N}$
95	$Z_{0.025} = -1.96$	$\bar{u}_k - 1.96\ \sigma/\sqrt{N}$
	$Z_{0.975} = 1.96$	$\bar{u}_k + 1.96\ \sigma/\sqrt{N}$
90	$Z_{0.05} = -1.64$	$\bar{u}_k - 1.64\ \sigma/\sqrt{N}$
	$Z_{0.95} = 1.64$	$\bar{u}_k + 1.64\ \sigma/\sqrt{N}$

We should note that $Z_{a/2}$ is the negative of $Z_{1-(a/2)}$. This is due to the symmetry of the normal distribution: The distances from the mean to the end points of the interval are equal but opposite in direction.

When a table of central areas such as Table 7.11 is available, one converts the confidence level directly to central areas and uses the entries for confidence factors provided in the table. When only a table of cumulative normal areas such as Table 7.6 is available, one converts the confidence level to an a value through $a = 1 -$ confidence level/100. The confidence factors $Z_{a/2}$ and $Z_{1-(a/2)}$ are determined from Table 7.6 by entering the table at one-sided areas corresponding to $a/2$ and $1 - (a/2)$. Figure 7.15 *(a)* and *(b)* illustrates the relationship among the central areas, one-sided areas, confidence levels, and a.

Example 7.17 (Confidence Intervals and Levels). A sample of 25 is taken from a universe of variance $\sigma^2 = 100$. The sample has a mean $\bar{u}_k = 60$. For what interval will we have a 95% confidence level that the constructed interval will contain the mean of the universe?■■

Solution: For a 95% confidence level we have $a = 0.05$. The confidence factors, defining the end points of the confidence interval, are $Z_{a/2} = Z_{0.025}$ and $Z_{1-(a/2)} = Z_{0.975}$. Using the tables we have,

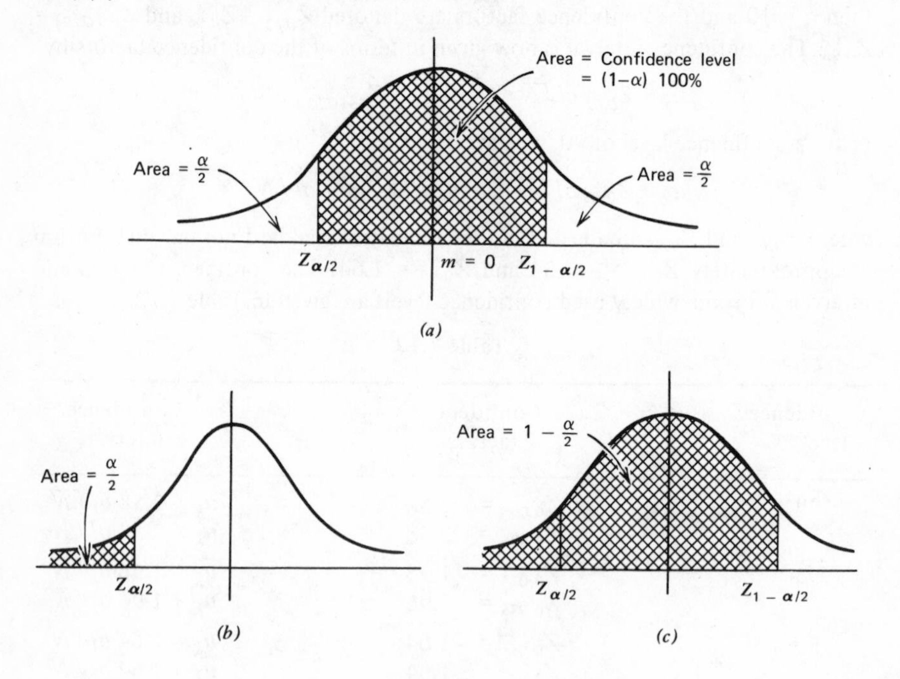

Figure 7.15 Confidence level and areas.

$$Z_{0.025} = -1.96$$
$$Z_{0.975} = 1.96$$

The confidence interval is given by

$$\bar{u}_k - 1.96\ \sigma/\sqrt{N} < m < \bar{u}_k + 1.96\ \sigma/\sqrt{N}$$

with $\sigma^2 = 100$, $\sigma = 10$; $N = 25$, $\sqrt{25} = 5$, and $\bar{u}_k = 60$, we obtain,

$$60 - 1.96\ 10/5 < m < 60 + 1.96\ 10/5$$
$$56.08 < m < 63.92$$

as the confidence interval for a 95% confidence level.■■

Exercise 7.16. From a sample of 50 college students it is determined that their average IQ is 115. It is known that IQs of college students have a variability of 16 units. Determine the IQ interval centered at 115 in which you can be 95% confident that the interval contains the mean IQ of all college students.■■

Exercise 7.17. If the confidence level sought is 80, what confidence index (a) should be employed?■■

Exercise 7.18. What confidence level is associated with the confidence factors $Z_{0.05}$ and $Z_{0.95}$?■■

Exercise 7.19. If $\bar{u}_k = 20$, $N = 25$, and $\sigma = 10$, what is the length of the confidence interval for the confidence factors of Exercise 7.18.■■

7.20 SAMPLE SIZE

From the inequality for the confidence interval we can see that the length of the confidence interval is dependent on the confidence level selected. This is indicated by the confidence factors $Z_{a/2}$ and $Z_{1-(a/2)}$ and the size of the sample N. Increasing the confidence level increases the size of the interval required. Increasing the sample size N decreases the size of the confidence interval for the same confidence level. Thus increasing sample size results in estimates of greater precision.

The length of a symmetric confidence interval is equal to twice the length from the sample mean to either end point. If we label the length of the upper half of the interval d, then,

$$d = Z_{1-(a/2)}\ \sigma/\sqrt{N}$$

We can fix the length of the interval for a given confidence level and by means of the above formula determine the sample size required to achieve the confidence level in the interval of length $2d$. To do this, we rewrite the equation as

$$N = \left\{ \frac{Z_{1-(\alpha/2)} \, \sigma}{d} \right\}^2$$

Example 7.18 (Sample Size v. Confidence Level). In Example 7.17 we had a confidence interval of length 7.84, a sample size of 25, and a confidence level of 95%. Let us keep the same confidence level but change the interval length to 4. We want to determine the increase in sample size required to compensate for the decrease in the length of the confidence interval while maintaining the same confidence level.■■

Solution: Half the new interval length is $d = 2$. The confidence level remains the same so $Z_{1-(\alpha/2)} = Z_{0.975} = 1.96$. The standard deviation σ remains at the value 10. The formula for the sample size N becomes

$$\begin{aligned} N &= \left\{ \frac{Z_{1-(\alpha/2)} \, \sigma}{d} \right\}^2 \\ &= \left\{ \frac{1.96 \cdot 10}{2} \right\}^2 \\ &= 96 \end{aligned}$$

Thus by increasing the sample size from 25 to 96, we can essentially double the precision in estimating the mean value of the distribution.

When measurements are taken in terms of physical units such as length, time, or mass, then the relationship between measurement precision and sample size assumes a higher level of relevance. This relevance is illustrated in Example 7.19.■■

Example 7.19 (Sample Size Determination). In the survey of large tracts of inaccessible areas, it was found that 68% of the length measurements had errors which did not exceed plus or minus 1 in., i.e., $\sigma = 1$ in. The surveyors had to survey a base line to an accuracy of plus or minus 1/2 in. Assuming the measurement errors follow a normal distribution, how many measurements of the base line should the surveyors take to have a 99% confidence that the measurements result in the necessary accuracy?■■

Solution: The standard deviation of the universe U of measurements is $\sigma =$ 1 in. The length of the confidence interval is 1 in. so that $d = 1/2$ in. The confidence level sought is 99%. This yields a confidence factor of $Z_{1-(\alpha/2)} = Z_{0.995} =$ 2.58. Using the formula for N, we have

$$\begin{aligned} N &= \left\{ \frac{2.58 \cdot 1}{1/2} \right\}^2 \\ &= 6.656 \cdot 4 \\ &= 27 \end{aligned}$$

The survey team should measure the base line 27 times and use the average of the measurements as the center of a 99% confidence interval of length equal to 1 in. Thus 99% of the time the true measurement will fall in such an interval. Note that in this example the value of the average measurement was not given nor needed. What was needed to determine the "error interval" to a 99% level of confidence was the variance with which measurements were accomplished.

The formula can be rewritten and used to determine the confidence level one will achieve with a given variance σ^2, interval size d, and sample size N. The formula is

$$Z_{1-(a/2)} = d\sqrt{N}/\sigma$$

Thus, if the survey team could only take 16 measurements, they would have to be willing to accept a lower confidence level in the accuracy of their measurements, i.e., there will be a greater chance that the resultant measurement will not be in the interval 1 in. in length. The confidence factor will be

$$\begin{aligned} Z_{1-(a/2)} &= \frac{1/2 \cdot 4}{1} \\ &= 2 \end{aligned}$$

The value $Z_{1-(a/2)} = 2$ corresponds to the one-sided area percentile of 0.9772. Since the error is two-sided and symmetric, we have

$$\begin{aligned} 1-(a/2) &= 0.9772 \\ (a/2) &= 1 - 0.9772 \\ a &= 2 \cdot 0.0228 \\ a &= 0.0456 \end{aligned}$$

and the confidence level is

$$\begin{aligned} (1-a)\ 100 &= (1-0.0456)\ 100 \\ &= 95.44\% \end{aligned}$$

So the reduction from 27 to 16 samples reduced the confidence level from 99 to 95.44.■■

7.21 ESTIMATION WHEN VARIANCE IS NOT KNOWN

There are many situations where the variance σ^2 of the distribution for the universe U is not known. The effectiveness of a medicine in removing the symptoms associated with a newly discovered virus, the reliability of a new product produced through new manufacturing techniques, and the *performance parameters* associated with newly captured and tested equipments are all examples of measurements that could come from universes with unknown variances. When the variance of the universe U is unknown, one expedient is to calculate the variance s^2 of the sample taken from U and use it in place of the variance σ^2 of universe U. This expedient is permissible, provided the following two changes in computational procedures in making estimates are implemented:

1. The variance s^2 of the kth sample $S_k(N)$ of N measurements from U is computed by the formula

$$s^2 = \frac{1}{N-1} \Sigma (x_i - u_k)^2$$

where $\bar{u}_k$ is the element (the kth sample mean) of U^* associated with the sample S_k, and

$$\bar{u}_k = (1/N) \Sigma_{i=1}^{N} x_i$$

The change in denominator (from N to $N-1$ in the formula for s^2) is needed to account for the fact that the average of the variances of all samples of size N taken from U is consistently less than the true variance σ^2 of the universe U.

Previously we had that the mean $\bar{u}$ of the sampling distribution U^* equals the mean m of the universe U. When the sampling distribution U^* of a parameter (mean, variance, median) has a mean which is equal to the population parameter, the estimate (statistic) is called an *unbiased estimate* of that parameter. From our previous discussions we can conclude that $\bar{u}$ is an unbiased estimate for m but $\sigma_{\bar{u}}^2$ is not an unbiased estimate for σ^2. If $\sigma_{\bar{u}}^2$ is replaced by s^2 as computed above then s^2 is an unbiased estimate of σ^2.

2. The confidence factors $Z_{a/2}$ and $Z_{1-(a/2)}$ may be changed to factors denoted by $t_{a/2}$ and $t_{1-(a/2)}$ which are derived from the t-distributions given in Table 7.13 rather than from the tables for the cumulative normal distribution. There is a t distribution for each sample size N. The t distributions are symmetrical but more dispersed than the normal distribution. For large N the t distributions are very close to normal.

7.22 THE USE OF t–STATISTICAL TABLES

Before we proceed to estimate the mean m of the distribution using the sample variance s^2 in place of σ^2, we illustrate the use of Table 7.13 in making these estimates. The factors $t_{a/2}$ and $t_{1-(a/2)}$ are obtained from the percentiles of the sampling distribution of

$$t = \frac{\bar{u} - m}{s/\sqrt{N}}$$

which has the same form as the standard normal variate with $s/\sqrt{N}$ replacing $\sigma_{\bar{x}}$ in,

$$Z = \frac{x - \bar{x}}{\sigma_{\bar{x}}}$$

However, t has the variable N in its denominator. As N increases, we can expect the distribution of t to become dispersed and more nearly normal in shape. To insure near normality of the variable t, it is assumed that the distribution for the universe U is normal.

To account for the sample size N, a new variable, called the *degree of freedom* and denoted *df*, is introduced; *df* is the number of measurements which may be varied independent of others when constraints are placed on the data. For example, in calculating the mean for N measurements, if an interval or range of values is specified for the mean, then only N–1 of the measurements may assume any value. The value of the remaining measurement is restricted to that range or value which will satisfy the condition that the mean be confined to a specific interval. For our use of the t distribution, *df* will be one less than the number N of samples, i.e.,

$$df = N - 1$$

The entries of Table 7.13 are arranged so that column headings are the t factors for percentiles corresponding to $1 - (a/2)$, the subscript of t. The rows are headed by the degrees of freedom, where $df = N - 1$. For a specified index a, say 5%, we have the confidence level of $(1-a)\ 100 = (1-0.05)\ 100 = 95\%$. The $t_{a/2}$ and $t_{1-(a/2)}$ factors are $t_{0.025}$ and $t_{0.975}$ which respectively give the 2.5 the 97.5 percentile points of the t distribution. Since the distribution is symmetric, only entries for the upper percentile are given. The table entries are the values for $t_{1-(a/2)}$. The interval end points $\bar{u}_k \pm t_{1-(a/2)}\ (df)\ s/\sqrt{N}$ yields an interval with $(1-a)\ 100\%$ confidence of containing the mean m of the universe U.

For a specific t factor with N–1 degrees of freedom, the table is entered at the row corresponding to N–1. Thus for $t_{0.975}(15)$ we read 2.131. For $t_{0.025}(15)$ we would read –2.131, i.e., if the subscripts of the two t factors are complementary (add to 1), then the table entry for $t_{a/2}$ is the negative of the table entry for $t_{1-(a/2)}$. When we substitute s for σ, $\bar{u}_k$ for $\bar{u}$, and the new confidence factors into the inequality for the confidence interval, we obtain

$$\bar{u}_k - t_{a/2}\ (df)\ s/\sqrt{N} < m < \bar{u}_k + t_{1-(a/2)}\ (df)\ s/\sqrt{N}$$

Table 7.13. Percentiles of the t distributions.

df	$t_{.60}$	$t_{.70}$	$t_{.80}$	$t_{.90}$	$t_{.95}$	$t_{.975}$	$t_{.99}$	$t_{.995}$
1	.325	.727	1.376	3.078	6.314	12.706	31.821	63.657
2	.289	.617	1.061	1.886	2.920	4.303	6.965	9.925
3	.277	.584	.978	1.638	2.353	3.182	4.541	5.841
4	.271	.569	.941	1.533	2.132	2.776	3.747	4.604
5	.267	.559	.920	1.476	2.015	2.571	3.365	4.032
6	.265	.553	.906	1.440	1.943	2.447	3.143	3.707
7	.263	.549	.896	1.415	1.895	2.365	2.998	3.499
8	.262	.546	.889	1.397	1.860	2.306	2.896	3.355
9	.261	.543	.883	1.383	1.833	2.262	2.821	3.250
10	.260	.542	.879	1.372	1.812	2.228	2.764	3.169
11	.260	.540	.876	1.363	1.796	2.201	2.718	3.106
12	.259	.539	.873	1.356	1.782	2.179	2.681	3.055
13	.259	.538	.870	1.350	1.771	2.160	2.650	3.012
14	.258	.537	.868	1.345	1.761	2.145	2.624	2.977
15	.258	.536	.866	1.341	1.753	2.131	2.602	2.947
16	.258	.535	.865	1.337	1.746	2.120	2.583	2.921
17	.257	.534	.863	1.333	1.740	2.110	2.567	2.898
18	.257	.534	.862	1.330	1.734	2.101	2.552	2.878
19	.257	.533	.861	1.328	1.729	2.093	2.539	2.861
20	.257	.533	.860	1.325	1.725	2.086	2.528	2.845
21	.257	.532	.859	1.323	1.721	2.080	2.518	2.831
22	.256	.532	.858	1.321	1.717	2.074	2.508	2.189
23	.256	.532	.858	1.319	1.714	2.069	2.500	2.807
24	.256	.531	.857	1.318	1.711	2.064	2.492	2.797
25	.256	.531	.856	1.316	1.708	2.060	2.485	2.787
26	.256	.531	.856	1.315	1.706	2.056	2.479	2.779
27	.256	.531	.855	1.314	1.703	2.052	2.473	2.771
28	.256	.530	.855	1.313	1.701	2.048	2.467	2.763
29	.256	.530	.854	1.311	1.699	2.045	2.462	2.756
30	.256	.530	.854	1.310	1.697	2.042	2.457	2.750
40	.255	.529	.851	1.303	1.684	2.021	2.423	2.704
60	.254	.527	.848	1.296	1.671	2.000	2.390	2.660
120	.254	.526	.845	1.289	1.658	1.980	2.358	2.617
∞	.253	.524	.842	1.282	1.645	1.960	2.326	2.576
df	$-t_{.40}$	$-t_{.30}$	$-t_{.20}$	$-t_{.10}$	$-t_{.05}$	$-t_{.025}$	$-t_{.01}$	$-t_{.005}$

When the table is read from the foot, the tabled values are to be prefixed with a negative sign. Interpolation should be performed using the reciprocals of the degrees of freedom.

Data are extracted from Table III of Fisher and Yates, *Statistical Tables*, with the permission of the authors and publishers, Longman Group Limited, Longman House, Burnt Mill, Harlow, Essex, England.

The mean value $\bar{u}_k$ is not to be confused with the mean u of the sampling distribution. Here $\bar{u}_k$ is the mean of the kth sample of N measurements being used in the current estimation. It is used as the estimator for the mean of the universe U; the $\bar{u}_k$ is but one measurement in the universe of sample means U^*. A few examples at this point should clarify the notation and usage of the tables.

Example 7.20 (*t* Statistic–Gas Mileage). A random sample of $N = 16$ compact cars is taken from the universe U. Measurements are made on the gasoline mileage each attains. The average gas mileage $\bar{u}_k$ of the sample is found to be 26 mpg. The variance s^2, as computed through the following formula is:

$$s^2 = \frac{1}{N-1} \Sigma_{i=1}^{16} (x_i - \bar{u}_k)^2 = 16$$

Determine the 95% confidence interval for the mean gasoline mileage m for the universe U. ■■

Solution: We first note that the degrees of freedom df for the statistic m is $N-1 = 15$. Second, the confidence level of 95% yields that $a = 0.05$, i.e., 95% = $(1-a)$ 100. Thus the confidence factors are

$$t_{a/2}\,(15) = t_{0.025}\,(15) \qquad \text{and} \qquad t_{1-(a/2)}\,(15) = t_{0.975}\,(15)$$

The confidence interval limits are

$$\bar{u}_k + t_{a/2}\,(df)\; s/\sqrt{N} \qquad \text{and} \qquad \bar{u}_k + t_{1-(a/2)}\,(df)\; s/\sqrt{N}$$

Obtaining the values for the confidence factors from Table 7.13 and substituting the values for s and N, we obtain

$$26 - 2.131\; 4/\sqrt{16} \leqslant m \leqslant 26 + 2.131\; 4/\sqrt{16}$$

or

$$23.87 \leqslant m \leqslant 28.13$$

■■

Example 7.21 (Sample Size versus Confidence Level). In the previous example, what increase in sample size N is required to achieve a confidence level of 98% for the same confidence interval length? ■■

Solution: We know that the length of the confidence interval is $2d = 2 \cdot 2.13$ so that $d = 2.13$. Therefore we have that

$$d = t_{1-(a/2)}\,(df)\; s/\sqrt{N}$$

or

$$N = \left\{ \frac{t_{0.99}\;(df)\; s}{2.13} \right\}^2$$

To simplify the problem, we evaluate the constant factor $(4/2.13)^2 = 16/4.537 = 3.52$, and substitute it back into the equation for N, obtaining

$$N = 3.52\ [t_{0.99}\ (df)]^2$$

In using the table for the t distribution, we are restricted to the column headed by $t_{0.99}$. We have to select an N which yields a confidence factor which when squared and multiplied by 3.52 is approximately N. We can try $N = 10$. For $N = 10$ the table yields (for $df = 9$) a percentile of 2.82. On substituting this value in the equation we have

$$10 < 3.52\ (2.82)^2$$

Hence we know that N must be larger. We try $N = 14$, and obtain

$$14 < 3.52\ (2.65)^2$$

and N is still not large enough. We next try $N = 25$, and obtain

$$25 > 3.52\ (2.492)^2 = 21.82$$

We now know that $N = 25$ is too high but it is getting closer to the desired value. Continuing the process we find that N should be between 22 and 23. Since N must be an integer, we select $N = 23$.■■

This last example illustrates the use of approximations in adjusting the values of two variables so that conditions on the statistic and its confidence interval and level are simultaneously satisfied. While the process is tedious, it is informative in the sense that one obtains a better feel for the effect the two variables, namely N and the confidence factor $t_{1-(a/2)}\ (df)$, have on each other. Knowing this, one can make educated guesses as to the sample size needed to achieve prespecified levels of confidence. The process could result in a considerable saving in time and labor, especially when new experiments and measurements are to be accomplished.

Exercise 7.20. A sample of 25 light bulbs is found to have an average life of 1000 hrs. The variance in the lifetime of the sample is computed using the formula

$$s^2 = \frac{1}{N-1}\ \Sigma_{i=1}^{25}\ (x_i - \bar{u}_k)^2$$

and is found to be 900. Determine the interval for which you are 90% confident that it will contain the average lifetime of the universe of light bulbs.■■

Exercise 7.21. If the sample size was increased to $N = 64$, what would be the resulting change in the length of the confidence interval?■■

7.23 BINOMIAL ESTIMATES

All the preceding estimates have dealt with normal distributions for the universe U or the sampling universe U^*. We will now deal with populations U whose members belong to one of K classes or intervals. In the simplest situation, we deal with 2 classes, i.e., $K = 2$. For example, we measure voters in terms of their being Democrats or Republicans. A treatment or medicine is measured as effective or ineffective. Students are measured in terms of pass and fail. Equipments are either reliable or unreliable. People are classed as either female or male. Whenever a universe of measurements is of the two-class variety, the resulting probability distribution is the binomial distribution.

In Section 7.2 we discussed some of the properties of the binomial distribution. We learned that the binomial probability distribution for the two classes, success and failure, expresses the probability of achieving r successes out of n trials, where the probability of success in any one trial is p. Thus p is the probability that a randomly selected element of the population belongs to one of two classes. The two classes, denoted U_1 and U_2, of U are disjoint and exhaustive. We can interpret p as the proportion of elements of the universe U which belongs to U_1. Our objective is to estimate p for the universe U from samples of size N from the sampling universe $S_u(N)$.

Let us consider a sample of size N. Suppose that N_1 elements of the sample belong to U_1. In this case the value $\hat{p} = N_1/N$ is an estimator for the proportion p of "successes" in the universe U. We make the following assertions about the sampling distribution U^* for $\hat{p}$. If the sample size N is large enough to satisfy $N\hat{p}(1-\hat{p}) > 9$, then the sampling distribution U^* of $\hat{p}$ is approximately normal with mean equal to p, the mean proportion of successes in the universe U. The variance of U^* is equal to $p(1-p)/N$. The number of standard normal units the estimator $\hat{p}$ is from the "expected" value p is given by the statistic

$$Z = \frac{\hat{p}-p}{\sqrt{p(1-p)/N}}$$

where Z is approximately normally distributed with mean 0 and variance 1. Since Z is normal, we can use the normal distribution to calculate confidence intervals for estimates on p. The $(1-a)$ 100% confidence interval for p is given by

$$\hat{p} + Z_{a/2}\,\sqrt{\hat{p}(1-\hat{p})/N} < p < \hat{p} + Z_{1-(a/2)}\,\sqrt{\hat{p}(1-\hat{p})/N}$$

where $Z_{a/2}$ and $Z_{1-(a/2)}$ are the confidence factors associated with the confidence interval desired, $\hat{p}$ is used in place of p in the formula $\sqrt{p(1-p)/N}$ for the variance, because p is not known and is the parameter under consideration.

Example 7.22 (Binomial Estimate). An optometrist, while gazing at the passing crowd in front of his office observed that 60 out of 100 adults wore glasses. Determine with 90% confidence the interval which covers the true proportion of adults wearing glasses. ■■

Solution: First we note that the point estimate for the proportion of adult glass wearers is $p = 60/100 = 0.6$. The size of the sample is $N = 100$. The confidence level is $90 = (1-a)\ 100\%$, so that $a = 0.1$. With $a = 0.1$, the confidence factors are $Z_{a/2} = Z_{0.05}$ and $Z_{1-(a/2)} = Z_{0.95}$. From Table 7.12 we read

$$Z_{0.05} = -1.64$$
$$Z_{0.95} = 1.64$$

The 90% confidence interval is

$$0.6 - 1.64\sqrt{0.6 \cdot 0.4/100} < p < 0.6 + 1.64\sqrt{0.6 \cdot 0.4/100}$$
$$0.6 - 1.64 \cdot 0.49/10 < p < 0.6 + 1.64 \cdot 0.49/10$$
$$0.52 < p < 0.68$$ ■■

Exercise 7.22. A political analyst randomly sampled 64 daily newspapers for political bias. The analyst reported that 48 were biased and 16 were neutral. Determine with 80% confidence the interval which covers the true proportion of biased papers. ■■

7.24 SUMMARY

The section on estimation considered the following three situations:

1. Estimate the mean m of the universe U knowing the variance σ^2 and the sample mean $\bar{u}_k$ of U and U^* respectively. No assumptions are made on U provided the sample size N is large, U^* was assumed to be normally distributed with mean $\bar{u}_k$ equal to the mean m of the universe U, and with standard deviation $\sigma_{\bar{u}} = \sigma/\sqrt{N}$.
2. Estimate the mean m of the universe U *not* knowing the variance σ^2 of U. The sample mean $\bar{u}_k$ is used to estimate m under the assumption that U^* is t-distributed with mean $\bar{u}_k$ equal to the mean m of the universe U, and with standard deviation $\sigma_{\bar{u}} = s/\sqrt{N}$, where
$$s^2 = \frac{1}{N-1}\ \Sigma_{i=1}^{N}\ (x_i - \bar{u}_k)^2$$
3. Estimate the proportion p of samples falling in one class of two mutually exclusive and exhaustive classes, U_1 and U_2 of U. The proportion $\hat{p}$ of a sample of size N falling in class U_1 is used as an estimator of p. The sampling universe of p, for large N, is assumed to be normally distributed with mean p and variance equal to $\sqrt{p(1-p)/N}$; $\hat{p}$ is used in place of p in the confidence interval estimates.

These are but a few of the many estimates of universe parameters that can be made through sampling. The field of estimation includes estimates for the variance, median, multiclass proportions, etc. What has been presented here is only an introduction to the subject and is required for the better understanding of tests of hypotheses, which will now be discussed.

7.25 TESTS OF HYPOTHESES

Estimates made on parameters associated with a universe are derived from information gained through samples taken from the universe. Now we are interested in making educated guesses about some properties associated with a universe U or its sampling universe U^*. We seek to verify our "guesses" by sampling the universe to obtain information for this purpose. Our guesses are our hypotheses, and the information gained from samples determine whether we should accept or reject our hypotheses. This aspect of statistics is known as *hypothesis testing* or *statistical inference.* Normally, in statistical inference, one starts with information gained from samples to make generalizations about properties of the universe from which the samples were drawn. These generalizations confirm or deny the stated hypotheses on the universe and we accept or reject them accordingly.

One may question the wisdom of using samples to make generalizations about universes. Why not use the entire universe? In most instances access to the entire population to determine its characteristics is not possible; the time and cost associated with using the entire universe as a sample may be excessive. As we shall see, samples of nominal size can be used in making generalizations about the entire universe. To achieve a high level of confidence in the generalizations, some simple expedients are available. By increasing the number of samples taken, one can achieve a higher level of confidence and yet sample but a very small percentage of objects of the universe. In this manner one can gain the information required at a considerable saving in effort. This is especially true when sampling is expensive such as when the items being sampled must be destroyed to achieve the appropriate measures. Bullets, tires, fuses, missiles, or light bulbs are but a few items which are destroyed or used when sampled.

Although there are several sampling procedures that can be followed, we shall not consider the sampling aspects of hypothesis testing in this text. We assume also that all our samples are drawn independently and randomly from universes large enough to discount the differences between sampling with and without replacement.

In statistical inference, the starting point is usually a question or an assertion such as the following: Will the Republicans win the next presidential election? Is the color blue favored over white by automobile buyers? Do more people watch NBC than ABC? It takes an average of 50 lb. of pressure to seal the hole properly. There is a television set in 80% of the American homes. Each of these questions or assertions states or implies the existence of a standard. For the

Republicans to win requires a majority of the electoral votes, i.e., more than 50%. White is favored over blue if more people buy white rather than blue automobiles, i.e., of the total buying, white and blue, more (>50%) buy white. The proportion P_1 of people watching NBC exceeds the proportion P_2 watching ABC. Will the hole be sealed properly if a pressure of less than 50 lb. is used?

Once the question or assertion is posed and the standard identified, the standard is related to some parameter or characteristic associated with the population or its samples. In some instances the standard may relate to the mean, median, or variance. In other instances it may be a probability, a proportion, a number, or a "shape." For example, NBC through previous polling may have determined that its proportion P_1 of viewers is 32% while ABC attracts $P_2 = 28\%$. This is the accepted standard stated in percentages. Thus, NBC may decide on a change in programs in the hope that it will attract more viewers. Thus, after the new programs are on for a while, NBC would like to determine whether its proportion of viewers, relative to ABC, has indeed increased as a result of the new programs. A new poll may find a new set of proportions as $P_1 = 35\%$ and $P_2 = 27\%$. Does NBC decide that its new programs are effective or are the observed differences due to chance?

Statements in logic or plausible reasoning differ from statistical statements (statistical hypotheses) in that the statistical hypotheses contain a measure or standard which normally can be related to the properties or parameters of well-defined distributions. For example, consider the statement,

S: 10 of 12 bombs dropped will hit the target.

In logic, this statement is true only if 10 bombs hit the target. It is false otherwise. In plausible reasoning, the statement is assigned a level of credibility based on human judgment and experience and need not depend on the availability of previous test or usage data. If sufficient data is not available, then, a credibility level is assigned. If sufficient data is available, then the statement can be considered as a statistical hypothesis and tested accordingly. The 10 of 12 could be viewed as a hypothesis and the probability of achieving this number of hits in an actual bombing is determined using the number of hits recorded in test samples or previous bombings. Thus, a statistical hypothesis is normally associated with distributions and their parameters. The hypothesis is accepted or rejected on the basis of the probability that it agrees with the information gained from the universe or sampling distribution. The determination of whether one should accept or reject a statistical hypothesis is based on testing procedures which state how and what samples are to be taken, identify decision rules through which the hypothesis is accepted or rejected, and relate the acceptance of the hypothesis to a level of confidence one can place in the decision.

Example 7.23 (Test of Hypothesis–Methods). On the strength of an automobile manufacturer's advertisement which states that its special compact car gets 24 miles to a gallon of gas, Joe purchased one. After driving it for a period of time, and checking his gasoline mileage, Joe determined that he was getting only 19.5 mpg. Joe accused the manufacturer of deceptive advertising. The manufacturer in return claimed that either Joe had a heavy foot or the car's carburetor required adjusting and suggested that Joe turn his car in for service to correct the latter defect, if it existed. After having his carburetor adjusted and using a lighter foot on the accelerator, Joe still received only 20 mpg. Not satisfied with his own experiences, Joe called the consumers' bureau and sought help in forcing the manufacturer to correct his poor gas mileage or having the manufacturer change the gas mileage claims in his advertising. The consumers' bureau listened to Joe's tale of woe and then started an investigation of its own. The consumer's bureau set itself the task of determining whether

1. Manufacturer's gas mileage (24) is correct and Joe's experience with his car is atypical, or
2. Manufacturer has employed deceptive advertising and Joe's experience gives further indication of this possibility.

There is little question as to the position Joe will take. In most cases he will take the second because he senses that the probability favors the advertising being deceptive or misleading. Is not overstatement one of the main techniques of effective advertising? But if the consumers' bureau takes this position, is it correct in taking it? There is also a real possibility that Joe's experience may be biased. Perhaps most of his driving is under severe traffic conditions, etc. How will the bureau know which position to take? Bluntly, the bureau can never be certain which position is correct. But through the techniques and procedures employed in testing of hypotheses, the bureau will be able to take the position which offers it a higher level of confidence in being right. The bureau should know that it will not be right all the time but it should also know the proportion of times that it will be right in assuming that the advertised gas mileage was overly optimistic. To decide on the position it should take, the consumer bureau sampled other car owners of the same model as Joe's and learned that 9 of them have kept accurate gas mileage records. For the 9 owners reporting, it was found that their average gas mileage was $\bar{x}_c = 23$ mpg. The standard deviation for this sample of 9 turned out to be $s_c = 4$ mpg. Statistically, the average gas mileage of the sample did not differ significantly from the manufacturer's advertised value. (We shall find out more about "statistical significance" shortly.) However, the consumer's bureau reasoned that at least 85% of all average drivers should realize at least the advertised gas mileage. They indicated to the car manufacturer that either 85% of its cars meet the advertised value, or the value should be changed accordingly. Naturally, the automobile manufacturer did not agree with the bureau's reasoning. They stated that if 50% of its model owners achieve the advertised gas mileage, then the advertisement is not deceptive or misleading. Here is an honest difference of opinion on what constitutes a fair level of acceptability. The consumers' bureau wants to reduce the buyer's risk in not achieving the advertised gas mileage. The manufacturer wants to attract as many buyers as possible by making the strongest claim without being cited for misleading advertisement. In terms of the distribution for gas mileage (assumed normal), the two standards are illustrated in Figure 7.16.

The manufacturer and consumers' bureau differ on two major points. Namely, the percentage of car owners that should achieve at least the advertised gas mileage, and what should be the average gas mileage. If the distribution is *normal* and has a standard deviation of $s_c = 4$ as indicated by the sample of 9, then the 85% *one-sided* confidence interval limit (with $\alpha = 0.15$) corresponding to the consumers' bureau g_c is

$$\begin{aligned} g_c &= \bar{x}_c + Z_\alpha\, s_c/\sqrt{N} \\ &= 23 - 1.04\ (4/3) \\ &= 21.62 \end{aligned}$$

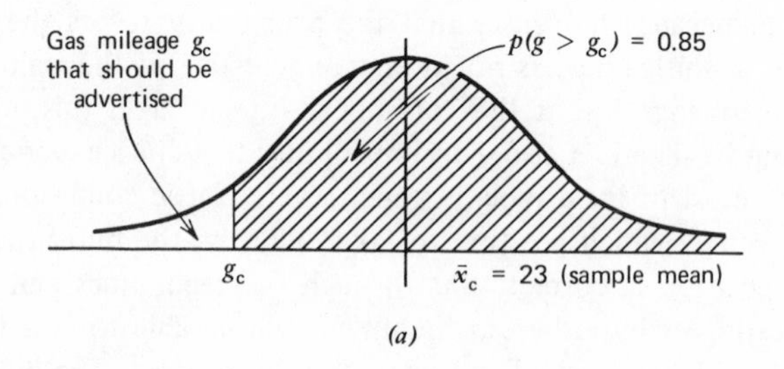

(*a*)

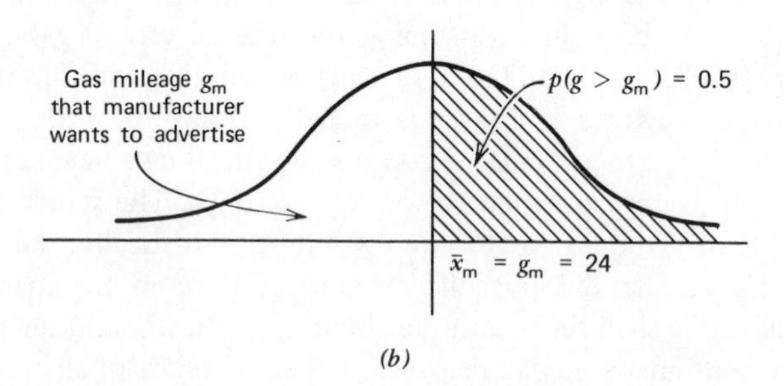

(*b*)

Figure 7.16 Viewpoint of (*a*) consumer's bureau and (*b*) manufacturer.

Table 7.14.

Sample size, N	Confidence level, $(1-a)$ 100%	Confidence factor, Z	Confidence interval limit
9	85	−1.04	22.7
16	85	−1.04	23.0
25	85	−1.04	23.2
64	85	−1.04	23.5
100	85	−1.04	23.6

Confidence limit $= \bar{x}_c + Z\, s_c/\sqrt{N}$ $\bar{x}_c = 24$, $s_c = 4$, $Z = 1.04$

The manufacturer's 50% *one-sided* confidence interval limit with $a = 0.5$ corresponding to the manufacturer's g_m is

$$\begin{aligned} g_m &= \bar{x}_m + Z_a\, s_c/\sqrt{N} \\ &= 24 + 0 \cdot s_c/\sqrt{N} \\ &= 24 \end{aligned}$$

The difference in value between the two confidence interval limits is 24 − 21.62 = 2.38. A difference of 1 is due to the difference in the sample mean of the consumers' bureau and the mean of the universe determined by the manufacturer. Now, how can the differences in average gas mileage confidence level be resolved? If neither is willing to compromise on the desired level, 85% by the consumers' bureau and 50% by the manufacturer, then the manufacturer can attempt the following to improve its position: Noticing that the consumers' bureau used a sample of 9, the manufacturer states that the sample is too small and that a much larger sample should have been used. The consumers' bureau, being statistically oriented, knows that the manufacturer's complaint is based on firm ground. They investigate what the resulting change in confidence interval limits would be for a set of sample sizes. The results are entered in Table 7.14 where a mean of 24 was used for all samples listed in the table.

Reviewing the table, the consumers' bureau realized that for samples of size $N = 16$, 85% of all sample means should fall in the gas mileage interval whose lower limit is 23 mpg. Also, further increases in sample size did not significantly change the confidence interval limit, at least in terms of round numbers. Hence the consumers' bureau proceeded to inform the manufacturer that an advertised gas mileage of 23 mpg. would be acceptable to them. The manufacturer accepted and closed the case.

Table 7.15.

Sample size	Confidence level	Confidence factor	Confidence interval limit
9	95	−1.64	21.8
16	95	−1.64	22.4
25	95	−1.64	22.7
64	95	−1.64	23.2
100	95	−1.64	23.4

There was still another alternative open to the consumers' bureau. Since it was asked to increase the sample size, it could have countered by asking for an increase in confidence level. Let us see what would happen if the confidence level is increased to 95%. The results are shown in Table 7.15.

Comparing Tables 7.14 and 7.15, we note that if the consumers' bureau is willing to increase its sample size to approximately 43, they could also improve the confidence level to 95% while retaining the advertised gas mileage at 23 mpg. Should the consumers' bureau use a larger sample size, it could then test the stated (hypothesized) gas mileage with greater accuracy. That the benefits derived from higher accuracy compensate for the additional effort required to process a larger sample is a question to be answered on the merits of the case. ■■

What we have just done is design an experiment and a test through which we could decide on (a) the gas mileage to be advertised, (b) the confidence level to be achieved, and (c) the sample size to be employed. There are many ways of assessing the design of an experiment. In the previous example, the consumers' bureau and the manufacturer had different viewpoints which each wanted the experiment and test to reflect. In this case, the acceptable experiment (if not the best) is the one that meets the needs of both parties, although each may have to bend a little to achieve a design acceptable to both. Even though an experiment may be "optimum," one must expect some variation when the experiment is repeated. A good design reduces the expected variations to tolerable or acceptable limits. For example, suppose at a later date the consumers' bureau checks on whether the manufacturers' automobile is still obtaining an average of 24 mpg. On sampling 16 automobiles, the average gas mileage turns out to be 20.4 mpg. What should the consumers' bureau do? Should it keep on testing, or should it tell the manufacturer to correct its product or reduce its advertised gas mileage. What if the sample mileage turned out to be 25.1? Should the consumers' bureau keep on testing, or should it allow the manufacturer to raise its advertised gas mileage? The experiment and test constructed by the consumers' bureau can be used to answer these questions. The accepted average for gas mileage was 24 mpg so as to have 85% of all samples of 16 having a mean equal to or greater than 23 mpg. Thus, if we accept that nothing is wrong, we should expect an average of 24 mpg. This is called the null hypothesis H_0. We write this as $H_0 : \bar{x}_c = 24$. What the consumers' bureau wants to protect against is having more than 15% of the owners obtaining gas mileage below 23 mpg. This is stated as the alternative hypothesis $H: \bar{x}_c < 24$. The consumers' bureau sets the advertised gas level at 23 mpg. knowing that an average of 85% of the owners will not complain. If many complaints are received, a random sample of 16 car owners can be taken to determine whether there has been a significant change in gas mileage. If the new sample has a mean of 24 or above, then the null hypothesis is accepted. If the sample mean is below 24, then the consumers' bureau considers the results critical. Hence, the interval or region determined by the values $\bar{x}_c < 24$ is called the *critical region.* Note that there is a probability of 0.50 that the statistic $\bar{x}_c$ will be in the critical region. The probability that the statistic $\bar{x}_c$ is in the critical region when the hypothesis is true is called the *level of significance.* In this example, the level of significance corresponds to the 0.50% critical region whose upper interval limit is $\bar{x}_c = 24$.

Thus in this test, if the new sample has a mean $\bar{x}_c$ which is not in the critical region we will accept the null hypothesis, i.e., nothing is wrong and the reports of low gas mileage are not significant. If the mean falls in the critical region $\bar{x}_c < 24$, we will reject the null hypothesis, i.e., something is wrong and the reports of low gas mileage are significant. We shall also say the *results are significant at the 0.50 level.* Note that the term "significant" applied to the rejection and not the acceptance of the null hypothesis.

There are many types of tests like the one described above. Such tests have many features in common. We shall describe a set of such nonparametric tests and illustrate the features which are common to most of these tests. We shall also emphasize that the tests selected are nonparametric due to their applicability to substantive analysis situations where parametric tests may be inapplicable.

7.26 SPECIFIC TESTS OF HYPOTHESES

The first test of hypotheses we shall consider is called the binomial test. It is used with populations which conceivably are made up of two mutually exclusive and exhaustive classes. Examples of the division of a population into two such classes are listed in the following table. The type of measurement used most often to distinguish the two classes in the population is also specified.

Classes	Measurements
Success–Failure	Ordinal
Male–Female	Nominal
Single–Married	Nominal
Slow–Fast	Interval
Thin–Fat	Ordinal
Short–Tall	Interval

In the discussion to follow, we shall distinguish between the two classes, success and failure. Success will mean membership in one class. Failure will identify members of the second class. Thus being a male, short, thin, democrat, etc., will be termed successes. Their complementary classes will be termed failures, or vice versa. What we call the classes is not important in binomial tests of hypotheses. What is important is the proportion in which members of the universe fall into each of the two classes. Knowing that one class has a proportion P of the universe yields that the complementary class has the proportion $Q = 1-P$ of the universe.

Now suppose we have a finite but large number of elements in a universe. We will assume that the proportion of successes and failures are P and $Q = 1-P$, respectively. If we draw a random sample from the universe, we should not expect the sample to be proportioned exactly as the universe. The proportions will vary from sample to sample. Small variations from the universe's proportions P and Q should be more probable than larger ones. The existence of sample variations is illustrated vividly in the more popular polls taken prior to elections. Although the several polls sample the same universe, the results have shown differences of 5 or more percentage points. If the sampling is done randomly, we assign to chance the differences observed between the proportions reported in the sample. If the sampling is stratified or based on judgment (as are some polls), we admit that factors other than chance may enter into the differences. We shall restrict our discussions to random sampling so these other factors will not be considered further.

7.27 BINOMIAL TEST

In the binomial test of hypotheses the null hypothesis may be that one class C_1 contains a proportion P of the universe. We can denote this as H_0:$P = P_0$ where H_0 designates the null hypothesis and $P = P_0$ is the value of the population parameter or characteristic under investigation. The probability that a sample of N

elements drawn from the universe contains r elements from class C_1 and $N-r$ elements from class C_2 is given by the binomial formula

$$p(r) = \begin{Bmatrix} N \\ r \end{Bmatrix} P^r \, (1-P)^{N-r}$$

where P is the proportion of elements in class C_1 and $Q = 1-P$ is the proportion of elements in class C_2. Recall that for a fixed value of N and P, the binomial probability distribution is obtained by evaluating the binomial formula for $r =$ 0, 1, ..., N. The binomial probability distribution changes in shape from a symmetric, normal distribution for $P = 1/2$ to a skewed shape distribution for the more extreme values for P (P close to 0 or 1). Recall that the normal distribution is a good approximation to the binomial distribution when $NP\,(1-P) \geqslant 9$. Thus, depending on what type tables are available, we can use either the binomial probability tables or their normal approximations in the calculations required in the following test. When $NP\,(1-P) \leqslant 9$, the use of the binomial formula or tables is advised if errors due to an inaccurate distribution approximation are to be avoided.

Example 7.24 (Binomial Test–Methodology). We shall use a binomial test of hypothesis to illustrate those features of tests which are common to virtually all tests of hypotheses. We test the hypothesis that voters in a specific county are split even along Republican-Democrat party lines. Let P = proportion of Republicans. We then want to test a hypothesis $H{:}P = 1/2$ against the alternative hypothesis H_1: $P > 1/2$. We have just decided on the first two features of the test:

1. State the null hypothesis $H{:}P = 1/2$.
2. State the alternative hypothesis $H_1{:}P > 1/2$.

To test the hypothesis a sample is required. The question is: What size sample is needed? For our purposes we shall test 25 voters. This gives the third feature of the test:

3. Determine the sample size. In binomial tests the sample size corresponds to the number of trials. In this example, $N = 25$.

The sample drawn may not be a representative one. We know that a chance exists for us to reject the null hypothesis when it is actually true. Thus we must state the percentage of times we are willing to accept the above type error (called a type I error). This is the level of significance and is denoted by a. Thus, if we are willing to accept an error of type I 5% of the time, then $a = 0.05$. We shall set a = 0.05, i.e., accept a type I error 5% of the time. This gives us the fourth feature of a test of hypothesis:

4. Select a level of significance, $\alpha = 0.05$ for the test of the null hypothesis H.

The next feature of a test of hypothesis is one of the most difficult to determine, especially for the analyst who is not well-versed in determining the types of distributions which describe the various sampling universes. In essence, one has to determine the statistic to be used in the test and its distribution. For our purposes, we will specify the statistic in all the tests considered and will make well-founded assumptions for the distribution of the statistic. For the current problem we can use either or both of two statistics. The first is the statistic

$$Z = \frac{\hat{p}-p}{\sqrt{p(1-p)/N}}$$

whose distribution for $Np(1-p) > 9$ approximates the binomial distribution with mean p and variance $p(1-p)$; Z is normally distributed with mean 0 and variance 1. The second is the binomial distribution given by

$$\begin{Bmatrix} N \\ r \end{Bmatrix} p^r \, (1-p)^{N-r} \qquad r = 0, 1, 2, \ldots, N$$

The binomial distribution is to be used when the inequality $Np\,(1-p) > 9$ does not hold. This distribution has a mean proportion p and variance $p(1-p)$.

In selecting either of the above, we have determined the fifth feature of a test of hypothesis. We state this feature as follows:

5. Determine the statistic to be employed in the test and its sampling distribution.

Having the statistic and its distribution, we can rely on tables of values for the statistic to specify regions for which we shall reject the hypothesis if the sample statistic falls in the specified regions. For the statistic

$$Z = \frac{\hat{p}-p}{\sqrt{p(1-p)/N}}$$

at the 0.05 level of significance, the Z factor is given as $Z = 1.64$. This yields the rejection region as depicted in Figure 7.17. Thus if the sample proportion $\hat{p}$ yields a value for the statistic Z which is 1.64 or greater, we shall reject the hypothesis $H{:}p = 1/2$. For the binomial distribution

$$\begin{Bmatrix} N \\ r \end{Bmatrix} p^r \, (1-p)^{N-r} \qquad r = 0, 1, 2, \ldots, N$$

the statistic is the number of successes (Republicans) r out of N trials. We write this as the cumulative probability $p(r \leqslant r_0) = \Sigma_{r=0}^{r_0} \begin{Bmatrix} N \\ r \end{Bmatrix} p^r (1-p)^{N-r}$. We must note that r takes on only integral values. Hence we seek the r_0 which is the first integer for which the cumulative probability $p(r \leqslant r_0)$ exceeds $1 - a$. Or, equivalently, we want $a > 1 - p(r \leqslant r_0) = p(r > r_0)$, where $p(r > r_0) = \Sigma_{r_0+1}^{N} \begin{Bmatrix} N \\ r \end{Bmatrix} p^r (1-p)^{N-r}$ is the area under the binomial probability distribution from the value $r_0 + 1$ to N. This area is illustrated in Figure 7.18.

For our example, we have

$$p(r>16) = \Sigma_{r=17}^{25} \begin{Bmatrix} N \\ r \end{Bmatrix} \cdot p^r \cdot (1-p)^{N-r} = 0.0539$$

$$p(r>17) = \Sigma_{r=18}^{25} \begin{Bmatrix} N \\ r \end{Bmatrix} \cdot p^r \cdot (1-p)^{N-r} = 0.0217$$

So, if the sample of 25 has 16 or fewer Republicans, we shall accept the null hypothesis. If there are 17 or more Republicans in the sample, then we shall reject it. We have just identified the critical region for the test under consideration. In this case we will reject the null hypothesis $H{:}p = 1/2$ if the statistic Z for the sample exceeds 1.64 or the number of successes r exceeds 17. This leads us to the sixth feature of a test of hypothesis:

6. Identify the critical region for the level of significance specified and identify it in terms of the statistic to be employed in the test. For our example,

   reject $H{:}p = 1/2$ if $Z \geqslant 1.64$

   or

   reject $H{:}p = 1/2$ if $r \geqslant 17$

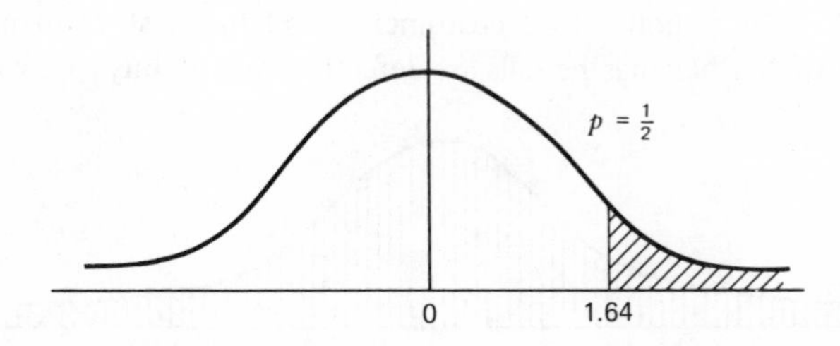

Figure 7.17 Critical region of a normal distribution.

Up to this point we have designed the experiment or statistical test. Now we have only to take the sample, compute the statistic, and determine whether it falls in the critical region. If it does, we reject the hypothesis. If it does not, we accept the hypothesis. We state this feature of the test as follows:

7. Compute the statistic (*s*) for the sample and accept or reject the hypothesis, i.e., state the statistical conclusion.

For our example, we shall assume that 15 out of 25 polled were Republicans. Hence $N_1/N = \hat{p}$ is the value to be used in the statistic Z. For the r statistic we use $r > r_0 = 15$ in the binomial formula

$$p(r > r_0) = \Sigma_{r=r_0+1}^{N} \begin{Bmatrix} N \\ r \end{Bmatrix} p^r (1-p)^{N-r}$$

Finally, we are left with one last feature of a test of hypothesis. We need to state the conclusion for the test or experiment. This feature is stated as follows:

8. State the conclusions drawn from the test or experiment.

In our example, the conclusion for the test is that at the 0.05 level of significance, we accept that the county's voters are equally divided between the Republican and Democratic parties.■■

Exercise 7.23. Select an individual from the class or department and design an experiment to test the hypothesis that he or she has ESP (extra sensory perception). Use a coin or playing card (2 colors) and limit the experiment to 25 trials. How many correct guesses out of 25 are required by the individual under test to provide you with a confidence of 95% that he or she has ESP?■■

Exercise 7.24. John believes that a die is loaded in favor of the side marked by three white dots. He is willing to roll the die 25 times. How many 3s should he get to be 90% confident in rejecting the null hypothesis that the die is a fair die, i.e., $p(3) = 1/6$?■■

Exercise 7.25. In response to a customer complaint, a store manager states that only 5% of the bananas he sells are defective. Mrs. B. buys a dozen bananas

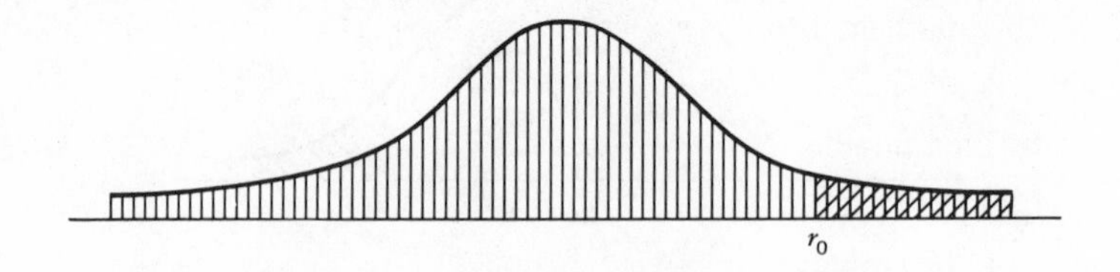

Figure 7.18 Binomial critical region.

and finds 5 to be defective. Does Mrs. B. have sufficient evidence to reject the store manager's claim and be 95% confident in her position? ■■

7.28 χ^2 TEST

The binomial test is designed to test hypotheses about populations whose measurements (objects) conceivably fall into either of two mutually exclusive classes. The χ^2 test is designed to test hypotheses about populations whose measurements (objects) conceivably fall into two or more mutually exclusive classes. In the binomial test, we tested the proportion of one class observed in the sample against a standard as stated in the null hypothesis. In the χ^2 test we deal mainly with frequencies rather than proportions and test whether the class frequencies observed in the sample differ significantly from an expected set of class frequencies. A typical χ^2 test would be the number of nations voting for, against, or not voting on an issue raised at the United Nations. The observed number in each of the three categories would be "tested" against the expected number as determined by previous votes on similar issues. Another example would be to test whether a sequence of characters was selected randomly from the alphabet or was formed by randomly permuting the characters from a page of English text. In the former case we expect all characters to appear with equal frequency. In the latter, we expect some characters like E, I, T, O, and N to have higher frequencies than X, Z, J, K, and Q. We use the χ^2 to test whether the text is random against its having some linguistic cohesion.

In a χ^2 test the null hypothesis (standard) is a statement on the expected frequencies of occurrence of objects or measurements in the categories of the population under investigation. If there are K categories in the population, then we shall denote the expected frequency for each category as E_i, where

E_i = expected number of objects or measurements in category i under the null hypothesis.

We shall draw a single sample of N objects or measurements from the population and separate them into the K mutually exclusive categories. The number that fall into each category i will be designated O_i, i.e.,

O_i = observed number of objects or measurements in category i for the sample drawn.

The statistic to be employed in testing the null hypothesis is given by,

$$\chi^2 = \Sigma_{i=1}^{K} (O_1 - E_i)^2 / E_i$$

Table 7.16 Critical values of chi square.[a]

df	Probability under H_0 that $\chi^2 \geqslant$ chi square													
	.99	.98	.95	.90	.80	.70	.50	.30	.20	.10	.05	.02	.01	.001
1	.00016	.00063	.0039	.016	.064	.15	.46	1.07	1.64	2.71	3.84	5.41	6.64	10.83
2	.02	.04	.10	.21	.45	.71	1.39	2.41	3.22	4.60	5.99	7.82	9.21	13.82
3	.12	.18	.35	.58	1.00	1.42	2.37	3.66	4.64	6.25	7.82	9.84	11.34	16.27
4	.30	.43	.71	1.06	1.65	2.20	3.36	4.88	5.99	7.78	9.49	11.67	13.28	18.46
5	.55	.75	1.14	1.61	2.34	3.00	4.35	6.06	7.29	9.24	11.07	13.39	15.09	20.52
6	.87	1.13	1.64	2.20	3.07	3.83	5.35	7.23	8.56	10.64	12.59	15.03	16.81	22.46
7	1.24	1.56	2.17	2.83	3.82	4.67	6.35	8.38	9.80	12.02	14.07	16.62	18.48	24.32
8	1.65	2.03	2.73	3.49	4.59	5.53	7.34	9.52	11.03	13.36	15.51	18.17	20.09	26.12
9	2.09	2.53	3.32	4.17	5.38	6.39	8.34	10.66	12.24	14.68	16.92	19.68	21.67	27.88
10	2.56	3.06	3.94	4.86	6.18	7.27	9.34	11.78	13.44	15.99	18.31	21.16	23.21	29.59
11	3.05	3.61	4.58	5.58	6.99	8.15	10.34	12.90	14.63	17.28	19.68	22.62	24.74	31.26
12	3.57	4.18	5.23	6.30	7.81	9.03	11.34	14.01	15.81	18.55	21.03	24.05	26.22	32.91
13	4.11	4.76	5.89	7.04	8.63	9.93	12.34	15.12	16.98	19.81	22.36	25.47	27.69	34.53
14	4.66	5.37	6.57	7.79	9.47	10.82	13.34	16.22	18.15	21.06	23.68	26.87	29.14	36.12
15	5.23	5.98	7.26	8.55	10.31	11.72	14.34	17.32	19.31	22.31	25.00	28.26	30.58	37.70
16	5.81	6.61	7.96	9.31	11.15	12.62	15.34	18.42	20.46	23.54	26.30	29.63	32.00	39.29
17	6.41	7.26	8.67	10.08	12.00	13.53	16.34	19.51	21.62	24.77	27.59	31.00	33.41	40.75
18	7.02	7.91	9.39	10.86	12.86	14.44	17.34	20.60	22.76	25.99	28.87	32.35	34.80	42.31
19	7.63	8.57	10.12	11.65	13.72	15.35	18.34	21.69	23.90	27.20	30.14	33.69	36.19	43.82
20	8.26	9.24	10.85	12.44	14.58	16.27	19.34	22.78	25.04	28.41	31.41	35.02	37.57	45.32
21	8.90	9.92	11.59	13.24	15.44	17.18	20.34	23.86	26.17	29.62	32.67	36.34	38.93	46.80
22	9.54	10.60	12.34	14.04	16.31	18.10	21.24	24.94	27.30	30.81	33.92	37.66	40.29	48.27
23	10.20	11.29	13.09	14.85	17.19	19.02	22.34	26.02	28.43	32.01	35.17	38.97	41.64	49.73
24	10.86	11.99	13.85	15.66	18.06	19.94	23.34	27.10	29.55	33.20	36.42	40.27	42.98	51.18
25	11.52	12.70	14.61	16.47	18.94	20.87	24.34	28.17	30.68	34.38	37.65	41.57	44.31	52.62
26	12.20	13.41	15.38	17.29	19.82	21.79	25.34	29.25	31.80	35.56	38.88	42.86	45.64	54.05
27	12.88	14.12	16.15	18.11	20.70	22.72	26.34	30.32	32.91	36.74	40.11	44.14	46.96	55.48
28	13.56	14.85	16.93	18.94	21.59	23.65	27.34	31.39	34.03	37.92	41.34	45.42	48.28	56.89
29	14.26	15.57	17.71	19.77	22.48	24.58	28.34	32.46	35.14	39.09	42.56	46.69	49.59	58.30
30	14.95	16.31	18.49	20.60	23.36	25.51	29.34	33.53	36.25	40.26	43.77	47.96	50.89	59.70

[a] Table is abridged from Table IV of Fisher and Yates: *Statistical tables for biological, agricultural, and medical research,* published by Longman Group Limited, Longman House, Burnt Mill, Harlow, Essex, England, by permission of the authors and publishers.

It is demonstrated in more advanced works in mathematical statistics that the sampling distribution of χ^2 under the null hypothesis is given by the χ^2 distribution. Like the t distribution, the χ^2 distribution is dependent on the degrees of freedom (df) associated with the sampling procedure. Table 7.16 gives critical values of χ^2 associated with the critical regions (column headings) and degrees of freedom (row headings). If we seek to test at the 0.05 level of significance with 10 degrees of freedom, we find that the critical value for the χ^2 is 18.31. If the sample yields a χ^2 which exceeds the critical value, then we would reject the null hypothesis at the 0.05 level of significance. Such values of χ^2 would lie in the critical region which contains 5% of the area under the χ^2 curve and to the right of the value $\chi^2 = 18.31$. Figure 7.19 illustrates the values and areas involved.

The degrees of freedom (df) with the χ^2 test is generally equal to one less than the number of classes in the population, i.e., $df = K - 1$. It is assumed that the null hypothesis H_0 includes a statement on the expected frequency E_i for each class. A typical null hypothesis for a sample of N elements from K classes is that N/K elements fall in each class. This assumes a uniform distribution across the classes of the sample, i.e.,

$$H_0{:}E_1 = N/K \qquad E_2 = N/K, \ldots, E_K = N/K$$

We illustrate the various features of the χ^2 test through two applications of the test.

Example 7.25 (χ^2 Test of Hypothesis). In 1850, a certain town in California had a ratio of 1 male to every 8 females, i.e., the proportion of males is $p = 1/9$. In a survey dated 1870, it was determined that a random sample of 450 citizens contained 68 men. Would we be justified in concluding at a 1% level of significance that the ratio had changed? ■■

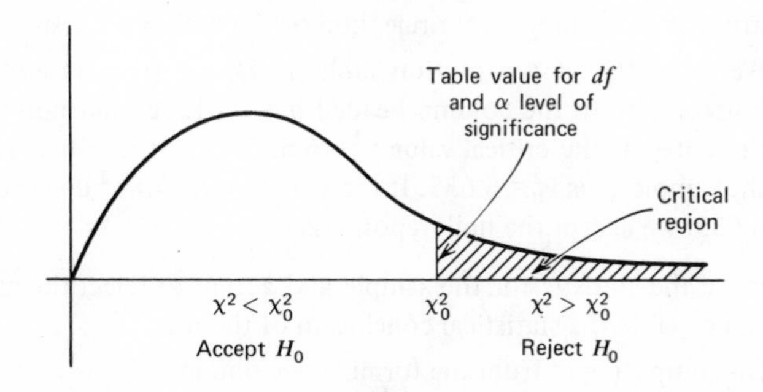

Figure 7.19 χ^2 critical region.

Solution: We shall follow and utilize a test of hypothesis as given in Section 7.26.

1. State the null hypothesis H_0.

   The null hypothesis can be stated in terms of proportions, i.e., male and female proportions are 1/9 and 8/9, respectively, or it can be stated in terms of frequency or expected number. We use the latter form of statement, i.e.,

$$H_0: E_1 = 450/9 = 50; E_2 = 8 \cdot 450/9 = 400$$

2. State the alternative hypothesis H_1.

   The alternative hypothesis is simply that the observed frequencies will not equal the expected frequencies.

$$H_1: E_1 \neq 50; E_2 \neq 400$$

3. Determine the sample size.

   In this example the sample size N is selected as 450.

4. Select a level of significance.

   In this example the level of significance is given as $\alpha = 0.01$.

5. Determine the statistic to be employed in the test of hypothesis and specify its sampling distribution.

   The statistic to be used is the χ^2 or "goodness of fit" given by

$$\chi^2 = \frac{(O_1 - E_1)^2}{E_1} + \frac{(O_2 - E_2)^2}{E_2}$$

   where O_1, O_2 are the sample values and E_1, E_2 are the expected values. The sampling distribution for the statistic is the chi-square with one degree of freedom, i.e., $df = K - 1 = 2 - 1 = 1$.

6. Identify the critical region for rejection of the null hypothesis.

   We enter the χ^2 distribution table at $df = 1$ (row 1) and proceed horizontally to the column headed $\alpha = 0.01$. At that point we read the entry as the critical value $\chi_0^2 = 6.635$. Thus the critical region for the sample χ^2 is $\chi_0^2 = 6.635$. If the sample value of χ^2 does not exceed 6.635, we accept the null hypothesis.

7. Compute the statistic for the sample and accept or reject the null hypothesis, i.e., state the statistical conclusion of the test.

   On computing χ^2 from the formula, we obtain

$$\chi^2 = \frac{(68 - 50)^2}{50} + \frac{(382 - 400)^2}{400} = 7.29$$

The value of χ^2 exceeds the critical value $\chi_0^2 = 6.635$ so we reject the null hypothesis.

8. State the conclusion drawn from the test or experiment.

   We conclude that the ratio of men to women has changed between 1850 and 1870.■■

Example 7.25 (χ^2 Test–Armaments). Country X reports in the SALT meetings that they are building four classes of aircraft in the ratio 9:3:3:1; Analysts from country Y reviewed reports on sightings of 800 aircraft of country X. There were 439, 168, 133, and 60 aircraft in the four aircraft classes in the order of the ratios given above. Is the difference between the stated ratio and observed numbers significant at the 0.05 level?■■

Solution: In this solution we shall go through the eight steps without identifying each of them except by number.

1. To state the null hypothesis, we convert the ratios into the expected frequency of occurrence of the four classes and sample of 800. These are:

   $$H_0{:}E_1 = 450,\ E_2 = 150,\ E_3 = 150, \text{ and } E_4 = 50.$$

2. The alternative hypothesis is:

   $$H_1{:}E_1 \neq 450,\ E_2 \neq 150,\ E_3 \neq 150, \text{ and } E_4 \neq 50.$$

3. The sample size was determined as $N = 800$.
4. The level of significance was selected at the 0.05 level.
5. The statistic to be employed is the "goodness of fit" or χ^2. The sample distribution is chi-squared with 3 degrees of freedom.
6. The critical value for the χ^2 test with $a = 0.05$ and $df = 3$ is obtained from Table 7.16. We read the critical value of χ^2 as $\chi_0^2 = 7.815$. We reject the null hypothesis if $\chi^2 > 7.815$.
7. The statistic is computed for the sample in the following table:

Aircraft type	1	2	3	4
O_i	439	168	133	60
$E_i = Np$	450	150	150	50
$O_i - E_i$	− 11	18	17	10
$(O_i - E_i)^2$	121	324	289	100

$$\chi^2 = \frac{(O_1-E_1)^2}{E_1} + \frac{(O_2-E_2)^2}{E_2} + \frac{(O_3-E_3)^2}{E_3} + \frac{(O_4-E_4)^2}{E_4}$$

$$= \frac{121}{450} + \frac{324}{150} + \frac{289}{150} + \frac{100}{50} = 6.36$$

The value for χ^2 is 6.36. This value does not exceed the critical value χ^2 so the null hypothesis is accepted.

8. The conclusion for the test is simply that country *Y* accepts country *X*'s stated ratio. Note that if country *Y* would be satisfied with a 10% level of significance, it would not accept country *X*'s stated ratio. The conclusion would then be that country *X* is falsifying its construction ratio in an attempt not to report the construction of some types of aircraft. ■■

7.29 LIMITATIONS OF THE χ^2 TEST

The χ^2 test should be avoided when certain conditions are not satisfied on the number of classes and the number of members which fall in these classes. When the degrees of freedom is equal to 1, i.e., we have two distinct classes, each class should have an expected frequency of 5 or more. When the degrees of freedom exceeds 1, i.e., we have more than two distinct classes, four-fifths of the classes should have expected frequencies of at least 5. In no case should an expected frequency be smaller than 1. In some cases classes with low-frequency counts can be combined with adjacent classes provided the combination is dimensionally compatible and meaningful. Normally adjacent categories on a nominal scale may not be combined. On an ordinal or higher scale, adjacent categories normally can be combined. If, after all combinations are completed the result is a two class population with one class having an expected frequency less than 5, then the use of the binomial rather than the χ^2 test is indicated.

Exercise 7.23. Using the voting preferences of members of your class as the sample, design an experiment to test the hypothesis that 60% of all college students are Democrats. Use $a = 0.05$. ■■

Exercise 7.24. In an accident investigation at five locations, the observed and expected (due to chance) number of accidents were found and recorded in the following table:

Location	Expected	Observed
1	44	59
2	82	75
3	112	87
4	106	132
5	54	35

Can the accidents at the five locations be considered as still occurring according to chance? At what level of confidence can you reject the chance hypothesis?■■

7.30 KOLMOGOROV–SMIRNOV (K–S) TEST

When the samples are few in number and adjacent categories need be combined to satisfy the requirements for applying the χ^2 test, the Kolmogorov–Smirnov test is applicable and generally provides a better test of agreement between the expected distribution under H_0 and the observed distribution. The K–S test is a one-sample goodness-of-fit test which determines whether the observed measurements could reasonably have been drawn from a population having the expected (theoretical) distribution.

In applying the K–S test, the theoretical frequency distribution is converted to a relative cumulative frequency distribution. The frequency distribution associated with the sample measurements is also converted to a relative cumulative frequency distribution. The two relative cumulative frequency distributions are compared point-by-point and the point where the two differ the greatest is observed. The sampling distribution of this difference has been previously determined and key values for this distribution are given in Table 7.17.

Figure 7.20 illustrates two frequency distributions, their relative cumulative frequency distributions and their differences. Figures 7.20 *(a)* and *(b)* are the graphs of the theoretical and observed frequency distributions, respectively. Figure 7.20 *(c)* contains the two associated relative cumulative frequency distributions. These are denoted $T(x)$ for the theoretical and $O_N(x)$ for the observed distributions. The subscript N indicates the dependence of this statistic on the sample size N. The dependency is also indicated in the sampling distribution as given by Table 7.17. In Figure 7.20 *(c)* the difference between the two relative cumulative frequency distributions is indicated at each interval end point. The difference is denoted by

$$|T(x) - O_N(x)|$$

Table 7.17 Critical values of D in the Kolmogorov-Smirnov one-sample test.[a]

Sample size, N	Level of significance for $D = \text{maximum } \lvert F_0(x) - S_N(X) \rvert$				
	.20	.15	.10	.05	.01
1	.900	.925	.950	.975	.995
2	.684	.726	.776	.842	.929
3	.565	.597	.642	.708	.828
4	.494	.525	.564	.624	.733
5	.446	.474	.510	.565	.669
6	.410	.436	.470	.521	.618
7	.381	.405	.438	.486	.577
8	.358	.381	.411	.457	.543
9	.339	.360	.388	.432	.514
10	.322	.342	.368	.410	.490
11	.307	.326	.352	.391	.468
12	.295	.313	.338	.375	.450
13	.284	.302	.325	.361	.433
14	.274	.292	.314	.349	.418
15	.266	.283	.304	.338	.404
16	.258	.274	.295	.328	.392
17	.250	.266	.286	.318	.381
18	.244	.259	.278	.309	.371
19	.237	.252	.272	.301	.363
20	.231	.246	.264	.294	.356
25	.21	.22	.24	.27	.32
30	.19	.20	.22	.24	.29
35	.18	.19	.21	.23	.27
Over 35	$\frac{1.07}{\sqrt{N}}$	$\frac{1.14}{\sqrt{N}}$	$\frac{1.22}{\sqrt{N}}$	$\frac{1.36}{\sqrt{N}}$	$\frac{1.63}{\sqrt{N}}$

[a]Adapted from Massey, F. J., Jr. 1951. The K–S test for goodness of fit. *J. Amer. Statist. Ass.*, 46, 70, with the kind permission of the author and publisher.

The K–S statistic concerns itself with the maximum of these differences. In Figure 7.20 *(c)*, the maximum occurs at $x = 6$ and the value of the differences at $x = 6$ is 0.35.

The theoretical relative cumulative frequency function $T(x)$ can be interpreted as the expected proportion of sample measurements with values less than

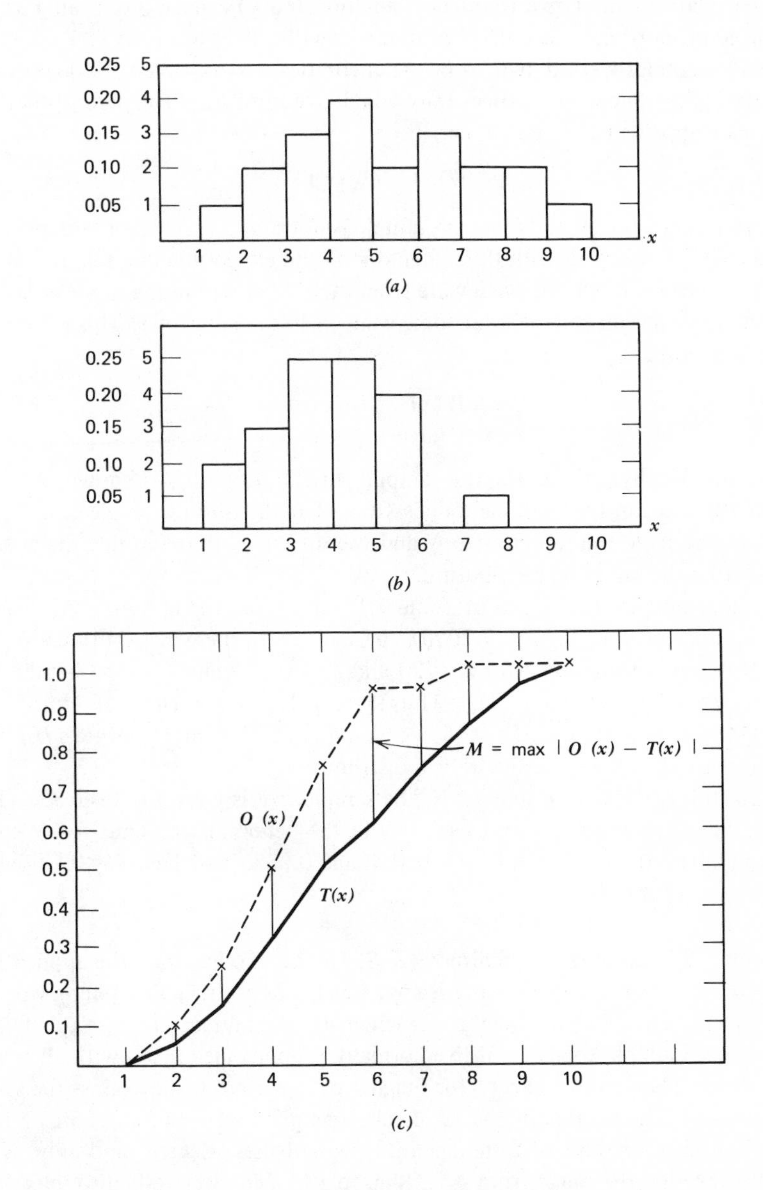

Figure 7.20 Differences of the Kolmogorov-Smirnov test: (*a*) theoretical distribution $T(x)$; (*b*) observed distribution $O_N(x)$; and (*c*) cumulative distributions with differences *D*.

x. The relative cumulative frequency function $O_N(x)$ can be interpreted as the observed proportion of sample measurements with values less than x.

If the samples were drawn from the theoretical (or expected) frequency distribution, then one would expect the two relative cumulative frequency functions to remain approximately equal, i.e.,

$$|T(x) - O_N(x)|$$

is close to zero for all x. When the sample is not drawn from the theoretical (or expected) frequency distribution, then one would expect the two relative cumulative frequency functions to deviate from each other within the range of x. The maximum deviation will occur at some value x_0 in the range of x. This maximum value is denoted

$$\max_x |T(x) - O_N(x)| = M$$

Under the null hypothesis H_0, the sampling distribution of M is known. A set of frequently encountered critical values from that distribution is given in Table 7.17. In the table the sample size N and the difference M determine the critical values from the sampling distribution.

As an example of the use of Table 7.17 in a typical situation, consider the distributions given in Figures 7.20 *(a), (b),* and *(c)*. In these distributions $N = 20$ and $M = 0.35$. Table 7.17 for $N = 20$ and $M = 0.35$ indicates a probability of occurrence of approximately 0.01. Thus if we had agreed on a level of significance $a = 0.05$, for $N = 20$ and $M = 0.35$, we would reject the null hypothesis H_0: The sample was taken from the theoretical distribution.

Note the last row of Table 7.17. The sample size is given as "over 35." The entries in the row are fractions with $\sqrt{N}$ in the denominator. Thus for $N = 49$, we would have 0.153, 0.163, 0.174, 0.194, and 0.233. If we have $M = 0.174$ with $N = 49$, we would read $a = 0.10$.

Example 7.27 (Kolmogorov–Smirnov (K–S) Test). To illustrate the application of the K–S test in a practical situation we test the hypothesis that young women show no preference in the height of their male companions. To test this hypothesis we obtain photographs of five normally proportioned males with all essential features blanked out except for height (photos were taken against the same background). The heights of the males photographed were chosen as 5 ft. 6 in.–6 ft. 2 in. in four steps of 2 in. Then 30 young ladies selected randomly, with heights covering the range from 4 ft. 8 in. to 5 ft. 7 in., were asked to select the male they would prefer to date from the five photographs. If the young ladies do not consider height in selecting their male companions, then one would expect that each of the males photographed, regardless of height, would be chosen with equal frequency. If there is a preference for height, then we would expect the taller or the shorter men being selected with higher frequency than the others.

We apply the K–S test to this experiment using the eight steps recommended for a test of hypothesis.

Step 1. State the null hypothesis H_0.

H_0: Height does not enter into the young ladies' choice of male companions.

Step 2. State the alternative hypothesis H_1.

H_1:Height enters into a young lady's choice of male companion.

Step 3. Determine the sample size

The sample size was selected as $N = 30$.

Step 4. Determine a level of significance.

We shall select a level of significance of $a = 0.05$, recalling that the K–S test is a two-sided test.

Step 5. Determine the statistic to be employed in the test of hypothesis and specify its sampling distribution.

The statistic to be employed is the maximum deviation M given by

$$M = \max_{x} |T(x) - O_N(x)|$$

The sampling distribution is known and critical values for this distribution are given in Table 7.17.

Step 6. Identify the critical region for rejection of the null hypothesis.

We enter the table of critical values for the K–S test at $N = 30$ and $a = 0.05$. There we find that $M = 0.23$. Thus the critical region for the statistic M is $M > 0.23$. If the sample value of M does not exceed 0.23, we accept the null hypothesis. Otherwise we reject it.

Step 7. Compute the statistic for the sample and accept or reject the null hypothesis, i.e., state the statistical conclusion of the test. The theoretical distribution is given in Figure 7.21. The observed distribution is given in Figure 7.22. The relative cumulative distributions with differences indicated

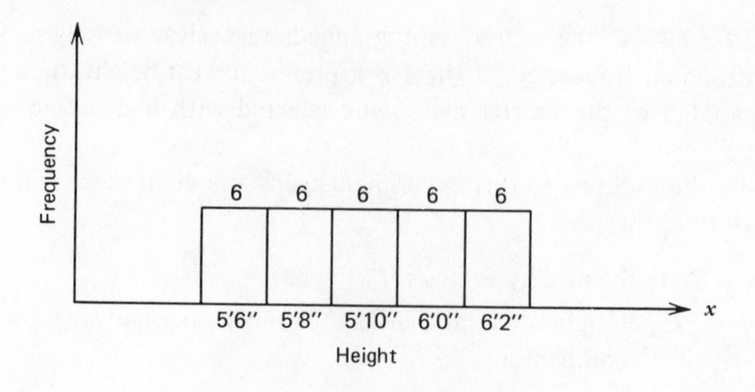

Figure 7.21 Theoretical distribution $T(x)$.

at interval end points are given in Figure 7.23. From Figure 7.23 we read that $M = 0.33$. This value exceeds the critical value $M = 0.23$ and we reject the null hypothesis.

Step 8. State the conclusion drawn from the test or experiment.

We conclude that height is a consideration in a young lady's selection of male companions. ■■

Exercise 7.25. Using the males (or females) in your class or department as a sample, test the hypothesis that the distribution of heights differs significantly (at the 0.05 level) from the expected normal distribution. Use intervals of two inches as the height measurement interval. ■■

Exercise 7.26. A comparison between the number of repairs a test automobile underwent in a 7 yr. period with the expected number experienced by a selected manufacturer produced the following table:

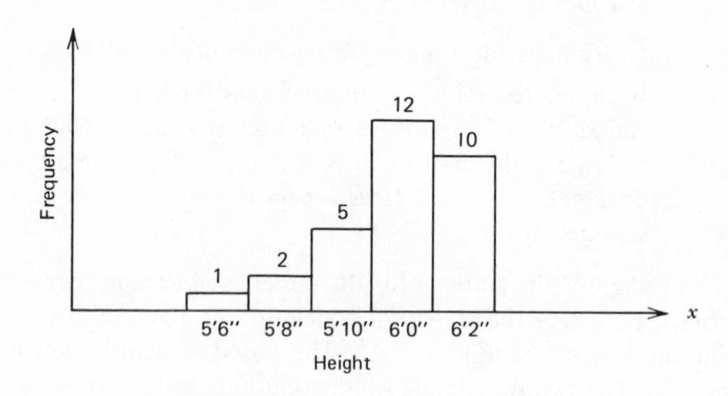

Figure 7.22 Observed distribution $O_N(x)$.

Year	Expected number	Actual number
1	2	4
2	3	6
3	5	9
4	4	6
5	5	12
6	8	9
7	12	14

Would you say the automobile under test is of a different make and be 90% confident in your statement? ■■

7.31 TEST FOR TWO RELATED SAMPLES

The tests considered up to this point were one-sample tests. Our sample was drawn from a population, a statistic obtained, and the hypothesis accepted whenever the value of the statistic remained out of a predetermined critical region for a prespecified level of significance. We will now consider a test in which two

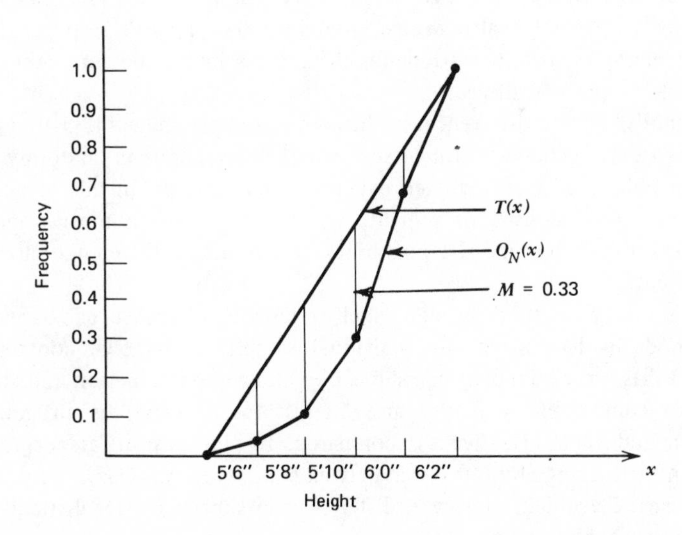

Figure 7.23 Relative cumulative frequency distributions with differences.

related samples are used to determine whether the results of two methods, treatments, or procedures are "better," "different," or "more effective." For example, two different treatments for the same illness may be compared as to effectiveness. Two teaching methods such as classroom lecture versus programmed instruction may be compared. The acceptance of a product with or without a new innovation (such as disc brakes on an automobile) may be evaluated. Or the reliability of a product before and after the introduction of a new manufacturing process or technology may be compared. In foreign relations it is possible to establish whether foreign aid, increased trade, or lowering of tariffs, results in significantly improved relations between the countries. The attitudes of the citizens of the country receiving aid toward the country giving aid is compared on a before and after basis. Another comparison could be between two countries, one country receiving aid, the other not receiving aid.

In applying a two related samples test, one must be careful not to attribute differences which exist to the methods, procedures, or treatments being tested when other factors contribute to the differences. For example, one may compare the effectiveness of two drugs in treating a specific illness. One group of patients is given one drug and the second group of patients is given the other drug. One group, by chance, may be younger, healthier, or generally in better condition than the second group to respond to the treatment by either drug. Hence, the differences noted in the results of the treatment are not due solely to the different effectiveness of the two drugs in treating the illness specified. Other factors such as general health or age enter into the patient's response to the drugs and hence contribute extraneous differences in the measurement of the effectiveness of the two drugs.

To avoid extraneous differences in the above example one could pair or match patients as to age, general health or severity of illness and then randomly assign the two members of each matched pair to the two groups. In this manner the two groups would consist of patients which are similar in most important respects and thus minimize the possibility of extraneous differences arising in the treatment.

There is another method through which extraneous differences can be avoided. The method involves using the individual samples as its own control. For example, a class may be taught one subject by classroom instruction and another subject (of comparable difficulty and interest to the class) by programmed instruction and their effectiveness compared. In this way differences due to motivation or learning abilities which may exist between two classes of different students can be avoided. However, differences in subject matter difficulty and interest must still be accounted for.

Generally, the method of using each sample as its own control is preferable to the matching and pairing procedure. Experimenters find it extremely difficult to match individuals or samples. Good matching requires full knowledge on all significant parameters which may affect the outcome of the test. In addition, even when the parameters are known, our measurements of these parameters do not possess the requisite accurate to establish good pairings or matches.

There are many nonparametrical statistical tests for the case of two related samples. We shall illustrate only one of them which is known as the McNemar test. We select the McNemar test because it can be used to test for the significance of changes where the parameters involved need to be measured only in the sense of a nominal scale. Thus, the parameter under test need not have an underlying continuous distribution. Nor need the measurement itself be continuous. The test can be used when the data is classified into categories which have no order relationship such as "greater than" among themselves. The test is equivalent to a test by a binomial distribution with $P = Q = 1/2$ and with N equal to the number of changes observed.

Other nonparametric tests are available but they impose conditions on the measurement scales or on the distribution of the samples. For example, one test requires that ordinal measurements between pairs is possible. Another requires an ordinal scale of measurement both within and between pairs. Still another requires that the distribution be both continuous and symmetrical. And finally, one test is applicable when the measurements can be made on at least an interval scale. However, all of these tests do not require the condition of normality which the parametric t test requires. In this sense each of the nonparametric tests should find wider application in the analysis of substantive information than the more widely used t test.

7.32 McNEMAR TEST

The McNemar test requires the establishment of an experiment in which measurements are taken on preferences, opinions or health state before and after an action, method, procedure, treatment, or technique is introduced. For example, polls check the popularity of presidents before and after an important domestic or international decision. Candidates give important speeches and wish to know whether the speech caused any significant changes in voter preferences. Health officials are interested in determining the effectiveness of a newly introduced health care program. The State Department and the United States Information Agency (USIA) would like to know whether pamphlets, broadcasts, or other means are effective in changing the attitudes of foreign nationals toward U.S. domestic and foreign policies.

To conduct a McNemar test one determines (a) the control group; (b) the action, technique, method, or treatment to be introduced; and (c) the measurement to be taken to determine the change before and after the action or treatment. For example, suppose that a candidate plans to make an important change in his platform and wishes to know its effect before springing it on the general public. This could be tested by obtaining a representative cross-section (sample) of the public and determining whether they favor the candidate prior to telling them of the change in the candidate's platform. Suppose that a sample of 200 voters is selected. Suppose also that the voters preference indicated that 82 favored the candidate and 118 opposed the candidate. At a later date, each member of the sample is told of the anticipated change in the candidate's platform. Each member is again asked if he favors the election of the candidate. At this time 101 persons indicate that they would favor the candidate and 99 would not. Note that the gross figures indicate a net change of 19. They do not show the direction in which the voters changed their voting preferences and those who did not change. A two by two square matrix which indicates voting preferences before and after the platform change is useful in this test. Figure 7.24 illustrates such a matrix where column and row headings N and F denote N = not favored, F = favored. The matrix indicates that 16 persons changed their votes from for (F) to against (N) and 35 changed their votes from against (N) to for (F), resulting in a net change of 19 votes or a total change in votes of 51; 66 voters remained favorable and 83 voters remained unfavorable, or did not change on the basis of the platform change.

The entries in the matrix indicate that the change in platform did cause some changes in voters preferences. If the platform change did not have any effect on voter preferences, we would expect that the number of changes from for-to-against would equal the number of changes from against-to-for. The resulting number would be the average of the numbers in the upper left and lower right cells of the matrix. We would expect $(16+35)/2 = 25\frac{1}{2}$ = either 25 and 26 changes to occur under the null hypothesis H_0: The platform will have no net effect on voter's preferences. To test this hypothesis we need introduce some notation in order to obtain an expression for the statistic to be used in testing the hypothesis. We relabel the matrix of Figure 7.25, where A and D signify the number of individuals (samples) which change from F to N and N to F,

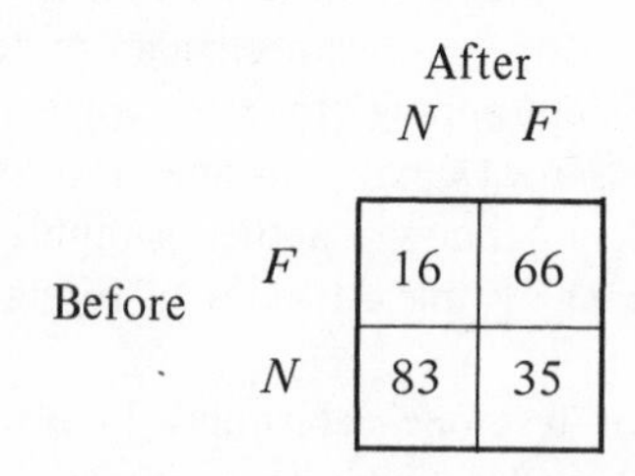

		After	
		N	F
Before	F	16	66
	N	83	35

Figure 7.24 Matrix of changes.

respectively, and B and C signify the number of individuals (samples) which remained F and N, respectively. At this point we can apply the χ^2 test to the two entries signifying changes A and D with each having an expected value of $E_i = (A + D)/2$. The result is

$$\chi^2 = \frac{(A-D)^2}{(A+D)}$$

where the degrees of freedom is $df = 1$. Thus we state that the sampling distribution used in the McNemar test for the significances of changes under the null hypothesis H_0 is distributed approximately χ^2 with 1 degree of freedom ($df = 1$).

We should note at this juncture that the binomial test could be used in place of the McNemar test. For the binomial test we would set $N = A + D$ and the number of success as r would be the smaller of A and D. The probability of success p is 1/2 in such a test. The binomial test is the better test to employ whenever the expected frequency $E_i = (A + D)/2$ is less than 5. Otherwise, the McNemar test should be used.

Example 7.28 (McNemar–Two Related Samples Test). We shall complete the example started in the discussion on the McNemar test.

Step 1. State the null hypothesis H_0.

H_0: Change in candidate's platform will cause an equal number of voters to change their votes in each direction, F to N and N to F.

Step 2. State the alternative hypothesis H_1.

H_1: Change in platform will affect voters' preferences favorably (one-sided).

Step 3. Determine the sample size.

In the example, a sample size of 200 was selected. The sample size, in this example, normally would be considerably larger.

Step 4. Determine a level of significance.

		After	
		N	F
Before	*F*	*A*	*B*
	N	*C*	*D*

Figure 7.25 Matrix of changes.

We shall select a level of significance to be 0.10 with $N = 200$. We must recall that the χ^2 statistics (to be used here) is two-sided and adjust the level of significance to obtain a one-sided test.

Step 5. Determine the statistic to be employed in the test of hypothesis and specify its sampling distribution.

The statistic to be employed is the χ^2 statistic given by

$$\chi^2 = \frac{(A-D)^2}{(A+D)} \qquad df = 1$$

The sampling distribution is approximately χ^2 with $df = 1$ and is given in Table 7.16. We could also use the binomial distribution with $N = A + D$ and the smaller of A and D as r, the number of changes.

Step 6. Identify the critical region for rejection of the null hypothesis.

Since $a = 0.10$ and Table 7.16 gives values for a two-tailed test, we take $a = 0.05$ and $df = 1$. From Table 7.16 we note that the critical region is $\chi^2 > 3.84$. Thus if the statistic

$$\chi^2 = \frac{(A-D)^2}{(A+D)}$$

yields a value which exceeds 3.84, we reject the null hypothesis. Otherwise, we shall accept the null hypothesis.

Step 7. Compute the statistic for the sample and accept or reject the null hypothesis, i.e., state the statistical conclusion of the test.

The matrix of voting changes from before-to-after the platform change was given as

		After N	After F
Before	F	16	66
Before	N	83	53

so that $A = 16, D = 53$. Hence

$$\chi^2 = \frac{(A-D)^2}{(A+D)} = \frac{(16-53)^2}{16+53}$$

$$= \frac{37^2}{69}$$

$$= 19.8$$

The value $\chi^2 = 19.8$ exceeds the critical value; so we reject the null hypothesis.

Step 8. State the conclusion drawn from the test or experiment.

We conclude that if the candidate's platform is changed, the result will be a net change in voter's preference which will favor the candidate.■■

Exercise 7.27. Using the class as its own control group, ask each student whether he or she believed this course would be of value to them in their careers prior to taking the course. Ask whether there has been a change in opinion, and its direction–better or worse–in the value of the course after its completion (with this exercise). Use the McNemar test to determine whether the course caused a significant change in the class opinion of its value at the 0.10 level of significance.■■

REFERENCES

Afifi, A. A., and Azen, S. P. *Statistical Analysis.* New York: Academic Press, 1972.

Dixon, Wilfred J. *Introduction to Statistical Analysis.* New York: McGraw-Hill, 1969.

Hoel, Paul G. *Introduction to Mathematical Statistics.* New York: John Wiley, 1966.

Siegel, Sidney. *Nonparametric Statistics for the Behavioral Sciences.* New York: McGraw-Hill, 1956.

Index

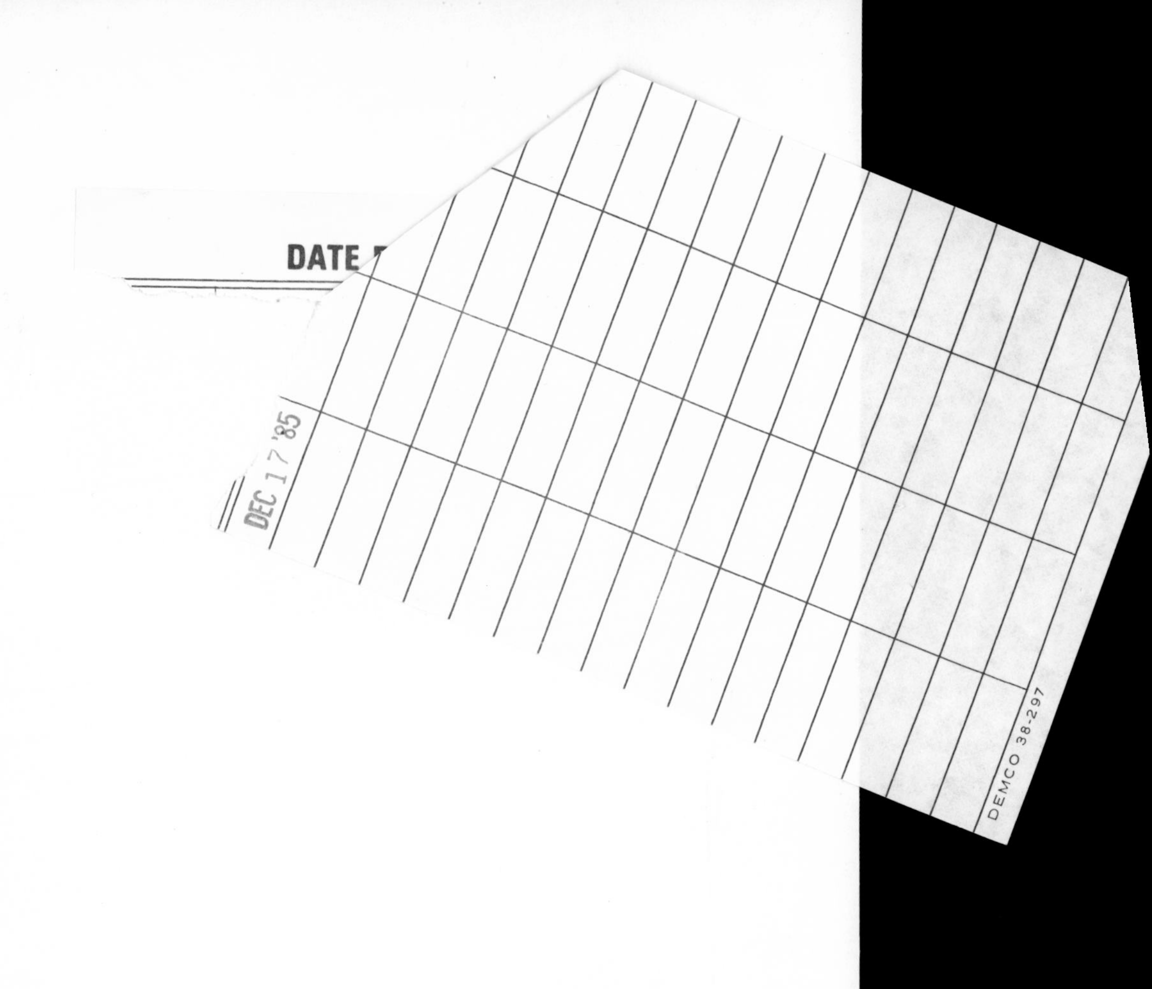
DATE
DEC 17 '85
DEMCO 38-297